The Enigma of Probability and Physics

Fundamental Theories of Physics

A New International Book Series on the Fundamental Theories of Physics: Their Clarification, Development and Application

The Enigma of Probability and Physics

by

Lazar Mayants

Department of Physics and Astronomy,
University of Massachusetts, Amherst, U.S.A.

D. Reidel Publishing Company

A MEMBER OF THE KLUWER ACADEMIC PUBLISHERS GROUP

Dordrecht / Boston / Lancaster

Library of Congress Cataloging in Publication Data

CIP

Mayants, Lazar, 1912–
The enigma of probability and physics.

(Fundamental theories of physics)
Bibliography: p.
Includes index.
1. Probabilities. 2. Quantum theory. 3. Statistical mechanics.
I. Title. II. Series.
QC20.7.P7M39 1984 530.1 84-16027
ISBN 90-277-1674-9

Published by D. Reidel Publishing Company,
P.O. Box 17, 3300 AA Dordrecht, Holland.

Sold and distributed in the U.S.A. and Canada
by Kluwer Academic Publishers,
190 Old Derby Street, Hingham, MA 02043, U.S.A.

In all other countries, sold and distributed
by Kluwer Academic Publishers Group,
P.O. Box 322, 3300 AH Dordrecht, Holland.

Printed in The Netherlands

To my wife ZILMA,
the dearest in the
world, with love

Table of Contents

Part II. Fundamentals of Probabilistic Physics

Foreword

Lazar Mayants is a recent Russian emigre noted for his work in theoretical physics. He was previously a professor at several universities of the Soviet Union and a distinguished member of the Academy of Sciences of the U.S.S.R, where he worked for about 30 years. In this book he presents a unique, extremely detailed, and embracive version of a subject that has suffered for a long time from numerous internal imperfections. His approach is new and original, the material covered features not only the foundations of the science of probability but also most of its applications, including statistical and quantum mechanics. The key methodological principle underlying the book is of extraordinary significance and deserves special attention.

The treatment excels in thoroughness of presentation, in its fulness of mathematical detail and the abundance of physical examples. The book is intended for a wide range of people interested in probability and its connection with modern science. It is written as a text for advanced students, and I predict that a reader who masters all its contents will become an expert in the subject of both probability and its physical implications, while enjoying its understanding and use.

HENRY MARGENAU

Veritas nihil veretur nisi abscondi (truth fears nothing except being hidden). Latin proverb

'What tremendously easy riddles you ask!' Humpty Dumpty growled out.

Lewis Carroll, *Through the Looking Glass*, Chap. 6.

Preface

The history of producing this book is rather complicated and not quite usual. When I worked at the Institute of Organoelemental Compounds of the Academy of Sciences of the U.S.S.R. (Moscow), I once learned from academician A. N. Nesmejanov's annual report about a strange chemical reaction that was supposed to be due to an intramolecular rearrangement. This was an incitement to me to begin pondering the problem of the possible existence of a criterion which would safely tell us whether or not a certain intramolecular rearrangement is feasible. And this started my series of researches in related fields.

As a result of my first search, I proposed a theory of kinetics of chemical and physical processes – a theory of transitional configuration – which contained, in particular, the criterion sought. The theory proved, however, to be in disagreement with the common views on what quantum physics is, and this made me scrutinize thoroughly the conventional foundations of the latter. I found them rather unsound and came to the conclusion that to comprehend what quantum physics really is, it is necessary, in the first place, to carefully examine what is the actual meaning of probability.

This research has resulted in a new approach to the consideration of probability and the construction of what I have at first called *complete probability theory*. This made it possible, in turn, to get a new approach, in particular, to the construction of both classical statistical mechanics and quantum mechanics, which turned out to be the interconnected domains of a unified science which I have termed *probabilistic physics*.

So, the logic of the researches has made me deal with probability in various domains of science. But, although I was engaged in this work in the U.S.S.R. from 1962, only here, in Amherst, where I came in 1980, did I fully realize that actually I have dealt with the creation of a specific science, namely, the science of probability, which I now term *probabilistics* (instead of complete probability theory), and with its applications to physics.

The usual course of forming a science may schematically be represented as follows. Somebody observes a phenomenon. Then he or somebody else tries to explore, maybe understand, and describe the phenomenon in terms of mathematics. If such an attempt proves successful, a new science arises, which may or may not be connected with some old ones. So, a science starts, as a rule, with an experiment – with the discovery of its basic phenomena. Description of the phenomena in terms of mathematics is the starting point of the theoretical part of the science. In order for a science to be successful, the primary notions related to it should be formulated distinctly enough and the subject of the science must be defined explicitly. The correctness of the theoretical part of any science is for it to be checked experimentally.

Such a way for emergence of a new science proved unrealizable for probabilistics. Instead of this unified science, basically two different disciplines, probability theory and statistics, have arisen and developed. The main reason probabilistics was not discovered as a specific science and did not exist until recent years may be thought to be as follows. The basic phenomenon of this science has not been formulated distinctly enough, although it was implicitly utilized both in statistics and in probability theory; and the subject of this science, namely probability, has not satisfactorily been defined either – in probability theory no attempt to define explicitly the notion of probability is made at all. The reason mentioned has been conditioned, in turn, by some causes of methodological and historical character.

The notion of probability goes far back in history. The beginning of a scientific approach to probability calculus can apparently be traced back to the sixteenth and seventeenth centuries. Particularly intensive developments of probabilistic methods, associated with the construction of the so-called axiomatic probability theory, began in the nineteen thirties.

Statistics is no less old, and the development of specific statistical methods, such as those used for analysis of statistical data, which are intimately connected with the concept of probability, often occurred in parallel, rather than simultaneously, with the development of probability theory. The relationship between statistics and probability theory was not organic enough; there has always been a keenly felt logical gap between the two disciplines.

The reader is certainly familiar with the wide use of probabilities in the most diverse fields of knowledge – both the past and recent times have seen the appearance of related new specific disciplines, such as informatics, cybernetics and others, which play an enormous part in modern life. But the question what probability is remained invariably open. All attempts to answer it failed for some reason or other. It was this circumstance that made mathematicians give up such attempts as a hopeless task and led them to construct probability theory purely axiomatically as a specific branch of mathematics – more narrowly, as a branch of the theory of measure. But this construction excludes any possibility of establishing a direct relationship between probability theory and statistics, an immediate transition from probability theory to statistics – no real probability problem can be solved within the framework of probability theory – a fact which is characterized in the axiomatic probability theory itself as the incompleteness of the theory.

Far from all scientists, even among the mathematicians, are reconciled with this state of affairs. Some mathematicians who do not specialize in probability theory believe, for example, that probability theory cannot be a purely mathematical discipline, if only for the fact that Bertrand's paradox (stating that there are a great number of answers to the question, 'What is the probability that the length of a randomly selected chord of a circle will exceed the side of a regular triangle inscribed in the same circle?', whereas the answer to this question must be unique) does not find its solution in this theory. On the other hand, many specialists, including most physicists, prefer using various statistical definitions of probability, which are not sufficiently rigorous from the mathematical standpoint.

Besides, ordinary probability theory completely overlooks the entire range of problems considered in quantum physics, which is known to make wide use of probabilistic terminology. The history of quantum physics is much shorter, but considerably more dramatic. The six decades since the establishment of quantum physics have seen two major rebellions (in the thirties and fifties) against the most widespread, semi-mystical Copenhagen interpretation of quantum phenomena, in which Einstein, Schrödinger, de Broglie, and Bohm, among others, took an active part. Both rebellions failed in the last analysis because the rebels did not propose anything that would introduce radical changes into the foundations of quantum theory, while this theory was seemingly 'working' well and was marvelously proved by experiment. Very many physicists, however, remain dissatisfied with the present state of quantum theory, which fact is exemplified by the incessant disputes on a number of its aspects.

This has led to a strange situation. Young physicists, who are just beginning to study quantum theory, can hardly gain a profound understanding of it under these conditions (as even Einstein could not 'understand' and accept it till the end of his life). At the same time they become convinced that its computational prescriptions are effective and lead to practically important results. Therefore, in solving some concrete problems they use these prescriptions pragmatically, making no attempt to get at the root of things. The current aphorism, 'Quantum physics cannot be understood – one must simply get used to it' protects them from any doubts and allows them to work quietly.

This state of things can hardly be considered satisfactory. It is not out of place to quote here, too, two of Dirac's remarks concerning this matter, which can be found in his well-known book *The Principles of Quantum Mechanics*, 4th edn. (Oxford University Press, 1958) and in its Russian edition (GIFML, Moscow, 1960). The first paragraph of his Preface to the Russian edition (dated 15 October, 1959) reads (in my translation from Russian): 'Our notions about the principles of the physical world are changing in the course of time by consecutive stages. We are now at a certain stage but there is no reson to think that this stage is the last one. We can anticipate further changes in the future, which will perhaps be as drastic as in the past.' The final paragraph of the book in the English edition says: 'It would seem that we have followed as far as possible the path of logical development of the ideas of quantum mechanics as they are at present understood. The difficulties, being of a profound character, can be removed only by some drastic change in the foundations of the theory, probably a change as drastic as the passage from Bohr's orbit theory to the present quantum mechanics.'

These words of Dirac are quite unambiguous and hardly need any comments.

The fact that probability has some bearing on quantum physics suggests that there must be a direct connection between quantum physics and probability theory. The unhappy situation in quantum theory and in probability theory may thus be intimately related. In the final analysis, the plight in quantum theory stems from its having been constructed 'with one's eyes closed', without sufficient penetration into its depths. Such construction of quantum physics does not naturally ensure the correctness of all its premises or the validity of all its conclusions. A similar situation prevails in contemporary probability theory.

If the basic phenomenon of probabilistics were formulated distinctly at the appropriate time and the notion of probability were defined satisfactorily enough, the development of any science related to probability could have proceeded in a quite different way. But in order to recognize this phenomenon clearly it is necessary to know and use one extremely significant methodological principle, without which that phenomenon could not explicitly be detected.

This key principle may be worded as follows: It is necessary to distinguish strictly between *concrete* and ***abstract*** objects. This premise, which plays a decisive role in the construction of probabilistics, should also be taken into consideration when forming any other sciences and especially social ones. One can hardly doubt that no social science can correctly be founded and developed without due allowance for this principle. So, it is this methodological principle which is the pivotal one for the construction of probabilistics and hence its applications to physics, that is, for forming probabilistic physics.

Making use of this principle one can clearly realize that the well-known basic fact of experimental statistics is a phenomenon which can now be worded as follows: When a very large number of random tests have been carried out on a set of concrete objects, the number of tests revealing concrete objects of each subset of this set is approximately proportional to the measure of this subset. It is precisely this phenomenon, which has always been invoked by statisticians in practice and used implicitly in the construction of axiomatic probability theory, but which has never before been formulated distinctly enough, that actually has been taken as the first axiom of probabilistics.

The present book thus contains basically the logical exposition of the fundamentals of probabilistics and probabilistic physics, proceeding from its basic phenomenon with permanent due allowance for the key methodological principle formulated above. This offers the reader an opportunity to see everything with his eyes open, to convince himself in particular that quantum theory is indeed tied most intimately with the probabilistic notions, and to comprehend all its apparent oddities (rather than getting 'used' to them!), proceeding from the correct understanding of what probability really is.

The book is written for a wide range of readers. It should be read from the very beginning and in the sequence in which the material is presented. But the reader may, of course, omit these sections which are not necessary for the understanding of the problems he is interested in that are considered further on in the text. For understandable reasons, we inevitably have employed the same terms, such as 'random variable', 'random function', and 'random field', as are used in ordinary probability theory, but these quantities are defined differently in probabilistics. Therefore, a person familiar with ordinary probability theory should be particularly attentive in reading the book to avoid misunderstanding.

The book consists of three parts. The first part exposes the fundamentals of probabilistics. The initial chapters contain the basic principles, including the elements of statistics. Furthermore, special material referring to random variables is presented in logical succession. In particular, random variables of a special type – 'Hamiltonian random variables' – are considered, and their specific, quantum peculiarities are discussed. Also, questions associated with joint probability distributions, in particular for functions of Hamiltonian random variables, are considered. This part is concluded with consideration of still more special material pertaining to random functions, in particular random fields. Much space is devoted to special-type random functions and fields which have been called *Hamiltonian*.

The second part describes the fundamentals of probabilistic physics, which follow logically from the fundamentals of probabilistics. In particular, the treatment of its special branch, quantum physics, happens to be actually completely covered in the first part of the book, where the peculiarities of Hamiltonian random variables are discussed. Moreover, separate questions of various degrees of complexity are considered, and a unified theory of the transformation kinetics of physical systems, embracing both decay-type physical transformations and chemical intramolecular rearrangements, is presented. The theory predicts, by the way, the possibility in principle of the decay of any particle and hence of a proton, too, which problem attracts at present, as is well known, the attention of many physicists, both theorists and experimenters. The second part also contains discussions of some specific features of the electromagnetic field and its relationship with photons; the related problem of the existence of zero-rest-mass particles is considered, too.

The third part covers some methodological problems of science in general and those relating to probability and physics in particular. Among other things, this part reveals the relationship between probabilistics, von Mises' theory, and ordinary probability theory; it explains some paradoxes of probability theory and physics and resolves many moot points of quantum physics, in particular the EPR, Bohm, and Bell paradoxes; the Dirac equation is also discussed in detail. It moreover formulates some new problems, bearing on microworld phenomena, which are still to be solved.

Being a single whole, the book encompasses problems of different degrees of generality and complexity. Therefore, its separate parts may be of interest to widely differing groups of readers. The initial sections, which purport to establish the difference and relationship between concrete and abstract objects, are designed for a more extensive range of readers. The delving into the difference and relationship between the concrete and abstract objects, which may be believed to be one of the basic principles of scientific cognition, should not be ignored by researchers in any domain of scientific knowledge, including the social sciences. Moreover, it may be thought that familiarization with this principle must be included in the mandatory high school curricula within the foreseeable future.

The initial chapters, which contain the basic premises of probabilistic, are aimed at a very wide and ever widening range of readers interested in probabilities and their practical applications and also at those who wish to acquaint themselves with the fundamentals of probabilistic physics. All these problems are therefore discussed most comprehensively. To facilitate the reading and understanding of the material by persons unfamiliar with the necessary elements of set theory, appropriate explanations are supplied where needed.

The remaining chapters of the first part and all the chapters of the second part are designed in particular for a broad range of readers, but primarily physicists and physical chemists interested in the fundamentals of statistical physics and quantum physics. Almost all these chapters are less comprehensive and require a knowledge of the elements of the corresponding branches of mathematics, including the fundamentals of linear algebra. A brief review of these, too, is given in the first part of the book.

The third part may be of particular value to specialists interested in methodological questions – problems of the philosophy of science. It may be believed, however, that this part will also be of great interest to a much wider circle of readers familiar with the paradoxes of both ordinary probability theory and quantum physics and with the stormy discussions in the latter field.

It is hard to find adequate words to thank Professor Henry Margenau of Yale University, a scientist and thinker of broad outlook, a man of high moral principles, a true humanist, whom I consider the 'Godfather' of my work on probability and without whose moral support and crucial help this work may not have achieved its present culmination.

I also greatly appreciate the permanent interest in and effectual support of my work on probability by Professor Alwyn van der Merwe of the University of Denver. I am especially thankful to him for his invariably favorable attitude towards the present book, his frequent and valuable advice, and his kind consent to be its Editor.

It is impossible to overestimate the role that my wife Zilma played in enabling me to pursue my work on probability and to complete this book. To her, in deep appreciation, I dedicate my book, which shall stand as a monument to her suffering and unwavering support.

I finally thank in advance all those who might care to send me comments and wishes concerning the book.

Amherst, Massachusetts, U.S.A. L. MAYANTS

Part I

Fundamentals of Probabilistics

Chapter 1

Principal Concepts

1.1. Concrete Objects

The starting point for the entire presentation is the concept of concrete objects. We regard this concept as a primary one, which cannot be reduced to any simpler concepts, and will only try to explain it. We perceive concrete objects in the most usual, trivial sense, that is, as quite definite, actually existing objects. It is precisely concrete objects that are the subject of any direct experimental study (observation, measurement, etc.).

Examples of concrete objects are the Earth we live on, the original sculpture of David by Michelangelo (each of its copies is also a concrete object, of course), each reader of this book, and the copy he has at his disposal, each coin in the purse of a definite person, a particle leaving a definite track in a definite cloud chamber or recorded by a definite Geiger counter, and so on.

In the simplest cases, in order to clearly specify the concrete object in question, it is sufficient to point it out in any arbitrary way (even with a finger!). In the general case, however, this can only be done by using the properties (features) of concrete objects. Thus, if two balls – say, white and black – are lying on a table, it will suffice to say 'the white ball' to specify clearly the concrete object implied. Here, one makes use of the 'color' property of the balls, and they are distinguished by the 'value' of this property, one ball being 'white', and the other 'black'. Subsequently we will always strictly differentiate between the properties of objects and the values of these properties. Different concrete objects possessing one and the same property may differ in values of this property, as in the above examples. Each property thus has some set of values.

If two balls are of the same color, but of different size, then in order to point out the concrete ball needed it is sufficient to use its 'size' property and say, for instance, 'the smaller ball'. The values of the 'size' property here are 'smaller' and 'larger', or 'small' and 'large', which is the same. Finally, even if two balls have the same values of all their own properties, they can be differentiated by their position in space (on the table, in our case) – a concrete ball can be singled out by specifying its space coordinates. Then one uses the 'spatial position' property of the ball, its values being the 'ball coordinates'. Different values of the 'time' property may also enable one to differentiate concrete objects. For instance, if several liquid drops fall out of a pipet, each concrete drop can be characterized by the 'time instant' of its fall.

In the above examples each property had a finite number of values, because we had a finite number of concrete objects. But this is not necessarily so. For instance, each concrete point of some concrete space differs from another concrete point of this space in the values of its coordinates (in the selected reference system). Then the 'spatial position' property of the space point has a nonenumerable set (that is, a set whose elements cannot be numbered with numbers of the natural series) of 'point coordinates' values. All this leads us to seemingly trivial, but actually extremely important conclusions, namely: (1) any concrete object possesses a definite set of properties; (2) each property has a definite set of values; (3) each concrete object has one, and only one quite definite value of each of its properties; (4) different concrete objects differ in the values of at least one of their properties. Assuming, for simplicity, that a set of properties is finite and taking advantage of mathematical language, we can write down all this as follows.

Each concrete object a has a set $T = \{t_0, t_1, t_2, \ldots, t_N\}$ of different properties (here, the set T is given by indicating all its elements). Each property t_j ($j = 0, 1, 2, \ldots, N$) has a set θ_j of different values f_j. To each concrete object there corresponds one, and only one value $f_j \in \theta_j$. Different concrete objects differ in the values $f_j \in \theta_j$, at least for one property t_j.

For different values of f_j, one could denote it by different superscripts or simply indicate the particular values. For instance, if we denote the 'color' property of a ball by t_1 and assume that the set θ_1 of the values of this property is $\theta_1 = \{\text{black, white}\}$, then for the black ball $f_1 = \text{black}$, and for the white one, $f_1 = \text{white}$. Further on, f_j will imply some one definite value of the property t_j. What value it is precisely, will be indicated separately, if required.

To any set of some number of properties $t_i, t_j, \ldots, t_k$, contained in the set T there also corresponds a property (combined) of a concrete object a. Denoting this property by $t_{ij\ldots k}$, we can write $t_{ij\ldots k} = (t_i, t_j, \ldots, t_k)$. The property $t_{ij\ldots k}$ has values $f_{ij\ldots k} \in \theta_{ij\ldots k}$, and $f_{ij\ldots k} = (f_i, f_j, \ldots, f_k)$, where $f_i \in \theta_i$, $f_j \in \theta_j$, $\ldots, f_k \in \theta_k$. The set $\theta_{ij\ldots k}$ of all the values $f_{ij\ldots k}$ is the so-called direct product of the sets $\theta_i, \theta_j, \ldots, \theta_k$, which can be written as $\theta_{ij\ldots k} = \theta_i \times \theta_j \times \cdots \times \theta_k$.

Explanations. Recording of the type $a \in A$, or $A \ni a$ means that a is an element of a set A (is contained in a set A). In an expression of the type $D = B \times C$, where B and C are arbitrary sets, the direct-product symbol '×' indicates that the set D consists of all the pairs of elements taken singly from the set B and the set C. Generalization for the case of an arbitrary number of sets multiplied is obvious – the direct product of n sets consists of all the manifolds of n elements taken singly from each of the sets multiplied.

Note. To each value of any property one can always juxtapose some proper number, and this can be done in an infinite number of ways, provided only that different values are associated with different numbers. Therefore sets θ_j can always be considered numerical, if required. Thus, for instance, by assigning the number 1 to the white color and 2 to the black one, we obtain $\theta_1 = \{2,1\}$ for the 'color' property in the above example with balls. We could equally make use of any other distinct numbers in place of 1 and 2.

1.2. Abstract Objects

In distinction to the 'concrete object' concept, selected as a primary one, the 'abstract object' concept, whose definition will follow, is a secondary one. Before proceeding to define it, we will consider some examples.

Example 1. Some concrete room contains 20 concrete persons, who differ primarily in their sex (t_1) and age (t_2). These persons can be divided into 4 groups in accordance with the values of the indicated properties: (1) 4 young women; (2) 3 aged women; (3) 7 young men; (4) 6 aged men. Besides, these persons may differ in the values of quite a number of other properties. We need not, however, attempt to enumerate these properties, because the only essential point is that the set of these properties, which is one (combined) property of each person, has different values for each concrete person in each of these groups. Taking advantage of the note at the end of the preceding section, we could, for instance, replace these values by different numbers and assign to each concrete person a definite ordinal number in the group to which he belongs.

Thus, the given room contains 20 concrete persons differing in the number of the group (corresponding to the value of the property t_{12} = (sex, age)) to which they belong and in their ordinal number in this group. Each of these concrete persons is a concrete object, which is, incidentally, dealt with by some other concrete persons, perhaps concrete teachers of different concrete subjects. If we detach (abstract) ourselves from the individual features of the concrete persons belonging, say, to group 1, and discard them, the remaining object will be 'a person in this room, who belongs to group 1', that is, 'a young woman in this room'.

'A young woman in this room' is the image of all the concrete persons in this room belonging to group 1, and since transition to it consists in abstracting oneself from the differences between the concrete persons of this group, it can be called 'an abstract object' corresponding to the concrete objects belonging to group 1. 'A man in this room' is the result of abstracting oneself from all the individual features (including the age) of those concrete persons in this room who belong to groups 3 and 4 and therefore is an abstract object corresponding to these concrete persons. And so on. Finally, 'a person in this room' is the result of abstracting oneself from all the differences between the concrete persons located in this room, and is therefore an abstract object corresponding to all these persons. By the way, all the concrete objects in this example have one and the same value of their common property, which determines their belonging to the same set, namely, the value 'a person in this room' of the property 'the person's location'.

To the property t_1 (sex) there corresponds the set of values: θ_1 = {woman, man}. To the property t_2 (age) there corresponds the set of values θ_2 = {young, aged}. The property 'the person's location', which will be denoted by t_0, has one and the same single value: f_0 = a person in this room for the entire set of the concrete persons under review (concrete objects); precisely this single value f_0 determines the fact that all these concrete persons belong to the same set.

To the property t_{12} = (sex, age) there evidently corresponds the set of values: θ_{12} = {a young woman, an aged woman, a young man, an aged man}, in accordance with which all the persons in this room were divided into 4 groups. There is still some other (combined, generally speaking) property t_3, whose different values differentiate concrete persons in each group. We have, however, agreed to replace these values by some numbers, say, by ordinals. Therefore, for instance, to a concrete person in group 1 with an ordinal number 1 there corresponds the following value $f_{0123} \in \theta_{0123}$ of the property t_{0123}: f_{0123} = (a person in this room, young, woman, 1).

'A person in this room' is the name of an abstract object corresponding to the whole set of concrete persons in this room. This abstract object can be denoted by the unique value f_0 of the property t_0, which indicates that concrete objects belong to the set in question. 'A young woman in this room' is the name of an abstract object corresponding to all the concrete persons having the enumerated values of the properties: f_{012} = (a person in this room, young, woman). And so on.

Example 2. Let us consider a set of concrete balls located in a definite urn (f_0 = a ball in this urn). The balls differ in size (t_1), color (t_2), and in some other individual features which distinguish them from one another. The property t_1 has two values: θ_1 = {large, small}; the property t_2 also has two values: θ_2 = {white, black}. 'A ball in this urn' is the name of an abstract object corresponding to the whole set of concrete balls in this urn. 'A large white ball in this urn' is the name of an abstract object corresponding to the concrete balls having the enumerated values of the properties: f_{012} = (a ball in this urn, large, white). And so on.

Example 3. Let us now consider a set consisting of all the points of a definite rectangle (f_0 = a point of this rectangle). The rectangle contains two quite definite circles (1 and 2), which have no points in common, either between themselves or with any side of the rectangle. For the property t_1 we select the position of a point of the rectangle relative to the circles. Then θ_1 = {inside circle 1, on circle 1, inside circle 2, on circle 2, outside the circles}. 'A point of this rectangle' is the name of an abstract object corresponding to the whole set of concrete points of this rectangle. 'A point of this rectangle inside circle 1' is the name of an abstract object corresponding to the concrete points having the enumerated values of the properties: f_{01} = (a point of this rectangle, inside circle 1). And so on, and so forth.

We can now proceed to the formal definition of abstract objects. We will always deal with a certain set A of concrete objects, and the fact that these concrete objects belong to set A is determined by their having one and the same value f_0 of some property t_0. This can be formulated as follows: a concrete object a belongs to A, that is $a \in A$, if, and only if it has the value f_0 of the property t_0. Besides, these concrete objects have at least n other properties t_j ($j = 1, 2, \ldots, n$), so that set T of their properties has at least the form: $T = \{t_0, t_1, t_2, \ldots, t_n\}$ ($n \leqslant N$). N is the number of all the properties of these concrete objects (cf. Sect. 1.1). Each of the properties t_j has a set θ_j of values f_j. As noted in Section 1.1, the concrete objects under consideration also possess the properties (combined) $t_{ij\ldots k} = (t_i, t_j, \ldots, t_k)$ (hereafter we assume that the subscripts $i, j, \ldots, k$ increase in that order, and $0 \leqslant i$). These properties have sets $\theta_{ij\ldots k}$ of values $f_{ij\ldots k} = (f_i, f_j, \ldots, f_k)$, where $f_i \in \theta_i$, $f_j \in \theta_j, \ldots, f_k \in \theta_k$.

Let us now choose some definite value $f_{ij\ldots k}$ of the property $t_{ij\ldots k}$ and single out in set A the concrete objects having this value of the indicated property. Then we obtain a definite subset of set A, which will be denoted by $A(f_{ij\ldots k})$. We will call this subset a class of concrete objects of A, which corresponds to the chosen value $f_{ij\ldots k}$ of the property $t_{ij\ldots k}$. We will subsequently bear in mind that $A(f_{0ij\ldots k}) = A(f_{ij\ldots k})$. Selecting in set A the concrete objects which have some other value $f'_{ij\ldots k}$ of the same property, we obtain another class $A(f'_{ij\ldots k})$ of concrete objects of A. Classes $A(f_{ij\ldots k})$ and $A(f'_{ij\ldots k})$ contain no concrete objects in common, since each concrete object has only one value of each of its properties and cannot have both the value $f_{ij\ldots k}$ and the value $f'_{ij\ldots k}$ of property $t_{ij\ldots k}$ simultaneously. On the other hand, each concrete object $a \in A$ unfailingly has one of the values $f_{ij\ldots k} \in \theta_{ij\ldots k}$. Therefore, selecting in set A concrete objects corresponding to each value $f_{ij\ldots k} \in \theta_{ij\ldots k}$, we obtain division of set A into classes $A(f_{ij\ldots k})$ of concrete objects. There is obviously a one-to-one correspondence between the set of these classes, which we denote by $A(t_{ij\ldots k})$ and the sets $\theta_{ij\ldots k}$ and $\theta_{0ij\ldots k}$, that is, to each class $A(f_{ij\ldots k})$ of concrete objects there corresponds, in the set $\theta_{0ij\ldots k}$ of values of the property $t_{0ij\ldots k}$, one definite element $f_{0ij\ldots k}$, which is called the image of the concrete objects of the class $A(f_{ij\ldots k}) \subseteq A$ in the set $\theta_{0ij\ldots k}$, and vice-versa, to each value $f_{0ij\ldots k} \in \theta_{0ij\ldots k}$ there corresponds one definite class $A(f_{ij\ldots k})$ of concrete objects in A. To the value f_0 of property t_0 there corresponds the class $A(f_0) = A$.

Since the class $A(f_{ij\ldots k})$ contains concrete objects of A, for which the property $t_{ij\ldots k}$ has the value $f_{ij\ldots k}$, that is, for which the property t_i has the value f_i, the property t_j the value $f_j, \ldots$, and the property t_k the value f_k, it consists of concrete objects common for the classes $A(f_i), A(f_j), \ldots$, and $A(f_k)$. Therefore we can write

$$A(f_{ij\ldots k}) = A(f_i) \cap A(f_j) \cap \ldots \cap A(f_k). \tag{1.1}$$

The following relation is evidently valid as well:

$$A(f_{ij\ldots k}) \subseteq A(f_l) \subseteq A \quad (l = i, j, \ldots, k; i \geqslant 0). \tag{1.2}$$

Further, since the class $A(f_{ij\ldots klm\ldots p})$ contains concrete objects of A, for which the property $t_{ij\ldots klm\ldots p} = (t_{ij\ldots k}, t_{lm\ldots p})$ has the value $f_{ij\ldots klm\ldots p} = (f_{ij\ldots k}, f_{lm\ldots p})$, the following relations hold true

$$A(f_{ij\ldots klm\ldots p}) \subset A(f_{ij\ldots k}) \tag{1.3}$$

$$A(f_{ij\ldots k}) = \bigcup_{\theta_{lm\ldots p}} A(f_{ij\ldots klm\ldots p}) \quad (0 \leqslant i < j < \cdots < k \leqslant n), \tag{1.4}$$

where $\theta_{lm\ldots p}$ is a set of values $f_{lm\ldots p} = (f_l, f_m, \ldots, f_p)$ of the property $t_{lm\ldots p}$; in particular,

$$A = \bigcup_{\theta_{ij\ldots k}} A(f_{ij\ldots k}) \tag{1.4a}$$

(for any $i, j, \ldots, k$).

Explanations. An expression of the type $B \cap C$ denotes the intersection of sets B and C, that is, a set consisting of all the elements belonging both to set B and set C. If two sets have no elements in common, their intersection is said to be empty, that is, it does not contain a single element. An empty set is often denoted by 0. Therefore the expression $B \cap C = 0$ means that sets B and C have no elements in common or, to put it differently, they do not intersect. From the foregoing it follows, for instance, that $A(f_{ij\ldots k}) \cap A(f'_{ij\ldots k}) = 0$. An expression of the type $B \subset A$ or $A \supset B$ means that B is a subset of set A (it is contained in set A, that is, all the elements of B are elements of A). If set B is a subset of A or coincides with it, this is recorded as $B \subseteq A$, or $A \supseteq B$. The expression $B \cup C$ means the union of the sets B and C, that is, a set, each element of which is contained in at least one of the sets B and C. For any number of sets B_k their union, which is denoted by $\cup_k B_k$, is a set, each element of which is contained in at least one of the sets B_k. Further on, we will have to use the concept of the difference of sets. The difference $A - B$ of the sets A and B is the set of all the elements of A which are not contained in B. If $B \subseteq A$, then $A - B$ is called the complement of set B (with respect to A).

Uniting all the concrete objects of A, which have one and the same value $f_{ij\ldots k} \in \theta_{ij\ldots k}$, into a single class, however, means abstracting oneself from the difference between the values of all the other properties of these concrete objects. Therefore, an abstract object corresponding to the class $A(f_{ij\ldots k})$ of concrete objects in A can be defined as the element $f_{0ij\ldots k}$ of set $\theta_{0ij\ldots k}$, that is, the image of the concrete objects of this class in set $\theta_{0ij\ldots k}$ of the values of the property $t_{0ij\ldots k}$. According to Eqn. (4a) set A can be divided into classes with respect to any property $t_{ij\ldots k}$ $(0 \leqslant i < j < \cdots < k \leqslant n)$, and since to each such class of concrete objects there corresponds an abstract object of its own – the image of the concrete objects of this class in the corresponding set $\theta_{0ij\ldots k}$, the general formal definition of the abstract objects is obtained from the following considerations. As noted above, there is a one-to-one correspondence between set $\theta_{0ij\ldots k}$ and set $A(t_{ij\ldots k})$ of the classes $A(f_{ij\ldots k})$. Therefore there also exists a one-to-one correspondence between set $\bigcup_{ij \ldots k} A(t_{ij \ldots k})$ $(i \geqslant 0)$ of all the classes of concrete objects of A and set $\theta = \bigcup_{ij \ldots k} \theta_{0ij \ldots k}$ $(i \geqslant 0)$ of all the values of all the properties (both basic and combined) of these concrete objects (note that $t_{00j \ldots k} = t_{0j \ldots k}$ and $\theta_{00j \ldots k} = \theta_{0j \ldots k}$). Each element of set θ is consequently the image of the concrete objects of a definite class, and it can be called (considering the method of transition to it from concrete objects) an abstract object corresponding to this class of concrete objects. Thus we obtain

DEFINITION 1. Abstract objects are elements of set θ.

This definition is the result of employing our conventional notions of what is abstracting oneself from some properties or other, which differentiate concrete objects, and therefore it fully corresponds to the conventional usage, as can be seen from the above examples. From the method of transition from concrete to abstract objects, it follows additionally that each abstract object also has definite values of some properties of concrete objects of A, namely those which figure in its denotation. Enumeration of these values fully determines the abstract object, not only in essence, but also in name (see above, Examples 1–3).

Abstract objects corresponding to different classes of concrete objects of A differ, generally speaking, in degree of their abstraction. Thus the abstract objects $f_{0ij \ldots k}$ are more abstract than $f_{0ij \ldots klm \ldots p}$. The least abstract objects are $f_{012 \ldots n}$, and the most abstract object is f_0 – the image of the whole set A of concrete objects, in other words, the abstract object corresponding to this whole set of concrete objects. In Example 1 the least abstract objects are: f_{012} = 'a young woman in this room'; f_{012} = 'an aged woman in this room'; and so on. The most abstract object in this example is: f_0 = 'a person in this room'. In Example 2 the least abstract object is, in particular, f_{012} = 'a large white ball in this urn', and the most abstract is, f_0 = 'a ball in this urn'. And so on. Recall that $f_{0ij \ldots k}$ denotes a single definite value of the property $t_{0ij \ldots k}$, and therefore a single definite abstract object. The identity of the abstract object is revealed by its name.

Transition from more abstract to less abstract objects consists of indicating the values of some other properties, which do not figure in the denomination of the initial abstract objects. One can therefore say that $f_{0ij \ldots klm \ldots p}$ is the abstract object $f_{0ij \ldots k}$ having the value $f_{lm \ldots p}$ of the property $t_{lm \ldots p}$; in particular, $f_{0ij \ldots klm \ldots p}$ is the abstract object f_0 having the value $f_{ij \ldots klm \ldots p}$ of the property $t_{ij \ldots klm \ldots p}$.

Note. None of the foregoing considerations will change essentially if we take some class $A(f_{ij \ldots k})$ of concrete objects of A in place of the initial set A. Then the property $t_{0ij \ldots k}$ of concrete objects of A is, for this class, that property t_0, whose unique value $f_0 = f_{0ij \ldots k}$ determines the new set $A(f_{ij \ldots k})$ of concrete objects and denotes the object which is the most abstract for this set. On the other hand, the initial set A itself may turn out (as is usually the case) to be a subset (class) of some broader set of concrete objects to which there corresponds a still more abstract object.

1.3. More About Concrete and Abstract Objects

An impatient reader may exclaim, 'How long can you harp on the same string? I am sick and tired of this empty philosophizing–concrete objects . . . abstract objects!! When will you get down to the brass tacks?' I can only retort politely, 'Don't fly into a rage, my dear reader, and don't skip over to the following parts of the book unless you are already well aware of the importance of the concepts of concrete and abstract objects. If you are really eager to understand what probability is and master the fundamentals of probabilistics (and this is the only way to get a clear view of probabilistic physics, and quantum physics in particular), do not move any further until you gain an insight into the relationship between concrete and abstract objects. Elucidation of this relationship and, in particular, realization of the absolute necessity of differentiation between concrete and abstract objects, is not empty philosophizing, but one of the most important principles of the study of nature and building up the various branches of science, and we will revert to this in the concluding, methodological part of the book. Meanwhile, here is a little more information on concrete and abstract objects.'

As has already been noted in Section 1.1, each of us (each concrete person) deals directly with concrete objects. Each of us eats concrete food (each time he does it). A concrete experimenter carries out concrete experiments with the aid of concrete devices on concrete objects and obtains concrete results each time. In a concrete laboratory, a concrete particle, colliding with a concrete screen, causes concrete scintillation observed by a concrete investigator. And so on.

At the same time, the general laws, which hold true for each concrete object – an element of a definite set of concrete objects – can be represented in the general form as laws valid for the corresponding abstract object. This lies at the basis of building up the various theoretical disciplines, which thus deal with abstract objects. The correctness of the theoretical conclusions following from the development of each theory is naturally checked by experimenting on the corresponding concrete objects. This well-known fact shows that the correlation of the properties and behavior of an abstract object and the related concrete objects, which actually exists on the strength of the definition of the abstract objects, is taken for granted (implicitly). But the necessity of strict differentiation of concrete and abstract objects is usually hardly realized clearly enough. At the same time, although confusion of a concrete object with an abstract one does not often cause misunderstanding, in some situations this confusion is entirely inadmissible, since it leads to paradoxes and absolutely erroneous conclusions, as will be seen below.

Therefore, in each particular case one must clearly understand whether concrete or abstract objects are meant. Without such clear understanding it is even impossible to formulate a problem correctly, still less to solve it properly. By way of example we quote the well-known question: What comes first, the chicken or the egg?

This question is usually considered insoluble – indeed, if we assume that the chicken comes first, there must be the egg from which it was hatched; so the egg comes first, but then it must be preceded by the chicken which hatched it, etc. The impression is that no solution can be obtained. In actual fact it is not so, the apparent paradox is due to the vague formulation of the problem, confusion of concrete and abstract objects. If the question refers to concrete chickens and eggs, the answer is unambiguous and very simple: each concrete chicken precedes the concrete egg it hatched, but succeeds the concrete egg it was hatched from. In the conventional formulation of this question it is usually implicitly assumed that it refers to an abstract chicken and an abstract egg, which is devoid of any sense and therefore leads to a paradox. Indeed, an abstract chicken is the image of all concrete chickens existing both some time ago and now, and also of those chickens which will exist in the future; thus, an abstract chicken cannot have a definite 'instant' value of the 'time' property, which is necessarily associated with the 'before' and 'after' concepts. The same is true of an abstract egg.

Another example. Let us pose the question: How long is an elephant's trunk? Thus formulated, the question is clearly senseless, since 'an elephant' is an abstract object corresponding to the whole set of concrete elephants, which lived before, are living now, and will live in the future, and in different countries to boot. Although an abstract elephant does have a trunk, since any concrete elephant has it, the trunk of an abstract elephant has no definite length, since concrete elephants have trunks of different lengths. The question of the length of an elephant's trunk becomes meaningful only when we speak of some concrete elephant and of a definite time of existence of this elephant.

Similarly, no sense is attached to the question: What is the speed of a train at this time instant? But this question would be quite meaningful if it referred to some concrete train, and not to an abstract object 'train'.

At the same time questions of, say, the average length of the elephant's trunk, or the average speed of the train, etc., are quite sensible. These questions are formulated for abstract objects 'elephant', 'train', etc., and to answer them one must know how many concrete elephants from among the whole set corresponding to the abstract elephant have a trunk of a given length, how many concrete trains from among the whole set corresponding to the abstract train move with a given speed etc. In other words, answering these questions requires a knowledge of what is called the trunk length distribution of concrete elephants, the speed distribution of concrete trains, etc.

Thus, when we speak of an abstract object $(f_{0ij \ldots k})$, there is no sense in the question: What are the values $(f_{lm \ldots p})$ of the properties $(t_{lm \ldots p})$ which do not figure in its name (its denotation)? because an abstract object has no definite values of these properties. At the same time it is quite sensible to ask a question about the distribution of a set of concrete objects corresponding to some abstract object with respect to the values of any property of these concrete objects. This and other related questions will be discussed in the following sections.

1.4. Measure of a Set of Concrete Objects

How many concrete objects of the whole set A have a definite value $f_{ij\ldots k}$ of the property $t_{ij\ldots k}$, that is, are contained in the class $A(f_{ij\ldots k})$? If we could answer this question, we would obtain, for set A, the distribution of the concrete objects with respect to the values of each of their properties, in other words the distribution of the concrete objects with respect to the corresponding abstract objects. In Example 1, in particular, this distribution was given from the very start, namely: to the abstract object f_{012} = 'a young woman in this room' there correspond 4 concrete objects (concrete persons); to the abstract object f_{012} = 'an aged woman in this room' there correspond 3 concrete objects, to the abstract object f_{012} = 'a young man in this room' 7 concrete objects, and to the abstract object f_{012} = 'an aged man in this room' 6 concrete objects. From the distribution of the persons in this room with respect to the values f_{12} of the property t_{12} it is easy to find their distribution with respect to the values f_1 of the property t_1 or with respect to the values f_2 of the property t_2. Indeed, simple arithmetic shows that to the abstract object f_{01} = 'a woman in this room' there correspond 4 + 3 = 7 concrete objects, and to the abstract object f_{01} = 'a man in this room', 7 + 6 = 13 concrete objects. It is equally easy to establish that to the abstract object f_{02} = 'a young person in this room' there correspond 4 + 7 = 11 concrete objects, and to the abstract object f_{02} = 'an aged person in this room', 3 + 6 = 9 concrete objects. To the abstract object f_0 = 'a person in this room' there correspond 7 + 13 = 20, or 11 + 9 = 20 concrete objects (concrete persons in this room), as should be the case. From consideration of Example 1 it is seen, incidentally, that it is sufficient to know the distribution of the concrete objects with respect to the least abstract objects in order to obtain their distribution with respect to any abstract objects corresponding to a given set A.

Example 2 contains no information as to the number of balls with different values of their property t_{12}. Therefore it is impossible at this juncture to give a direct answer to the question posed at the beginning of the section. It is clear, however, that the answer exists, since a definite urn must contain a definite number of balls of each size and color from among those indicated in this example, and therefore the question is posed correctly.

The situation is worse in Example 3. Indeed, there we deal with a set of points in a definite rectangle, inside a definite circle, etc. None of these sets, however, can be characterized by figures. But the question 'how many'? requires the answer 'so many', that is, an indication of some definite number, which a priori cannot be done in this example. Therefore we must consider that for this example the question posed at the beginning of the section is formulated incorrectly, and it should be restated. To do this, it is necessary, first of all, to introduce such an appropriate quantitative characteristic of a set of concrete objects which would also be suitable for nonenumerable sets of such objects. A measure of a set, defined in some way or other, answers this requirement. In what follows, the measure of set A will be denoted by mA. In the example at hand, for the measure of a set of points of any part of the rectangle we can choose the total area of this part equal to the sum of the areas of all the domains of the rectangle determined by their envelopes, of which this part of the rectangle consists.

Similarly, for the measure of a set of all the points on a line segment we may take the length of this segment; for the measure of a set of all the points in some continuous domain of a three-dimensional space, the volume of this domain, etc. For the measure of a set of concrete points in some continuous domain of an n-dimensional Euclidean space we can thus take the n-dimensional volume of this domain. We may, however, come across sets of concrete points of a more complex structure than those mentioned so far. Among them is, for instance, a set of points of a three-dimensional space whose Cartesian coordinates are irrational. Therefore, in a more general case one should use a more general measure, namely, the so-called Lebesgue measure, which is suitable for quantitative characterization of sets of concrete points, not only those fully occupying continuous domains of the n-dimensional Euclidean space, but also more complex ones, in particular those of the type just indicated. The rigorous definition of the Lebesgue measure can be found in the appropriate math text-books (see Suggested References at the end of the chapter). We, however, are interested at the moment exclusively in the fact that the Lebesgue measure, which is concisely designated as the measure (L), for each continuous domain of the n-dimensional Euclidean space (often denoted by R_n), coincides with the volume of this domain, and for a finite or denumerable set of concrete objects (corresponding to zero-dimensional space) can be taken equal to the number of objects in this set; the properties of the measure (L) are the same as for the volumes of continuous domains of the space R_n, namely: (1) the measure (L) is nonnegative, that is, $mA \geqslant 0$ for any A; (2) the measure (L) of a union of non-intersecting sets is equal to the sum of measures (L) of these sets – $m \cup_j A_j = \Sigma_j mA_j$, if $A_j \cap A_k = 0$ $(j \neq k)$.

Sets which have a measure (L) are called measurable (L). Hereafter we will deal exclusively with such sets.

We can now re-state the question posed at the beginning of the section, namely: What is the measure (L) of a set of concrete objects of A which have a definite value $f_{ij\ldots k}$ of the property $t_{ij\ldots k}$? This formulation is evidently correct for Example 3 as well. Answers to the question for this example have the following formal pattern: $mA(f_1$ = inside circle 1) = area of circle 1; $mA(f_1$ = on circle 1) = area of circumference 1 = 0; etc. (See note at the end of this section.)

In the general case of a measurable (L) set A of concrete objects, with due allowance for the properties of the measure (L), from Eqn. (3) follows

$$mA(f_{ij\ldots klm\ldots p}) < mA(f_{ij\ldots k}) \tag{1.5}$$

and since sets $A(f_{ij\ldots klm\ldots p})$ corresponding to different values of $f_{lm\ldots p}$ do not intersect, Eqn. (4) leads to

$$mA(f_{ij\ldots k}) = \sum_{\theta_{lm\ldots p}} mA(f_{ij\ldots klm\ldots p}) \quad (i \geqslant 0). \tag{1.6}$$

An important particular case of equality Eqn. (6) is the following equality, which stems from Eqn. (4a) for $f_{ij\ldots k} = f_{12\ldots n}$,

$$mA = \sum_{\theta_{12\ldots n}} mA(f_{12\ldots n}). \tag{1.6a}$$

To the equality (6a) there correspond: in Example 1, the equality of the number of all the persons in this room to the sum of the numbers of the persons in each of the four groups into which they were divided; in Example 2, the equality of the number of all the balls in this urn to the sum of the numbers of balls which have different values of at least one of the properties indicated there ('size', 'color'); in Example 3, the equality of the area of the entire given rectangle to the sum of the areas of both circles (1 and 2) and of the remaining part of the rectangle (the areas of the two circumferences can be neglected, because they are equal to zero).

Note. A strict distinction must be made between the measure (L) of some set of concrete objects in itself, that is, in the space of the same number of dimensions as for the set under consideration, and the measure (L) of the same set in the space of a larger number of dimensions. Thus, the measure (L) of any finite or denumerable set in itself is equal to the number of elements of this set. At the same time the measure (L) of an identical set of points in a one-dimensional space (R_1) is equal to zero, since the length is the measure (L) of a set of points in this space, and the length of a finite, or even a denumerable set of points is equal to zero. Similarly, in a two-dimensional space (R_2) the area is the measure (L) of a set of points, and therefore the measure of a finite, denumerable or one-dimensional set of points in this space is equal to zero, etc. We can thus assert that in an n-dimensional space (R_n) the measure (L) of any k-dimensional set is equal to zero, provided $k < n$. A set whose measure is equal to zero is called a zero-measure set. It turns out that a union of even a denumerable set of zero-measure sets is also a zero-measure set. Therefore in the equality (6) we can omit summation over all the values $f_{lm\ldots p} \in \theta_{lm\ldots p}$, for which $mA(f_{ij\ldots klm\ldots p}) = 0$.

1.5. Experimental Determination of the Measures of Sets of Concrete Objects

How can one find the measures of different subsets of some set of concrete objects? The answer to this question is quite simple if the set is specified well enough. Then these measures can be found by simple calculation, thus, in Example 1 we are given the number of persons in the room contained in each of the four groups corresponding to different values of the property t_{12} = (sex, age). Therefore, the number of persons in any union of these groups can be calculated directly by Eqn. (6). If we knew, in Example 2, the number of balls in the urn which have different values of the property t_{12} = (size, color), it would be a cinch to calculate the measure of any subset of balls in this urn by using Eqn. (6). Similarly, if we knew, in Example 3, the dimensions of the rectangle and the radii of both circles, simple calculations with the use of Eqn. (6) would enable us to find the measure of any subset of points of this rectangle, provided this subset is a union of the classes of points of the rectangle which correspond to different values f_1 of the property t_1. If we are interested, say, in the measure of the set of the points of the rectangle which are outside circles 1 and 2, we must, according to Eqn. (6), deduct the sum of the areas of both circles (1 and 2) from the area of the rectangle.

All this looks trivial enough. Of less triviality, however, is the question of finding the measures of different subsets of a set of concrete objects by experimenting. First, it may happen that the question can be resolved in more ways than one; secondly, in resolving this question, problems of various degrees of difficulty may arise; thirdly, for most sets of concrete objects this question is altogether insoluble, in principle, when thus stated.

In Example 1 the experimenter may obtain the answer to this question, say, by direct observation if he is in the same room, or by questioning over the telephone, etc. In Example 2 the experimenter may find the number of balls in the subsets he is interested in by direct counting, spilling them out, on the table for instance, or withdrawing them one by one, etc. In Example 3, it will suffice, say, to measure the lengths of the sides of the rectangle and the radii of both circles in order to find the measures of any subsets (of the above-indicated type) of the set of points of this rectangle.

In this example one can, however, make use of another technique for finding the measures of these subsets experimentally. Let us divide the rectangle into N_1N_2 identical rectangles with the aid of straight lines parallel to its sides. By calculating the number of the little rectangles which completely or almost completely fit inside each domain of the initial rectangle, we obtain approximate measures of all these domains – true enough, these will not be absolute, but relative measures (with an accuracy to a factor which is the area of a scaled-down rectangle). The smaller the area of the scaled-down rectangle, that is, the larger are N_1 and N_2, the more accurate is the result obtained. To determine the measures we are interested in, it will now suffice to find the area of the scaled-down rectangle by measuring its sides and to multiply it by the number of these little rectangles in each domain of interest. If we were interested, not in the measures, but just in their ratios, we could do without measuring the area of the scaled-down rectangle.

The above examples were simple enough as regards resolving the question posed. It is, however, possible to quote an indefinite number of examples where resolving this question is either extremely difficult or even entirely impossible in principle. How, for instance, can we determine experimentally the population of some concrete country and the measures of the sets of persons in it which differ in the values of age, sex, etc., at some definite time instant? The solution of this problem evidently (1) involves great difficulties (they are greater, the larger the country and its population) and (2) cannot, in principle, be obtained with an accuracy, say, to one person, despite the fact that here we deal with a finite set of concrete objects, citizens of a given country. Experimental determination, at a certain given time instant, of the measures of subsets of concrete molecules, whose velocities lie within given ranges, from among the whole set contained in a concrete gas sample, which fills an enclosed vessel, appears (at least now) entirely impossible, in principle, although here, too, we deal with a finite set of concrete objects.

Thus, the problem of experimental determination of the measures of different subsets of some set of concrete objects is by no means always soluble. The problem of experimental determination of the ratios between these measures is simpler, however, as follows from the above consideration of Example 3. When the initial rectangle is divided into N_1N_2 identical rectangles with the aid of straight lines parallel to its sides, the ratios between the numbers of these small rectangles, which completely or almost completely fit into the domains of interest, are evidently approximately equal to the ratios between the measures of these domains, and with an increase in N_1 and N_2 the former ratios progressively approach the latter. The technique used here cannot, of course, result in some universal method for experimental determination of the ratios between the measures of different subsets for any set of concrete objects. There is, however, a method for experimental determination of the ratios between the measures of different subsets, which is suitable, in principle, for an extremely broad class of sets of concrete objects and can therefore be considered universal enough. We are now proceeding to consider this method, which can be called the statistical method.

1.6. The Statistical Method

Before we go any further it should be emphasized that the statistical method enables one to experimentally determine ratios between the measures of subsets of some set of concrete objects only approximately. The essence of the method is that a set A of concrete objects is subjected to a great number of random tests, that is, tests by which some concrete object belonging to A is revealed accidentally. Then a statement is used, which is taken as the primary principal premise of probabilistics, and which can be formulated as follows:

AXIOM 1. If we have a very large number of random tests, the number of tests revealing concrete objects of each subset of the set A is approximately proportional to the measure of this subset.

This statement, which is seemingly self-evident, is actually based on a huge body of statistical data at the disposal of experimental statistics, and therefore must be considered substantiated by statistical experiment. Of course, for Axiom 1 to be applicable, the tests must actually be of a random nature, that is, they must actually lead to accidental results. Indeed, if the tests were carried out so that some subsets of A were given preference over the others, the approximate proportionality between the measures of different subsets and the numbers of tests revealing concrete objects of these subsets would be disturbed. Therefore the main condition for the application of the statistical method is ensuring the randomness of the tests.

Thus, it is precisely the randomness of the tests that is essential to the statistical method, since it ensures the possibility of experimental determination of the sought-for ratios between the measures. At the same time these ratios themselves are quite definite quantitative characteristics of the structure of the set A of concrete objects subjected to the tests, and bear no relation to randomness.

What is 'randomness' after all? Without going into the philosophic niceties of this question, it should be stressed here that 'randomness' is introduced into probabilistics as a primary concept, which is not to be defined, but which has the most usual, trivial, rather everyday-life meaning of something which is not predetermined, which may or may not happen. Since the 'randomness' concept is not defined, neither is the 'random test' concept. It is, however, possible to indicate in each particular case what techniques should be used to ensure the randomness of the tests and hence the applicability of the statistical method.

Thus, in Example 2 the statistical method can be realized, for instance, by repeated withdrawal of a single ball from the urn and placing it back (after each withdrawal). To ensure the randomness of the appearance of each ball, the balls must, of course, be thoroughly mixed prior to each withdrawal. (Thorough mixing of the balls is precisely the technique ensuring the randomness of the tests in this example.) Then, on the strength of Axiom 1, with a sufficiently large number of tests (ball withdrawals) the number of appearances of a ball with definite values of the properties is approximately proportional to the measure of the corresponding subset of balls in the urn.

In Example 3, to realize the statistical method it is necessary to have some device which would ensure accidental revelation of sufficiently small parts of the rectangle so that none of them has any advantages over the others. Many such devices can be proposed. Suppose, for instance, that the part of the rectangle which is located outside the circles is made of wood, and the circles themselves, of different metals. Then one can use, say, a sufficiently small steel bead freely falling onto a horizontally positioned plane of the rectangle (from a sufficiently large height and from an unfixed position in space), and determine by ear which parts of the rectangle the bead hits. If the number of tests is sufficiently large, the number of strokes of each tone is, on the basis of Axiom 1, approximately proportional to the areas of the corresponding domains of the rectangle, that is, to the measures of the corresponding subsets of the points of the rectangle.

In these two examples the random tests were made directly by the experimenter. In Example 1 the statistical method can be applied by using, say, the following technique, which allows for the fact that the experimenter deals with living persons, who also take part in the realization of this experiment. The experimenter is outside the room and communicates with them by telephone. On each of the experimenter's calls the phone is taken by one of the persons in the room, precisely the one who is to take the phone according to the result of any appropriate random sampling. The experimenter enquires only about the sex and age of the speaker (a man, a woman, young, aged). With a sufficiently large number of such calls the experimenter can, by using Axiom 1 and on the basis of the answers received, determine approximately the ratios between the numbers of persons in the room who have different values of the properties he is interested in (provided, of course, that the answers are true). In this example the randomness of the tests is ensured by the random sampling of the person taking the phone. Naturally, the indicated way of realization of the statistical method in this example is not unique, either, and those who wish can propose various other techniques. But, as an example of how the statistical method can be realized in practice, it is of some interest, being something of a model of correct 'population interviewing'.

In the examples discussed it was necessary to resort to artificial measures in order to ensure the randomness of the tests. But whenever an experimenter applies the statistical method in studying natural phenomena, Nature itself 'takes care' to ensure the randomness of the tests. In these cases the role of the experimenter, who may not even clearly realize that he is dealing with the statistical method (with a statistical experiment), amounts merely to arranging the experiment and recording its results (by observing, measuring, etc.), while the accidental character of these results is ensured by Nature itself, as indicated above.

An example of this is experimental determination of the molecular velocity distribution of some gas in an enclosed vessel at a definite temperature. No matter how the experiment is arranged, the randomness of the value of the velocity of a molecule escaping from the enclosed vessel, say, through a narrow slit, is ensured by the randomness of the effects exerted on this molecule by a practically infinite number of accidental causes. Indeed, the walls of any indefinitely hermetic vessels, in which the gas may be placed, themselves consist of molecules, and therefore, even if these walls do not directly pass accidental foreign particles (although in fact this is by no means the case), they transmit their effect; the motion of gas molecules inside any hermetically sealed vessel changes, in particular, on their collisions with the molecules of which the vessel walls consist, and these changes depend on the motion of the latter (molecules), which, in turn, changes accidentally as a result of collisions with random particles arriving from any parts of the Universe.

In some cases the experimenter has at his disposal the whole set A of concrete objects at a time, and the whole set is investigated by the statistical method. It is, however, possible to quote an indefinite number of examples where the experimenter deals with separate concrete objects, and not with the entire set, which is inaccessible to him as a whole. This is the case, for instance, in cosmic ray investigations. But here, too, there is not the shadow of a doubt that the separate concrete objects accidentally revealed in experiment belong to some set A determined by some value of some one of their properties. These sets A, however, may (depending on the conditions of the statistical experiment) prove to be different subsets of one and the same broader set of concrete objects.

Note. Since the statistical method enables one to determine approximately only the ratios between the measures of different subsets of some set A of concrete objects, it can, in principle, be modified as follows. Assume that there are N

independent identical sets A and each of them can be subjected to random tests. Because of the independence of these sets the results of the random tests they are subjected to are also independent. Therefore the statistical data obtained from a sufficiently large total number of random tests made on all the N identical sets A must approximately coincide with those obtained from a sufficiently large number of random tests made on any separate set A. It follows from this that if the number N is very large, it will suffice to make even a single random test on each of N identical sets A to obtain approximately the sought-for ratios between the measures.

1.7. Probability. Preliminary Consideration

When using the statistical method one cannot, of course, predict the result of each separate random test. But if the ratio of the measure (mA') of some subset $A' \subset A$ to the measure (mA) of set A is close to unity, we can assume with considerable confidence that a single random test will reveal precisely some one of the concrete objects $a \in A'$, since in this case, even with a very large number of random tests, only a small number of them can, on the strength of Axiom 1, reveal concrete objects of the subset $A - A'$. The larger the ratio mA'/mA, the greater is the confidence that the above assumption is correct; the less this ratio, the less is the confidence. These considerations bring us right to the definition of probability as the degree of confidence in the correctness of the assumption that a single random test made on a set A of concrete objects will reveal some concrete object of the subset $A' \subset A$, and it is natural to estimate the probability understood in this sense by a number equal to mA'/mA.

But with a sufficiently large number of random tests the frequency of revealing concrete objects of A', that is, the ratio of the number of the tests revealing concrete objects of A' to the number of all the tests carried out, must be approximately equal just to mA'/mA as follows from Axiom 1. Therefore probability, as understood in the indicated sense, approximately coincides numerically with the frquency of revealing the corresponding concrete objects (provided that the number of random tests is sufficiently large). For this reason many people actually understand by probability the frequency of revealing concrete objects of the corresponding subset of the entire set in a very large number of random tests (so-called statistical probability).

The question of probability arises when there is no complete confidence as to what some 'accidental result' will be (that is how we will sometimes term, for brevity, the result of a random test, irrespective of how it is realized – the participation of the observer, the experimenter, in obtaining accidental results is not mandatory at all). The statement of the question in the following form is typical: What is the probability of this or that? Hence, one always speaks of the probability of something. But what does this question refer to? The probability of what is implied? To be able to answer these questions we will consider some examples of the statement of the probability question.

In Example 2 one can, say, pose the question: What is the probability that a ball withdrawn at random from a given urn will be white? Here the term probability evidently again implies the degree of confidence that the indicated accidental result will occur. But the urn can yield exclusively what it contains – only a ball located in it can be withdrawn (further on we will always keep in mind this circumstance, namely that only a concrete object which is an element of the set being tested can be revealed). Therefore the question can be re-stated as follows: What is the probability that a ball in this urn is white? But this question clearly refers to 'a ball in this urn', that is, to an abstract object corresponding to the entire set of concrete balls in this urn; it concerns the probability that for this abstract object the 'color' property has the value 'white'. Similarly, one can pose the question: What is the probability that a black ball in this urn is large? Then the question refers to 'a black ball in this urn', that is, to an abstract object corresponding to a class of balls in this urn, for which the 'color' property has the value 'black'; it concerns the probability that for this abstract object the 'size' property has the value 'large'. And so on.

In Example 1 we can pose, say, the question: What is the probability that a person in this room is a man? This question obviously refers to an abstract object 'a person in this room'; it concerns the probability that for this abstract object the 'sex' property has the value 'man'. And so on, and so forth.

In Example 3 we can pose, say, the question: What is the probability that a point of this rectangle lies inside the circles? The question refers to an abstract object 'a point of this rectangle'; it concerns the probability that for this abstract object the 'position in space' property has the value 'inside circle 1', or 'inside circle 2'. And so on.

From the examples just considered it follows (and is confirmed by any other examples) that the question of probability can always be referred to some abstract object corresponding to some set (class) of concrete objects, and one can enquire about the probability that for this abstract object some property has some definite set of values. In its usual formulation the question implies enquiry about the probability that a random test will reveal that the values of this property are contained in some definite set, that is, the question concerns the probability of an accidental result. But with either formulation of the question the probability must evidently be expressed by one and the same number. Therefore the probability that for an abstract object corresponding to the whole set A of concrete objects some one of its properties has some definite set of values corresponding to a subset $A' \subset A$ of these concrete objects must also be equal to mA'/mA, that is, to the relative measure of the subset A'. This circumstance lies at the basis of the formal definition of probability as an objective quantitative characteristic of the structure of the set A of concrete objects.

At this point the reader may experience various doubts. Some people may ask: 'Is it permissible to use the term "probability" to denote any objective value, while the word "probability" itself obviously contains an intrinsic meaning of a degree of confidence (or doubt), which has a shade of subjectivity? . . .' (Webster: '*probability* (1) likelihood; chance stronger than possibility but falling short of certainty.') Others may ask: 'Is it necessary to define probability at all? The contemporary, axiomatic, probability theory (In the following text we will simply call it the ordinary probability theory) does without answering this question, it is based on the known properties of probability.'

I will have to answer the second question only briefly, for the time being: It is precisely the lack of a direct definition of probability that leads to the incompleteness of the ordinary probability theory – it is constructed as a purely mathematical discipline, which does not require the knowledge of the object in question. Here, we present the fundamentals of a concrete science – probabilistics (science of probability), in which, as in any other concrete science, the concrete subject of its study must necessarily be defined.

This problem is discussed in more detail in the concluding, methodological part of the book. As for the first question, I can answer it as follows: Yes, the term 'probability', as applied to an objective quantity, does sound rather inappropriate. It should have been replaced here by the correct term 'the relative measure of a subset of concrete objects'. But the word 'probability' is so firmly entrenched in the literature that we also have to use it out of respect for the tradition, always bearing in mind its new objective content, however.

Proceeding to the formal definition of probability, we will start with consideration of subsets $A' \subseteq A$ of concrete objects, which we mentioned before. Any such subset consists of concrete objects, which are, in the general case, elements of a union of some classes, or, more precisely, elements of a union, in some broader class of concrete objects, of some narrower classes. Thus, in Example 2, if subset A' consists of large and small white balls, the concrete balls of this subset are included in the union of classes of large white balls and small white balls. It can, therefore, be represented as a union, in the white-ball class, of the classes of these balls with different values of the 'size' property: 'large' and 'small'. In this example the union obtained coincides with the whole class of white balls, since the 'size' property has only the indicated two values. In the general case, however, such a union is merely a subset of a class, as would be the case if, for instance, the 'size' property had at least one more value, say 'medium'.

Thus, in the general case we deal with subsets of the type

$$A'(f_{ij\ldots k};\theta'_{lm\ldots p}) = \bigcup_{\theta'_{lm\ldots p}} A(f_{ij\ldots klm\ldots p}) \tag{1.7}$$

consisting of the concrete objects contained in the class $A(f_{ij\ldots k})$, for which the property $t_{lm\ldots p}$ has a set $\theta'_{lm\ldots p} \subseteq \theta_{lm\ldots p}$ of values $f_{lm\ldots p}$. To each class $A(f_{ij\ldots klm\ldots p})$ of concrete objects there corresponds an abstract object $f_{0ij\ldots klm\ldots p}$. Therefore, to the set $A'(f_{ij\ldots k};\ \theta'_{lm\ldots p})$ there corresponds a set of abstract objects, which will be designated as $f_{0ij\ldots k}(\theta'_{lm\ldots p})$. In particular, to the set $A' = A(f_0;\ \theta'_{12\ldots n})$ of concrete objects there corresponds a set of abstract objects $f_0(\theta'_{12\ldots n}) = \theta'_{012\ldots n}$.

Recall now that $f_{0ij\ldots klm\ldots p}$ is an abstract object $f_{0ij\ldots k}$, which has the value $f_{lm\ldots p}$ of the property $t_{lm\ldots p}$ (see end of Sect. 1.2). Therefore

$mA'(f_{ij\ldots k};\theta'_{lm\ldots p})$ is the measure of a set of concrete objects corresponding to the abstract object $f_{0ij\ldots k}$, which has a set $\theta'_{lm\ldots p}$ of values $f_{lm\ldots p}$ of the property $t_{lm\ldots p}$, and $mA(f_{ij\ldots k})$ is the measure of the whole class of concrete objects corresponding to the abstract object $f_{0ij\ldots k}$ itself. In particular, $mA' = mA'(f_0;\theta'_{12\ldots n})$ is the measure of a set of concrete objects corresponding to the abstract object f_0, which has a set $\theta'_{12\ldots n}$ of values of the property $t_{12\ldots n}$, while mA is the measure of the whole set of concrete objects corresponding to the abstract object f_0 itself. Further, according to the note at the end of Section 1.2, any class $A(f_{ij\ldots k})$ can be taken as a new initial set of concrete objects. All this considered, probability (as an objective quantitative characteristic of the structure of a set of concrete objects) can be defined as follows.

DEFINITION 2. By the probability that for an abstract object $f_{0ij\ldots k}$ a value $f_{lm\ldots p}$ of the property $t_{lm\ldots p}$ belongs to a set $\theta'_{lm\ldots p} \subseteq \theta_{lm\ldots p}$ is meant the ratio of the measure of the set $A'(f_{ij\ldots k};\theta'_{lm\ldots p})$ to the measure of the set $A(f_{ij\ldots k})$.

According to this definition probability refers to abstract objects and bears no relation to chance. But, as noted above, statement of probability problems is meaningful only in connection with the possibility of accidental revelation, as a result of a random test, of some concrete object in some subset, that is, in connection with statistical manifestations of probability. (We emphasize once again that realization of a random test by no means requires mandatory participation of an observer, experimenter.)

On the other hand, the statistical method can be considered as a method for direct measurement of probabilities. On the strength of Axiom 1 the frequency of revelation of concrete objects of $A' \subseteq A$ (that is, the ratio of the number of random tests revealing concrete objects of $A' \subseteq A$ to the number of all random tests made on a set A, provided this number is large enough) is approximately equal to the corresponding probability. Thus, the frequency must be considered the measured value of the probability. The entire set of random tests can then be regarded as a single measurement of the probabilities corresponding to all subsets $A' \subseteq A$ of concrete obejcts. On the strength of the note at the end of the preceding section this single measurement can also be realized in a different way, namely by carrying out only one random test on each of a sufficiently large number of identical sets A of concrete objects.

Definition 2 is evidently suitable only for cases where the set A of concrete objects is given, and it is possible to introduce a measure (L) into it. But in most cases with which probabilistics is concerned, none of these assumptions is fulfilled, although the set T of properties (together with their values), and hence the set θ of abstract objects, is given each time. For all such cases Definition 2 is clearly unsuitable, and it must be appropriately modified, which will be done later on by using the concept of mutual statistical adequacy of sets of concrete objects.

1.8. Mutually Adequate Sets

Let us consider the problem of the probability of a coin turning up 'head' on a single toss. What should be considered concrete objects here? To be able to answer this question we must recall how a random experiment is made in this case. A coin (some concrete coin, of course) is tossed. Landing on some smooth horizontal surface (the floor or a table) the coin may show the picture on its face ('head') or the denomination on its back ('tail'). Since a single random test on the coin reveals (opens up) its upper side, it is the upper sides of coins that must be considered concrete objects. But (and this is essential) the set A of these concrete objects is not given, since, though each concrete coin as such has two sides, neither of them is a priori the upper or the lower one; the sides can be classified only as a result of tests. The property t_1 is the 'image', which has two values: 'head' and 'tail'. The abstract objects are: f_{01} = 'the upper side of the coin with the head'; f_{01} = 'the upper side of the coin with the tail'.

If the coin on which random tests of the indicated type are made can be considered 'good', then for set A of concrete objects one could choose a set consisting of two concrete objects: the upper side of the coin the 'head'; the upper side of the coin the 'tail'. In the general case set A in this problem depends both on the specific properties of the coin itself and on the conditions of the random tests. If, for instance, random tests of a coin of nonuniform density are made by tossing in the presence of an atmosphere and gravitational forces, and in their absence, the sets A for these two cases will be different. If the coin is magnetized, set A will depend on whether the random tests are made in the presence of a magnetic field or in its absence. And so on. This is associated with the fact that the coin is actually only part of the physical system subjected to random tests, and these physical systems are different in the different cases considered.

It may also happen (as indicated above) that set A is given, but the measure (L) cannot be introduced directly into this set. This is the case, for instance, with a set of chords in a given circle or with a set of concentric circles within the same circle. In more general terms, this is surely the case when the necessary condition of the possibility of introducing the measure (L) is not fulfilled; this condition can be formulated as follows: the measure (L) can be introduced only into sets which are subsets of points of an n-dimensional Euclidean space (R_n).

In all cases where the measure (L) cannot be introduced into set A (in particular, when it is not given), two alternatives are possible: either the statistical method is applicable (that is, this set of concrete objects can be subjected to random tests) or it is inapplicable in principle. In the latter case, statement of probabilistic problems hardly makes any sense at all. In the former case, the initial set A under test can be replaced by a (statistically) adequate concrete set $A^{(a)}$. Before proceeding to the formal definition of such a set we will consider this question in more detail.

According to the note at the end of Section 1.1 any set of values (even qualitative ones) of some property of concrete objects of A can be replaced by a corresponding set of numbers, and in an infinite number of ways. Moreover, the initial concrete objects themselves can also be replaced by some other concrete objects, even those possessing different properties. On any such replacement one obtains, in place of the initial set of concrete objects, a set of some other concrete objects. But under certain conditions all these sets may prove to be statistically adequate to the initial set, and therefore to each other as well. What does this mean? Assume that on such replacement the set of values of each initial property goes over to an equivalent set of values of the property which corresponds to this property in the new set of concrete objects; then the initial set of abstract objects also goes over to an equivalent set of new abstract objects. If the frequencies corresponding to the initial abstract objects are approximately equal to frequencies, and hence to probabilities (in the sense of Definition 2) corresponding to the equivalent new abstract objects (of course, we assume that this new set of concrete objects is measurable (L)), then such a new set will be considered adequate to the initial one.

EXPLANATIONS. Two sets are called equivalent if it is possible to establish one-to-one correspondence between their elements. The elements of these sets which correspond to each other will also be called equivalent.

From the foregoing it follows, first of all, that for each set of concrete objects, a collection of any number of identical sets is adequate to it. Therefore, in Example 1, to the set of persons in the given room is adequate a set of $20N$ persons, where N is any whole number, if $4N$ of them are young women, $3N$ aged women, $7N$ young men, and $6N$ aged men. Further, to this set is adequate any set of $20N$ (N being an arbitrary whole number) concrete objects possessing at least some two properties t_1 and t_2, each of which has two values ($\theta_1 = \{f_1', f_1''\}; \theta_2 = \{f_2', f_2''\}$), if the distribution of these concrete objects by classes has the form: in class $A(f_{12} = (f_1', f_2')) - 4N$, in class $A(f_{12} = (f_1', f_2'')) - 3N$, in class $A(f_{12} = (f_1'', f_2')) - 7N$, and in class $A(f_{12} = (f_1'', f_2'')) - 6N$ concrete objects. Similarly one can obtain mutually adequate sets in Examples 2 and 3. In particular, the set of balls in Example 2 may (if the above conditions are fulfilled) prove adequate to the set of persons in Example 1.

We will now give a formal definition of a set $A^{(a)}$ adequate to some set A of concrete objects.

DEFINITION 3. Let some set $T^{(a)}$ of properties of concrete objects, elements of a measurable (L) set $A^{(a)}$ be equivalent to a set T of properties of concrete objects, elements of a set A. Let, further, for any $i, j, \ldots, k$, a set $\theta_{ij\ldots k}^{(a)}$ of values $f_{ij\ldots k}^{(a)}$ of the property $t_{ij\ldots k}^{(a)}$ be equivalent to a set $\theta_{ij\ldots k}$ of values $f_{ij\ldots k}$ of the property $t_{ij\ldots k}$. Let, finally, for any values of all the suffixes, the ratio of the frequency corresponding to the abstract object $f_{0ij\ldots klm\ldots p}^{(a)}$ to that corresponding to the abstract object $f_{0ij\ldots k}^{(a)}$, and hence the ratio of the measure of the class $A^{(a)}(f_{ij\ldots klm\ldots p}^{(a)}) \subseteq A^{(a)}$ to the measure of the class $A^{(a)}(f_{ij\ldots k}^{(a)}) \subseteq A^{(a)}$ be approximately equal to the ratio of the frequency corresponding to the abstract object $f_{0ij\ldots klm\ldots p}$ to the frequency corresponding to the abstract object

$f_{0ij\ldots k}$. Then we will say that set $A^{(a)}$ is adequate to set A. If set $A^{(a)}$ is adequate to sets A_1 and A_2, all these sets are mutually adequate.

The definition given differs somewhat from the initial one stated by the author in [5]. Note that the requirement of measurability (L) of set $A^{(a)}$ is necessary only to the extent that we wish to make use of this set later on to arrive at the general definition of probability. Although in Definition 3 we refer to frequencies, in many cases a set $A^{(a)}$ can readily be found from theoretical reasoning. Consider, for instance, a set A of concrete chords in a given circle of radius R, choosing 'the length' as the property t_1. The set θ_1 of the values $f_1 \equiv l$ of this property is nonnumerable and has the form $\theta_1 = [0, 2R]$. Since a set of chords is not a set of points in some Euclidean space R_n, the measure (L) cannot, a priori, be introduced into it. But a set $A^{(a)}$, which is adequate to it, can readily be constructed, using the following considerations.

Suppose that in performing a random test on a given circle in some way or other we reveal some concrete chord and measure its length. How can we identify this chord? This can be done in different ways. The simplest one is to identify the chord by the position of the point in the circle which bisects the chord. Each concrete point in the circle, except its center, uniquely determines the concrete chord whose midpoint it is, and vice versa, each concrete chord uniquely determines the concrete point in the circle which bisects the chord. A random test revealing some concrete chord simultaneously reveals the corresponding point in the circle – the midpoint of the chord. Consequently, the frequency of revealing chords whose lengths fall within a certain range of their values coincides with the frequency of revealing the points in the circle which bisect these chords and is approximately equal to the relative measure of the corresponding subset of these points.

Therefore set $A^{(a)}$ of concrete points in a given circle is adequate to set A of concrete chords in it. The length l of each chord is uniquely related to the distance r from its midpoint to the center of the circle by the expression

$$l = 2(R^2 - r^2)^{1/2}. \tag{1.8}$$

The set $\theta_1^{(a)} = (R, 0)$ is equivalent to set $\theta_1 = (0, 2R)$. An abstract object $f_{01}^{(a)}$ = 'a point of a given circle at a distance r from its center' is equivalent to an abstract object f_{01} = 'a chord of length l in this circle'. The measure of the subset $A^{(a)'}(f_0^{(a)}; \theta_1^{(a)'} = (r, r + dr))$ of the points of the circle r to $r + dr$ distant from its center is equal to $2\pi r\, dr$. The relative measure, $mA^{(a)'}/mA^{(a)}$, of this subset is equal to $2R^{-2} r\, dr$, due account being taken of $mA^{(a)} = \pi R^2$. Further, from (8) it follows that $|r\, dr| = 4^{-1}|l\, dl|$. Therefore the frequency of revealing chords whose lengths fall within the range from l to $l + dl$ is approximately equal to $2^{-1}R^{-2}|l\, dl|$.

Thus we have found one set $A^{(a)}$ adequate to the set of concrete chords in the given circle – the set of concrete points in this circle – and determined the approximate value of the frequency corresponding to the set $f_0(\theta_1' = (l, l + dl))$ of abstract objects – 'chords of given length'. Proceeding from this we can construct other sets $A^{(a)}$ adequate to the set of concrete chords in the given circle, which are characterized by the single property, 'the length'. If some variable q is related to l one-to-one by expressions $q = f(l)$ and $l = \varphi(q)$, then

$$l \, dl = \varphi(q)\varphi'(q) \, dq. \tag{1.9}$$

A set of points on a plane, whose Cartesian coordinates are q and ρ and which is bounded by the abscissa $\rho = 0$, a curve $\rho(q) = 2^{-1}R^{-2}|\varphi(q)\varphi'(q)|$, and straight lines $q = f(0)$ and $q = f(2R)$, is also adequate to the set of chords under review. Indeed, the set $\theta_1^{(a)} = [f(0), f(2R)]$ of values $f_1^{(a)} \equiv q$ of the property $t_1^{(a)}$ 'abscissa of the point' is equivalent to set $\theta_1 = [0, 2R]$, while the relative measure of the subset of these points located between $q = q'$ and $q = q' + dq$, with an allowance for (9), is equal to $2^{-1}R^{-2}|\varphi(q')\varphi'(q') \, dq| = 2^{-1}R^{-2}|l \, dl|$ and, in virtue of the foregoing, is thus approximately equal to the frequency corresponding to the set $f_0(\theta_1' = (l', l' + dl))$ of abstract objects.

A particularly important case is $q = l$, it corresponds to a set of points bounded by the abscissa, a straight line $\rho(l) = 2^{-1}R^{-2}l$, and straight lines $l = 0$ and $l = 2R$. By Definition 2 the probability that an abstract object – 'a point of this set' – has a set of values of the 'abscissa' property in the range from l to $l + dl$ is equal to

$$\rho(l) \, dl = 2^{-1}R^{-2}l \, dl. \tag{1.10}$$

Therefore $\rho(l)$ can be called the density of the probability that 'a point of this set' has an 'abscissa' value equal to l. But this quantity is clearly approximately equal to the density of the frequency of revealing chords of length l in random tests, and therefore coincides with the probability density of the corresponding accidental result.

Other examples of sets of concrete objects for which adequate sets can be obtained from theoretical considerations will appear further on. For the present, we will proceed to formulate the general definition of probability as an objective quantity.

1.9. Probability. General Definition

In Section 1.7 probability was actually defined as the relative measure (L) of the corresponding subset of the set A of concrete objects. In the general case, however (as has already been noted), the measure (L) cannot be introduced into a set of concrete objects, either because this set itself is not given, or because, although being given, it is not a subset of points of some space R_n. Therefore such a definition of probability is inapplicable to the general case, as indicated above. Then the question arises: Is it possible to modify the definition of probability so that it should suit the general case as well? This question is answered in the affirmative, for the following reasons.

Assume that the initial set A of concrete objects is measurable (L). Then Definition 2 holds good for it. Some other sets A may be adequate to this initial set, and for all these mutually adequate sets, including A, the relative measures of any subsets of concrete objects corresponding to equivalent sets of abstract objects are equal to each other and are approximately equal to the frequencies of the corresponding accidental results. Therefore, for set A, in the case at hand, the probability can be computed, proceeding not only from this set itself, but from any other set $A^{(a)}$ adequate to it. This circumstance enables us to define the probability for any one of mutually adequate sets of concrete objects as the relative measure of the corresponding subset of concrete objects of some one of these sets. Such a definition of probability can be extended for the general case as well.

However, this requires, first, that the 'frequency' concept be meaningful and, secondly, that there exist at least one set $A^{(a)}$ adequate to the set A. For the 'frequency' concept to make sense it is necessary, in turn, that in performing any arbitrary number of series of a sufficiently large number of random tests the frequencies of the corresponding accidental results in all these series be approximately equal (stability of statistical results). If this condition is fulfilled, we can assume that the random tests are actually performed on some definite set of concrete objects. And conversely, if the random tests are carried out on some definite set of concrete objects, this condition can hardly fail to be satisfied. Therefore the stability of statistical results can be considered a necessary and sufficient condition for these results to refer to some definite set of concrete objects.

On the other hand, the existence of at least one set $A^{(a)}$ adequate to the set A of concrete objects has to be postulated; we formulate the second (and the last) basic principle of probabilistics as follows.

AXIOM 2. For any set A of concrete objects which can be subjected to random tests there exists at least one adequate set $A^{(a)}$.

This assertion, being an axiom, requires no proof, but some substantiation would certainly be desirable. We will supply it a little later; now, taking advantage of this axiom, we will formulate the general definition of probability.

DEFINITION 2a. By the probability that for an abstract object $f_{0ij\ldots k}$ a value $f_{lm\ldots p}$ of the property $t_{lm\ldots p}$ belongs to a set $\theta'_{lm\ldots p} \subseteq \theta_{lm\ldots p}$ is meant the ratio of the measure of the set $A^{(a)\prime}(f^{(a)}_{ij\ldots k}; \theta^{(a)\prime}_{lm\ldots p})$ to the measure of the set $A^{(a)\prime}(f^{(a)}_{ij\ldots k})$, where $A^{(a)}$ is any one of the sets adequate to A.

Denoting this probability by $P(f_{0ij\ldots k}; \theta'_{lm\ldots p})$, we write Definition 2a as

$$P(f_{0ij\ldots k}; \theta'_{lm\ldots p}) = mA'(f_{ij\ldots k}; \theta'_{lm\ldots p})/mA(f_{ij\ldots k}) \tag{1.11}$$

(for simplicity, the superscripts 'a' on the (right-hand side) of (11) are omitted, although implied).

We now revert to Axiom 2. If set A is given and measurable (L), this axiom is satisfied explicitly, since A is adequate to itself (Definition 2a then coincides with Definition 2). On the other hand, in the general case this axiom refers to some set A of concrete objects, which can be subjected to random tests. We either know beforehand that we are dealing with a definite set A of concrete objects, and therefore are sure of the stability of the corresponding statistical results, or, on the contrary, we are sure that we are dealing with a definite set A of concrete objects precisely because these statistical results are stable. In either case it is significant that the statistical results for set A exhibit stability, which suggests the existence of definite probabilities for them, and hence also for the corresponding abstract objects. But if such probabilities do exist, then, according to Definition 2a, there must also exist at least one set $A^{(a)}$ adequate to the initial set A, with the aid of which these probabilities can be determined. Moreover, for any set A of concrete objects, for which the probabilities corresponding to each abstract object are known, it is possible to indicate even a general method for constructing a set $A^{(a)}$ adequate to it. We will first consider this method using the above example of the set of chords in a given circle.

In this case one of the sets $A^{(a)}$ adequate to the set at hand could be obtained by constructing a curve $\rho = \rho(l)$, where $\rho(l)$ is the probability density of the corresponding accidental result, and hence the probability density of the value l of the chord length in this circle as well; we can do the same in any other case. In accordance with the note at the end of Section 1.1 the set of values of any property of concrete objects can be considered numerical. Therefore the value $f_{12\ldots n}$ of the property $t_{12\ldots n}$ of these objects can be represented as a point in space R_n, then the probability density ρ (see Chapter 3) corresponding to the abstract objects $f_{012\ldots n}$ can be represented as an n-dimensional surface in the corresponding space R_{n+1}, on whose $(n + 1)$th axis ρ is laid off. The set of points in this space R_{n+1} bounded by an n-dimensional plane $\rho = 0$, an n-dimensional surface $\rho =$ probability density, and also by the other bounding surfaces, which are determined by the boundary values of the various properties of concrete objects, is, according to Definition 3, actually set $A^{(a)}$ adequate to the initial set A of concrete objects. If the values of some properties are discrete, then the $n + 1$-dimensional set $A^{(a)}$ is degenerated to a less-dimensional set, which fact does not prevent it from remaining adequate to the initial set A of concrete objects. Thus in the case of $n = 1$, if the property t_1 has a finite number of values $\theta_1 = \{l_1, l_2, \ldots, l_k\}$ and to the value $f_1 = l_j$ there corresponds the probability p_j, then in the two-dimensional space (the plane $l0\rho$) set $A^{(a)}$ is degenerated to a collection of k line segments determined by equations $l = l_j$ $(j = 1, 2, \ldots, k)$ and bounded by points $\rho = 0$ and $\rho = p_j$ $(j = 1, 2, \ldots, k)$. The one-dimensional set of the points of this collection of segments is, in the case under study, actually set $A^{(a)}$ adequate to the initial set of concrete objects (but it is constructed in a two-dimensional space).

We have thus shown that the assertion of the existence of at least one set $A^{(a)}$ adequate to the set A of concrete objects is tantamount to the assertion of the existence of probability. The existence of at least one set $A^{(a)}$ ensures the existence of probability (Definition 2a), and vice versa, the existence of probability ensures the existence of at least one set $A^{(a)}$ (knowledge of the probability corresponding to each abstract object enables one to construct this set).

The concept of mutually adequate sets is of prime importance to probabilistics.

Solution of various probabilistic problems reduces in some cases to finding at least one set adequate to the set of concrete objects actually figuring in the problem. We will illustrate this with the Bertrand problem mentioned in the Preface, formulating it as follows: What is the probability that the length of a chord in a given circle of radius R exceeds the side of a regular triangle inscribed in it, that is, $R3^{1/2}$? Before we proceed to solve the problem, we will note that its formulation here is considerably simplified as compared with the previous, rather ponderous wording. This is done to simplify the style, assuming at the same time that the reader has already reached sufficient understanding of the essence of the problem: we are dealing with a 'chord of a given circle', that is, with an abstract object, and we ask about the probability that the value of the 'length' property for this abstract object is larger than $R3^{1/2}$. Henceforth we will avoid using the earlier heavy formulations unless it is absolutely necessary.

From Eqn. (8) it follows that if $l \geqslant R3^{1/2}$, then $r \leqslant 2^{-1}R$. Using the set of points in the circle as $A^{(a)}$ and applying Eqn. (11), we get

$$P[f_0; \theta_1' = (R3^{1/2}, 2R)] = \pi(2^{-1}R)^2/\pi R^2 = \tfrac{1}{4}. \tag{1.12}$$

The same value is obtained (as should be the case) by integrating the right-hand side of (10) with respect to l from $R3^{1/2}$ to $2R$, which corresponds to choosing for $A^{(a)}$ an appropriately bounded set of points under the straight line $\rho(l) = 2^{-1}R^{-2}l$. But this coincidence does not confirm the correctness of construction of the two utilized sets $A^{(a)}$, because they are not independent. It is easy, however, to construct, in an independent way, one more set A adequate to the set of chords in the circle.

A concrete chord can also be identified by the length of the intercepts of the line, on which it lies, on the axes of some fixed Cartesian system of coordinates X0Y, whose origin coincides with the center of the circle. We denote the intercepts by x and y, respectively. Then a set of points satisfying the equation

$$x^{-2} + y^{-2} = r^{-2} \tag{1.13}$$

in this system of coordinates is equivalent to the set of chords of length l, where r has the former meaning and is related to l by Eqn. (8). Therefore the set of points bounded by the coordinate axes and the curve $x^{-2} + y^{-2} = R^{-2}$ is also a set $A^{(a)}$ adequate to the set of chords in the circle. The measure of this set is evidently infinite, as is the measure of the subset of points we are interested in, which is bounded by the coordinate axes and the curve Eqn. (13) at $r = R/2$. But the ratio of these measures is finite. Indeed, the measure of any subset of a similar type can be written as $\iint_{S(r)} \mathrm{d}x\, \mathrm{d}y$, where $S(r)$ is an r-dependent integration range bounded by the coordinate axes and the curve (13) at some fixed value of r. Introducing new variables u and v with the aid of relations $x = ru$, $y = rv$, we find $\iint_{S(r)} \mathrm{d}x\, \mathrm{d}y = r^2 \iint_S \mathrm{d}u\, \mathrm{d}v$, where S is an integration range no longer dependent on r and bounded by the coordinate axes and the curve $u^{-2} + v^{-2} = 1$. Making use of this circumstance and applying Eqn. (11), we again obtain $P = \frac{1}{4}$ as solution to Bertrand's problem.

It is worth noting here that the circumstance discovered by Bertrand and consisting in the dependence of the result obtained for the above problem on the method of its solution is only an apparent paradox. It arises from the fact that to solve this problem Bertrand used different sets, of which only one – namely, the set of points in the circle – is adequate to the set of chords, whereas all the others are not adequate to it. Bertrand's problem is formulated quite correctly and unambiguously, and it must have a unique solution, which can be checked by statistical experiment. Therefore the assertion that all the solutions obtained by Bertrand are equally valid (see, for instance, [4]) is radically wrong.

In the case of Bertrand's problem the correctness of our choice of each of the two independent sets $A^{(a)}$ is well substantiated and is confirmed by the agreement between the results of calculations using both these sets. In the general case the correctness of choosing a set $A^{(a)}$ also is subject to some kind of independent verification – theoretical, if possible, or experimental, based on analysis of the results of the corresponding statistical experiment. We will revert to this question, in particular, when considering the Buffon problem, which (like many other probabilistic problems) is solved in the ordinary probability theory, also with the use of mutually adequate sets (to which there corresponds the introduction of so-called geometrical probabilities), although one does it without being conscious of that.

As follows from Definition 2a, probability exists regardless of whether it is measured or not, that is, whether or not the corresponding random tests are made. But the probability of appearance of the corresponding accidental result on a single random test has the same numerical value and is therefore also expressed as the right-hand side of Eqn. (11). The appearance of an accidental result is usually called a random event, and one speaks of the probability that some random event will occur on realization of some definite complex of conditions. Proceeding from the form of the left-hand side of (11), we can denote the corresponding random event by $(f_{0ij \ldots k}; \theta'_{lm \ldots p})$ or, using the denotation of the corresponding set of abstract objects, by $f_{0ij \ldots k}(\theta'_{lm \ldots p})$. The complex of conditions which is usually meant, singles out some definite set of concrete objects under test and hence the corresponding abstract object. In the indicated denotation of a random event the abstract object $f_{0ij \ldots k}$ corresponds to this complex of conditions which determines $A(f_{ij \ldots k})$, and the set $\theta'_{lm \ldots p}$ of values $f_{lm \ldots p}$ of the property $t_{lm \ldots p}$, to the random event of interest.

Considering the aforesaid, we will sometimes, for fluency sake and to bring the text closer to everyday speech, interpret the left-hand side of (11) as the probability that a single random test will result in the event indicated there. Some other points of contact between probabilistics and the ordinary probability theory will be discussed as need arises, but the interrelation of the two disciplines will be treated in more detail only in the final part of the book.

Problems

1.1. A family A consists of 6 concrete people: the couple John and Mary, their son Stan and his wife Lola, and the latter couple's children – Bob and Ann. Their ages are: John – 59, Mary – 54, Stan – 33, Lola – 29, Bob – 12, and Ann – 8. Divide family A into classes in accordance with the properties t_1 = 'sex' and t_2 = 'age', taking (conditionally) as values of age $f_2^{(a)}$ = 'child' (below 10), $f_2^{(b)}$ = 'adolescent' (between 10 and 20), $f_2^{(c)}$ = 'adult' (between 20 and 35), and $f_2^{(d)}$ = 'mature' (over 35).

1.2. How will the division of family A into classes change in 5 years?

1.3. Are the concrete families A in problems 1 and 2 the same or different?

1.4. Verify that Eqns. (1)–(3) and (4a) are fulfilled in problems 1 and 2.

1.5. What is the complete set of abstract objects in problems 1 and 2 and what are their names?

1.6. Which one of the abstract objects f_{012}^{ad} and f_{02}^{d} is more abstract in problems 1 and 2? Which abstract object is the most abstract and which abstract objects are the least abstract in these problems?

1.7. What are the measures of all the classes in problems 1 and 2?

1.8. Verify that Eqns. (5)–(6a) are fulfilled in both problems 1 and 2.

1.9. Form at least two sets A^a adequate to family A in each of problems 1 and 2.

1.10. What are the probabilities that a female member of family A is an adolescent in problems 1 and 2?

1.11. What is the probability that a pair of members of family A (in problems 1 and 2) is a couple?

1.12. What is the probability that a point in a circle of radius R is inside the concentric circle of radius R/n?

1.13. A separate molecule of gas is in a closed vessel of volume V. What is the probability that it is in a fixed volume V' of the vessel?

1.14. Put problems 10–13 in terms of random events.

Suggested References

1. Bertrand, J.: 1889 (1st edn.) and 1907 (2nd edn.), *Calcul des Probabilites*, Gauthier-Villars, Paris; reprint of the 2nd edn. by Chelsea, New York, 1972. The book contains the original formulation of Bertrand's paradox.
2. Cramer, H.: 1946, *Mathematical Methods of Statistics*, Princeton U.P., Princeton. This excellent book contains sufficient information on the theory of sets, including the definition of Lebesgue's measure.
3. Kamke, E.: 1956 (1st edn.) and 1960 (2nd edn.), *Das Lebesgue–Stieltjes-Integral*, Taubner, Leipzig. This wonderful book can be used by readers who wish to gain a deeper insight into some aspects of the mathematical apparatus of probabilistics.

4. Kendall, M. G., and P. A. P. Moran: 1963, *Geometrical Probability*, C. Griffin and Co., London. Here one can find the formulation of Bertrand's and Buffon's problems. However, the explanation of Bertrand's paradox is erroneous.
5. Mayants, L. S.: 1973, *Found. Phys.* 3, 413. This paper formulates for the first time the fundamental principles of probabilistics, which is referred to as the *complete probability theory*.

Chapter 2

Main Theorems

2.1. Sum and Product of Events

The event $Q = (f_0; \theta'_{lm \ldots p})$ will be called the sum (or union) of the events $Q_r = (f_0; \theta'^{(r)}_{l_r m_r \ldots p_r})$ if $\theta'_{lm \ldots p} = \bigcup_r \theta'^{(r)}_{l_r m_r \ldots p_r}$. We write the sum of the events as $Q = \Sigma_r Q_r$ if the set of events Q_r is finite or denumerable, or as $Q = \bigcup_\alpha Q_\alpha$ if it is nonenumerable. (We use the term 'event' in place of 'random event' for the sake of brevity, except when this may cause misunderstanding.)

The event $Q = (f_0; \theta'_{lm \ldots p})$ will be called the product (or intersection) of the events $Q_r = (f_0; \theta'^{(r)}_{l_r m_r \ldots p_r})$ if $\theta'_{lm \ldots p} = \bigcap \theta'^{(r)}_{l_r m_r \ldots p_r}$. The product of events will be written as $Q = \Pi_r Q_r$.

Since an event is the appearance, in a single random test, of a result corresponding to some subset of concrete objects of their set A, the event $Q = \Sigma_r Q_r$, which is the appearance of an accidental result corresponding to a subset $A' = A'(f_0; \theta'_{lm \ldots p}) \subseteq A$, occurs if this accidental result corresponds to at least one of the subsets $A'(f_0; \theta'^{(r)}_{l_r m_r \ldots p_r})$, whose union is A', that is, the event Q occurs if at least one of the events Q_r takes place.

Similarly, the event $Q = \Pi_r Q_r$ occurs only when the accidental result of a single test corresponds at once to all the subsets $A'(f_0; \theta'^{(r)}_{l_r m_r \ldots p_r})$, whose intersection is A', that is, if all the events Q_r take place.

As follows from the definition of the sum and product of events, event Q is representable (and, generally speaking, more than in one way) both as the sum of some events and as the product of some other events. Exceptions are events of the type: (1) $Q = (f_0; \theta'_l)$ and (2) $Q = (f_0; \{f_{12 \ldots n}\})$, where θ'_l is a set of values of some one property t_l, and $\{f_{12 \ldots n}\}$ is a set consisting of one element – a single definite value of the property $t_{12 \ldots n}$. Events of the first type can evidently be represented only as the sum, but not the product, of events, while events of the second type, only as the product, but not the sum, of events. Of special interest are second-type events, since it follows from the above, with due allowance for (1.7) and (1.4), that any event can be represented as the sum of such events. Therefore these type of events will be called elementary events. Elementary events correspond to the least abstract objects for a given set A of concrete objects (see Sect. 1.2), while to the set of elementary events, which we denote by U, there is equivalent the set $\theta_{12\ldots n}$ of the least abstract objects. Any event Q can thus be represented as

$$Q = (f_0; \theta'_{12\ldots n}), \tag{2.1}$$

where $\theta'_{12\ldots n} \subseteq \theta_{12\ldots n}$.

In a single random test performed on a set A of concrete objects some accidental result must necessarily appear, that is, some event must take place unfailingly. An event which must necessarily occur can naturally be called the certain event. Since it must contain any elementary event as a summand, it coincides with the set U of elementary events, and we designate it by U. We can thus write

$$U = (f_0; \theta_{12\ldots n}). \tag{2.2}$$

The nonappearance of any result at all in a single random test, in other words the appearance of a result belonging to an empty set, is impossible, and we call such an event the impossible event. Denoting such an event by 0, we can evidently write $0 = (f_0; 0)$.

If the product of two events Q_1 and Q_2 is the impossible event, that is, $Q_1Q_2 = 0$, these events are incompatible. Indeed, writing for brevity $Q_1 = (f_0; \theta'_{(1)})$, $Q_2 = (f_0; \theta'_{(2)})$, we obtain $Q_1Q_2 = (f_0; \theta'_{(1)} \cap \theta'_{(2)})$, and therefore $\theta'_{(1)} \cap \theta'_{(2)} = 0$. But this means that if the result of a single random test belongs to $\theta'_{(1)}$, it cannot belong to $\theta'_{(2)}$, and vice versa, that is, these two events cannot occur jointly.

Event $Q = (f_0; \theta'_{12\ldots n})$ will be called the difference of the events $Q_1 = (f_0; \theta'^{(1)}_{12\ldots n})$ and $Q_2 = (f_0; \theta'^{(2)}_{12\ldots n})$ and written as

$$Q = Q_1 - Q_2 \tag{2.3}$$

if $\theta'_{12\ldots n} = \theta'^{(1)}_{12\ldots n} - \theta'^{(2)}_{12\ldots n}$ is the complement of the set $\theta'^{(2)}_{12\ldots n}$ with reference to the set $\theta'^{(1)}_{12\ldots n}$ (see Explanation in 1.2). Since $(\theta'^{(1)}_{12\ldots n} - \theta'^{(2)}_{12\ldots n}) \cap \theta'^{(2)}_{12\ldots n} = 0$, then $QQ_2 = 0$, that is, the events $Q_1 - Q_2$ and Q_2 are incompatible.

The event

$$\bar{Q} = U - Q \tag{2.4}$$

will be called opposite to Q. This name stems from the fact that the occurrence of this event means the nonoccurrence of event Q. If $Q = (f_0; \theta'_{12\ldots n})$, then $\bar{Q} = (f_0; \theta_{12\ldots n} - \theta'_{12\ldots n})$. Since $\bar{\bar{Q}} = U - (U - Q)$, then $\bar{\bar{Q}} = Q$. The following relations are self-evident: $\bar{U} = 0; \bar{0} = U; UQ = Q$.

Example. In a single tossing of a coin the certain event is the appearance of either the 'head' or the 'tail'. The impossible event is the nonappearance of either the 'head' or the 'tail'. The appearance of the 'head' and that of the 'tail' are opposite events.

Note. As distinct from the sum and product of events, which are meaningful for any events, as seen from their definition, the difference of events is defined only for cases where $\theta'^{(2)}_{12\ldots n} \subseteq \theta'^{(1)}_{12\ldots n}$, that is, where the event Q_1 can be represented in (3) as a sum of events, one of which is Q_2. From $Q = Q_1 - Q_2$ it clearly follows that $Q_1 = Q + Q_2$ and $Q_2 = Q_1 - Q$. But from $Q = Q_1 + Q_2$, in the general case, follows neither $Q_1 = Q - Q_2$ nor $Q_2 = Q - Q_1$. Indeed, let Q_1 correspond to $\theta'_{(1)}$ and Q_2, to $\theta'_{(2)}$. Then Q corresponds to $\theta' = \theta'_{(1)} \cup \theta'_{(2)}$. In the general case $\theta'_{(1)} \cap \theta'_{(2)} \neq 0$, and therefore $Q_1 Q_2 \neq 0$. Thus, in the general case $\theta'_{(1)} \cup \theta'_{(2)} - \theta'_{(1)} = \theta'_{(2)} - \theta'_{(1)} \cap \theta'_{(2)}$, and $\theta'_{(1)} \cup \theta'_{(2)} - \theta'_{(2)} = \theta'_{(1)} - \theta'_{(1)} \cap \theta'_{(2)}$. Hence, in the general case $(Q_1 + Q_2) - Q_1 = Q_2 - Q_1 Q_2$ and $(Q_1 + Q_2) - Q_2 = Q_1 - Q_1 Q_2$.

2.2. Addition Theorem

Using (1), we can write: $Q = (f_0;\, \theta'_{12\ldots n})$; $Q_r = (f_0;\, \theta'^{(r)}_{12\ldots n})$. By definition $Q = \Sigma_r Q_r$ if $\theta'_{12\ldots n} = \cup_r \theta'^{(r)}_{12\ldots n}$. From (1.11) follows

$$P\left(\sum_r Q_r\right) = P(f_0;\theta'_{12\ldots n}) = mA'(f_0;\theta'_{12\ldots n})/mA. \tag{2.5}$$

Further we obtain, with due regard for (1.7):

$$A'(f_0;\theta'_{12\ldots n}) = \bigcup_{\theta'_{12\ldots n}} A(f_{12\ldots n}) = \bigcup_r \bigcup_{\theta'^{(r)}_{12\ldots n}} A(f_{12\ldots n})$$

$$= \bigcup_r A'(f_0;\theta'^{(r)}_{12\ldots n}). \tag{2.6}$$

If the events Q_r are pairwise incompatible, then $\theta'^{(r)}_{12\ldots n} \cap \theta'^{(s)}_{12\ldots n} = 0$ $(r \neq s)$. By virtue of the additivity of the measure (L) (see Sect. 1.4)

$$mA'(f_0;\theta'_{12\ldots n}) = \sum_r mA'(f_0;\theta'^{(r)}_{12\ldots n}). \tag{2.7}$$

Substituting (7) into (5) and using (1.11) again, we find that pairwise incompatible events obey *the addition theorem*:

$$P\left(\sum_r Q_r\right) = \sum_r P(Q_r) \quad (Q_r Q_s = 0; r \neq s). \tag{2.8}$$

A particular case of (8) is

$$P(Q_1 + Q_2) = P(Q_1) + P(Q_2). \tag{2.8a}$$

Now, from (3) and taking into account $QQ_2 = 0$ and (8a), we get

$$P(Q_1 - Q_2) = P(Q_1) - P(Q_2). \tag{2.9}$$

From (2) and (1.11) immediately follows

$$P(U) = 1. \tag{2.10}$$

Then from (4) and (9) we obtain for the opposite event

$$P(\bar{Q}) = 1 - P(Q). \tag{2.11}$$

In the particular case $\bar{U} = 0$

$$P(0) = 0. \tag{2.12}$$

Equations (12) and (10) set up the range of probability of any event. As follows immediately from (1.11) and from the properties of the measure (L), probability satisfies the relation: $0 \leqslant P(Q) \leqslant 1$.

The addition theorem proved above refers to pairwise incompatible events. Let us now consider the case of two arbitrary events Q_1 and Q_2. As was established in the preceding section (see Note)

$$(Q_1 + Q_2) - Q_1 = Q_2 - Q_1 Q_2. \tag{2.13}$$

Applying Eqn. (9) to both sides of this equality, we get

$$P(Q_1 + Q_2) - P(Q_1) = P(Q_2) - P(Q_1 Q_2), \tag{2.14}$$

whence

$$P(Q_1 + Q_2) = P(Q_1) + P(Q_2) - P(Q_1 Q_2). \tag{2.15}$$

If Q_1 and Q_2 are pairwise incompatible events, then (15) obviously goes over to (8a). In the general case, however, as can be seen from (15), the expression for the probability of the sum of two events contains the probability of the product of events, which will be considered in the next section. But before we proceed any further it will be appropriate to read the following:

Very Important Note. All the events figuring in Eqn. (8) belong to one and the same set A of concrete objects and, hence, to one and the same abstract object f_0. In terms of the ordinary probability theory it means that all the events in Eqn. (8) belong to one and the same fixed complex of conditions which produces a single definite and fixed probability distribution. Therefore this equation is inapplicable when different events belong to different condition complexes (and, hence, to different sets of concrete objects and, accordingly, to different abstract objects). The importance of this note is due to the fact that the neglect of this circumstance is usually utilized to prove that the rules of addition of probabilities in quantum physics differ from those expressed by (8). This problem will be considered in more detail in the methodological part of the book.

2.3. Multiplication Theorem

Let $Q = (f_0; \theta'_{12\ldots n})$, $Q_1 = (f_0; \theta'^{(1)}_{12\ldots n})$, and $Q_2 = (f_0; \theta'^{(2)}_{12\ldots n})$. By definition $Q = Q_1 Q_2$ if $\theta'_{12\ldots n} = \theta'^{(1)}_{12\ldots n} \cap \theta'^{(2)}_{12\ldots n}$. From (1.11) follows

$$P(Q_1 Q_2) = P(f_0; \theta'_{12\ldots n}) = mA'(f_0; \theta'_{12\ldots n})/mA. \tag{2.16}$$

If neither Q_1 nor Q_2 is equal to 0 (and it makes sense to consider only this case), the right-hand side of (16) can also been written thus:

$$\frac{mA'(f_0; \theta'_{12\ldots n})}{mA} = \frac{mA'(f_0; \theta'_{12\ldots n})}{mA'(f_0; \theta'^{(1)}_{12\ldots n})} \cdot \frac{mA'(f_0; \theta'^{(1)}_{12\ldots n})}{mA}, \tag{2.17}$$

$$\frac{mA'(f_0; \theta'_{12\ldots n})}{mA} = \frac{mA'(f_0; \theta'_{12\ldots n})}{mA'(f_0; \theta'^{(2)}_{12\ldots n})} \cdot \frac{mA'(f_0; \theta'^{(2)}_{12\ldots n})}{mA}. \tag{2.18}$$

According to the definition of probability the first term on the right-hand side of (17) has the meaning of the probability that for the abstract object f_0 the value $f_{12\ldots n}$ of the property $t_{12\ldots n}$ belongs to set $\theta'_{12\ldots n}$, provided the set of all the values of this property is $\theta'^{(1)}_{12\ldots n}$. This can be written

$$P(f_0; \theta'_{12\ldots n}/\theta'^{(1)}_{12\ldots n}) = mA'(f_0; \theta'_{12\ldots n})/mA'(f_0; \theta'^{(1)}_{12\ldots n}). \tag{2.19}$$

Since set $\theta'_{12\ldots n}$ corresponding to event $Q_1 Q_2$ is a subset of set $\theta'^{(1)}_{12\ldots n}$ corresponding to event Q_1, the left-hand side of (19) can be interpreted as the conventional probability of the event Q_2 relative to event Q_1. Denoting this probability by $P(Q_2/Q_1)$, we can consequently write

$$P(Q_2/Q_1) = P(f_0; \theta'_{12\ldots n}/\theta'^{(1)}_{12\ldots n}). \tag{2.20}$$

Using (1.11) again, we find from (17), with due regard for (19) and (20), that for any events Q_1 and Q_2 *the multiplication theorem*

$$P(Q_1 Q_2) = P(Q_1)P(Q_2/Q_1) \tag{2.21}$$

is valid.

Obviously, (18) yields

$$P(Q_1 Q_2) = P(Q_2)P(Q_1/Q_2), \tag{2.21a}$$

where

$$P(Q_1/Q_2) = P(f_0; \theta'_{12\ldots n}/\theta'^{(2)}_{12\ldots n}). \tag{2.20a}$$

If the event Q is represented as

$$Q = \sum_j QQ_j, \tag{2.22}$$

where events Q_j are pairwise incompatible, then from (8) and (21) follows

$$P(Q) = \sum_j P(Q_j)P(Q/Q_j). \tag{2.23}$$

Representation of some event Q in the form of (22) is possible, of course, only when the relation $\Sigma_j Q_j = U$ is satisfied.

The event Q_2 is said to be independent of the event Q_1 if $P(Q_2/Q_1) = P(Q_2)$. Since in this case, as follows from (20), (19), and (1.11), $mA'(f_0; \theta'_{12\ldots n})/mA'(f_0; \theta'^{(1)}_{12\ldots n}) = mA'(f_0; \theta'^{(2)}_{12\ldots n})/mA$, then also $P(Q_1/Q_2) = P(Q_1)$, that is, events Q_1 and Q_2 are mutually independent. For independent events Eqn. (21) acquires the form

$$P(Q_1Q_2) = P(Q_1)P(Q_2). \tag{2.24}$$

If events Q_1 and Q_2 are independent, then events Q_1 and $\bar{Q}_2$, Q_2 and $\bar{Q}_1$, and $\bar{Q}_1$ and $\bar{Q}_2$ are also independent. Indeed, for independent events Q_1 and Q_2, from $Q_1 = Q_1Q_2 + Q_1\bar{Q}_2$ and Eqns. (23) and (24) follows $P(Q_1) = P(Q_1)P(Q_2) + P(\bar{Q}_2)P(Q_1/\bar{Q}_2)$, whence $P(Q_1/\bar{Q}_2) = P(Q_1)$. The independence of the other indicated pairs of events follows immediately from this.

We illustrate the application of some of the equations obtained above on Buffon's problem, which we formulate as follows: A needle of length $2l$ is dropped at random onto a horizontal plane with parallel lines drawn on it at intervals of $2a$ $(l < a)$. What is the probability that the needle will cross any one of the lines? To solve this problem we consider a set of possible positions of the needle relative to the line nearest to its midpoint. Each concrete position of the needle relative to this line can be identified by the distance (t_1) from the needle midpoint to the line and the angle (t_2) between the needle and the line. The set of values $f_1 \equiv x$ of the property t_1 has the form: $\theta_1 = (0, a)$, and the set of values $f_2 = \varphi$ of the property t_2, $\theta_2 = (0, \pi)$. The elementary event has the form: $(f_0; f_{12})$, where $f_{12} = (x, \varphi)$. The event consisting in the fact that x is in the interval between x' and $x' + \Delta x$, and φ is in the interval between φ' and $\varphi' + \Delta\varphi$ is a union of elementary events over Δx and $\Delta\varphi$. We can write this event as: $(f_0; (\Delta x, \Delta\varphi))$ (for simplicity we omitted the indication of the values x' and φ' from which the respective intervals are laid off). The distance from the needle midpoint to the line and the angle it makes with the line are not related in any way. Therefore the corresponding events can be considered mutually independent. Since all the values of x, as well as all the values of φ, are equivalent, we can write, on the basis of (24), the following expression for the probability that x is in the interval between x' and $x' + \Delta x$, and φ is in the interval between φ' and $\varphi' + \Delta\varphi$:

$$P(f_0; (\Delta x, \Delta\varphi)) = \Delta x a^{-1} \Delta\varphi\pi^{-1}. \tag{2.25}$$

The events we are interested in, that is, crossings of a line by the needle, and the events corresponding to x or φ are no longer mutually independent, because, for instance, at some definite value of φ the needle can cross the line only when the following condition is satisfied

$$x \leqslant l \sin \varphi. \tag{2.26}$$

The probability that the needle will cross the line at an angle φ in the interval $\Delta\varphi$ has the following form, due account being taken of (8) and (26):

$$\frac{\Delta\varphi}{\pi} \int_0^{l \sin \varphi} \frac{dx}{a} = \frac{\Delta\varphi}{\pi} \frac{l \sin \varphi}{a}. \tag{2.27}$$

Now using Eqn. (23), we obtain for the probability that the needle will cross the line at any value of φ

$$P = \frac{l}{\pi a} \int_0^{\pi} \sin \varphi \, d\varphi = \frac{2l}{\pi a}, \tag{2.28}$$

which is the final solution of Buffon's problem.

The above method for solving Buffon's problem corresponds to the usual method for solving it with the aid of 'geometrical probabilities', which actually consists in selecting, as a set $A^{(a)}$, adequate to the set of concrete positions of the needle relative to the line, a set of points bounded by the axes of a Cartesian system of coordinates 0φ (abscissa) and $0x$ (ordinate) and by lines $x = a$ and $\varphi = \pi$. The set of needle positions in which it crosses the line is representable then by the set of points located between the abscissa and the curve $x = l \sin \varphi$. The above reasoning would seem to be airtight, and therefore such a choice of set A should not cause any doubt. However, if we apply similar reasoning to the consideration of the problem of the probability that a point chosen at random within a circle of radius R is located within the circle of half-radius with the use of polar coordinates, the answer will be erroneous, namely 1/2 in place of 1/4 (incidentally, this problem is adequate to that of Bertrand); due to this the reasoning given above in solving Buffon's problem cannot be considered fully convincing. Therefore, as noted at the end of Section 1.9, in order to check the correctness of the solution, even of Buffon's problem, that is, to verify the correctness of the choice of the corresponding adequate set in this problem, one should make use of some independent method. This is best done by using a statistical experiment. For Buffon's problem such experimental verification has been carried out more than once, by Buffon himself, in particular [1], and the solution given by (28) proved correct.

2.4. Sequence of Random Tests

So far, speaking of the probability of events, we meant the probability of the appearance of some accidental result in a single random test on a set A of concrete objects. In a statistical experiment, however, we deal, not with a single test, but wth a manifold of such tests, which may be, in particular, a sequence of random tests. Therefore, the question naturally arises as to the probability associated with a sequence of random tests, that is, the probability relating to a sequence of random events. This problem is solved very easily, since a sequence of random tests can always be represented as a single random test performed on the corresponding manifold of sets of concrete objects. We will first consider the case of two successive tests.

Suppose there are two arbitrary sets of concrete objects A_1 and A_2. We agree that a single random test performed on the manifold of these two sets is the manifold of two single tests performed on each of these sets separately. As a result of each such single test, some pair of concrete objects – each one from set A_1 and set A_2 – is revealed. This means that a single random test on the manifold of sets A_1 and A_2 is equivalent to a single random test performed on set $A = A_1 \times A_2$ of pairs of concrete objects (one each from A_1 and A_2), which is the direct product of these sets (see Explanation in Sect. 1.1). The result of some single test performed on set A is some event Q, which is a pair of events $Q_1^{(1)}$ and $Q_2^{(2)}$, the former belonging to set A_1 and the latter to set A_2 (the parenthetical superscripts indicate to which sets the corresponding events belong). Let us write this down as

$$Q = Q_1^{(1)} \times Q_2^{(2)}. \tag{2.29}$$

If set A_2 has arisen from set A_1 as a result of the fact that the single random test performed on set A_1 led to event $Q_1^{(1)}$, then the single random test performed on set $A = A_1 \times A_2$ is equivalent to the sequence of two random tests performed on set A_1 (provided the first of them led to event $Q_1^{(1)}$). Equation (29) defines event Q consisting in successive appearance of event $Q_1^{(1)}$ and event $Q_2^{(2)}$ in these tests.

We will now proceed to calculate the probability of event Q. If to set A_1 there corresponds an abstract object $f_0^{(1)}$ and to set A_2 an abstract object $f_0^{(2)}$, the abstract object f_0 corresponding to set A is a pair of abstract objects $(f_0^{(1)}, f_0^{(2)})$. Since this cannot cause any misunderstanding, we agree to write f_0 as $f_0 = f_0^{(1)} f_0^{(2)}$ (it would be correct to write $f_0 = f_0^{(1)} \times f_0^{(2)}$, but this would complicate the subsequent expressions). If $Q_1^{(1)} = (f_0^{(1)}; \theta'^{(1)}_{12\ldots n'})$ and $Q_2^{(2)} = (f_0^{(2)}; \theta'^{(2)}_{12\ldots n''})$, then $Q = (f_0^{(1)} f_0^{(2)}; \theta'_{12\ldots n})$, where $\theta'_{12\ldots n} = \theta'^{(1)}_{12\ldots n'} \times \theta'^{(2)}_{12\ldots n''}$.

Let $A_1^{(a)}$ be one of the sets adequate to A_1 and $A_2^{(a)}$ be one of the sets adequate to A_2. Then one of the sets adequate to A is obviously the set: $A^{(a)} = A_1^{(a)} \times A_2^{(a)}$. Omitting the indicies 'a' again for simplicity, we obtain the following expression for event Q of the type (29) with the use of (1.11)

$$P(Q) = mA'(f_0^{(1)} f_0^{(2)}; \theta'_{12\ldots n})/mA. \tag{2.30}$$

But since $\theta'_{12\ldots n} = \theta'^{(1)}_{12\ldots n'} \times \theta'^{(2)}_{12\ldots n''}$, evidently $A'(f_0^{(1)} f_0^{(2)};\ \theta'_{12\ldots n}) = A'(f_0^{(1)};\ \theta'^{(1)}_{12\ldots n'}) \times A'(f_0^{(2)};\ \theta'^{(2)}_{12\ldots n''})$. Taking into account that all the sets considered here are measurable (L) and also that the measure (L) of the direct product of sets is equal to the product of their measures (L), we get for the right-hand side of (30)

$$\frac{mA'(f_0^{(1)}f_0^{(2)};\theta'_{12\ldots n})}{mA}=\frac{mA'(f_0^{(1)};\theta'^{(1)}_{12\ldots n'})}{mA_1}\cdot\frac{mA'(f_0^{(2)};\theta'^{(2)}_{12\ldots n''})}{mA_2}. \tag{2.31}$$

Now (30) with an allowance for (1.11) and (29), finally gives

$$P(Q)=P(Q_1^{(1)}\times Q_2^{(2)})=P(Q_1^{(1)})P(Q_2^{(2)}). \tag{2.32}$$

Generalization of the obtained result for the case of the sequence of any number n of tests is obvious:

$$P(Q)=P(Q_1^{(1)}\times Q_2^{(2)}\times\cdots\times Q_n^{(n)})=\prod_{i=1}^{n}P(Q_i^{(i)}). \tag{2.33}$$

Equation (33) thus determines the probability that in performing n successive tests on the initial set A_1 of concrete objects there arises the sequence of events: $Q_1^{(1)}$, $Q_2^{(2)}, \ldots, Q_n^{(n)}$ (the parenthetical superscript means the test number and, hence, that of the new tested set; the subscripts distinguish the events). In this equation the sequence of the events is essential, because event $Q_{i+1}^{(i+1)}$ belongs to set A_{i+1}, which is obtained from set A_i as a result of the fact that a single random test performed on it yielded event $Q_i^{(i)}$ (if it had yielded some other event, then set A_{i+1} would have to be different, generally speaking).

If the sequence of tests is such that the probability of any event is independent of the test number, that is, of which event took place in the preceding test, such a sequence is called a sequence of independent tests. For a sequence of independent tests Eqn. (33) goes over to

$$P(Q_1\times Q_2\times\cdots\times Q_n)=\prod_{i=1}^{n}P(Q_i). \tag{2.34}$$

We will illustrate the application of (34) on the usual example of the so-called Bernoulli scheme. This scheme assumes that as a result of a test some one event Q may or may not occur, in other words, there may occur an event Q or the event $\bar{Q}$, and the probability of event Q (and hence of $\bar{Q}$) is the same for all the tests. It is required to determine the probability that in n random tests event Q will occur m times. Denote $P(Q)=p$ and $P(\bar{Q})=1-p=q$. According to (34) the probability of some definite sequence of n events consisting of m events Q and $n-m$ events $\bar{Q}$ is equal to p^mq^{n-m}. Since the total number of sequences of this type is equal to $\binom{n}{m}$, by virtue of (8) the sought for probability, which we denote by $P_n(m)$, has the form

$$P_n(m)=\binom{n}{m}p^mq^{n-m} \tag{2.35}$$

(it is easy to ascertain that the conditions under which the equality (8) is valid, actually hold true in the case at hand).

As is well known, the probability distribution (35) bears the name of the binomial distribution. The importance of the Bernoulli scheme and, hence, of Eqn. (35) is due to the fact that in many cases the statistical experiment is realized precisely according to this scheme. In particular, this scheme is followed in solving Buffon's problem with the aid of a statistical experiment. Therefore it would be appropriate to dwell here on Eqn. (35) in more detail.

Since events Q_m ($m = 0, 1, \ldots, n$), consisting in the realization of event Q m times in n tests, are incompatible, the following equality holds:

$$\sum_{m=0}^{n} P(Q_m) = \sum_{m=0}^{n} P_n(m) = \sum_{m=0}^{n} \binom{n}{m} p^m q^{n-m} = 1, \tag{2.36}$$

which, incidentally, follows immediately from $(p + q)^n = 1$. It is easy to see that $P_n(m + 1)/P_n(m) = (n - m)p/(m + 1)q$. Therefore $P_n(m)$, as a function of m, increases while $(n - m)p - (m + 1)q = np - m - q$ remains positive and begins to decrease when this expression becomes negative (it is assumed, of course, that the condition $np - q > 0$ is fulfilled; even if p is very small, this condition must always be fulfilled beginning with some sufficiently large n). Thus, for any $p \neq 0$ and a sufficiently large n, $P_n(m)$, as a function of m, reaches the maximum value at some $m = k$, which satisfies the approximate equality $np \cong k$. This equality, which obviously cannot, in principle, be exact at any value of n and can only be approximate, determines the so-called most probable value of m. But the probability of this most probable value is found to be very low and reduces progressively as n increases. To ascertain this we have to resort to approximate calculations, since direct calculations by Eqn. (35) are practically impossible.

At large values of l the well-known Stirling formula

$$l! = (2\pi l)^{1/2} l^l \mathrm{e}^{-l} \tag{2.37}$$

is approximately valid. No matter what the value of $p \neq 0$, it is always possible to take a sufficiently large n such that $k \cong np$ is also sufficiently large. Using (37), we obtain for $\binom{n}{k}$ the approximate expression

$$\binom{n}{k} = n!\, k!^{-1} (n - k)!^{-1} = \left(\frac{n}{2\pi k(n - k)}\right)^{1/2} \left(\frac{n}{k}\right)^k \left(\frac{n}{n - k}\right)^{n-k}. \tag{2.38}$$

Substituting (38) into (35) for $m = k$, we find

$$P_n(k) = \left(\frac{n}{2\pi k(n - k)}\right)^{1/2} \left(\frac{np}{k}\right)^k \left(\frac{nq}{n - k}\right)^{n-k}. \tag{2.39}$$

In approximate calculation of the right-hand side of (39) we take into account that from $np \cong k$ follows $nq \cong n - k$. Introducing now

$$np/k = 1 + x; \qquad nq/(n - k) = 1 + y, \tag{2.40}$$

where x and y are obviously small numbers, and calculating $(1 + x)^k(1 + y)^{n-k}$ with an accuracy to terms of the second order of smallness, we get, after simple rearrangements and with an allowance for the relation stemming from (40), $y = -k(n - k)^{-1}x$,

$$(1 + x)^k(1 + y)^{n-k} = 1 - knx^2/2(n - k). \tag{2.41}$$

Substituting (41) into (39) and replacing k by np and $n - k$ by nq, we find, using (40) again

$$P_n(k) = (2\pi npq)^{-1/2} [1 - (2npq)^{-1}(np - k)^2]. \tag{2.42}$$

Since np differs from k by not more than unity, the second term in brackets is a small number, which is the less, the larger is n, and the probability of k (that is, the probability of the most probable value of m), which is approximately equal to $(2\pi npq)^{-1/2}$, is indeed very small for large values of n (for a large number of tests) and reduces as n increases.

This fact, however, does not contradict the statement, following from Axiom 1 and from the definition of probability, that for a sufficiently large number of random tests the frequency of realization of some event is approximately equal to its probability. Qualitatively, this can be ascertained as follows: With a sufficiently large number of tests the frequency m/n of realization of the event Q is approximately equal to p, not only for $m = k$, but also for all the values of m lying in the interval between $k - \Delta m$ and $k + \Delta m$, where Δm and the deviation of the frequency from p, which will be designated by ϵ, are related by the self-evident equation $\epsilon = \Delta m/n$. With a prescribed approximation of the frequency to the probability, that is, with a prescribed ϵ, $\Delta m = \epsilon n$ is proportional to n. Therefore, although the probability of the most probable value of m reduces with increasing n, the number of values of m, for which the probability is close to the maximum probability, increases with n and, as will be shown in the next section, for any indefinitely small prescribed ϵ the probability that the absolute magnitude of the deviation of the frequency from the probability is less than ϵ tends to 1 with increasing n.

2.5. Law of Large Numbers

Equation (42) holds good when k is replaced by m, provided m lies in the above indicated interval. To estimate the right-hand side of (42) for $k = m$, we write the following expression, taking into account the smallness of the second term in brackets at the values of m we are interested in:

$$P_n(m) = (2\pi npq)^{-1/2} \exp[-(np - m)^2/2npq]. \tag{2.43}$$

To calculate the probability that the frequency deviates from p not more than by ϵ, we must sum (43) over all the values of m lying in the corresponding interval. Introducing the variable η with the aid of the relation $m - np = n\eta$, we replace the summing over m by integrating with respect to η in the range from $-\epsilon$ to ϵ. Then we obtain for the sought-for probability, which we designate as $P(|m/n - p| < \epsilon)$:

$$P(|m/n - p| < \epsilon) = n^{1/2}(2\pi npq)^{-1/2} \int_{-\epsilon}^{\epsilon} \exp\,[-n\eta^2/2pq]\; d\eta. \tag{2.44}$$

Replacing the variables: $n^{1/2}\eta(2pq)^{-1/2} = z$, we get from (44)

$$P(|m/n - p| < \epsilon) = \pi^{-1/2} \int_{-z_0}^{z_0} \exp(-z^2)\, dz, \tag{2.45}$$

where $z_0 = n^{1/2}(2pq)^{-1/2}\epsilon$. No matter how small ϵ, it is always possible to choose such a large n that z_0 will be larger than any preassigned number N. Therefore, the following equality holds true for any ϵ:

$$\lim_{n \to \infty} P(|m/n - p| < \epsilon) = \pi^{-1/2} \int_{-\infty}^{\infty} \exp(-z^2)\, dz = 1. \tag{2.46}$$

This expresses one of the so-called laws of large numbers, namely Bernoulli law of large numbers (Bernoulli's theorem).

It would be appropriate here to say a few words about the meaning of Eqn. (46) (that is, about the meaning of the law of large numbers). Since this equation expresses the probability of some definite event, it relates, not to concrete objects (not to concrete events, in this particular case), but to the corresponding abstract object (that is, to the corresponding abstract event). Therefore it cannot be considered (as is usually done) that the law of large numbers establishes a relationship between probability theory and statistics – indeed, statistics deals with concrete objects only (we imply experimental statistics, of course). This problem will be considered in more detail in the conclusive part of the book. Here we will only note that the possibility of approximate estimation of the probability of an event by the frequency of its appearance in a sequence of a very large number of independent random tests is merely an experimental fact. At the same time the law of large numbers agrees with Axiom 1, as should be the case.

2.6. Limiting Cases of Binomial Law

As already noted in 2.4, direct calculations by Eqn. (35) are very difficult, and even practically impossible when n are large. Therefore it is desirable to replace this equation for practical application by some other expressions, perhaps approximate, but easier to handle. One such equation is obtained by introducing $(np - m)/(npq)^{1/2} =$

v into (43) and regarding v as a continuous variable. Then the summing over m is replaced by integrating the function

$$\rho(v) = (2\pi)^{-1/2} \exp(-v^2/2) \tag{2.47}$$

with respect to v.

The probability distribution law, determined by (47), is called the normal law. (More correctly, Eqn. (47) describes a particular case of the normal law; the general form of the law will be indicated in Chapter 3.) From the derivation of (43) it follows that the normal law best approximates the binomial one, the larger is pq, that is, the closer is p to 0.5, and the larger is n.

In practical work, however, one often encounters cases where p is very small and n is not too large. Then it is impossible to use Stirling's formula and, hence, to approximate (35) with the aid of the normal law. If, as before, $np - q > 0$, then the most probable value of m, which is approximately equal to $np = \lambda$, is relatively small, and, according to (35), only a small number of the first terms corresponding to the small values of m may have an appreciable magnitude. Considering the foregoing, we rewrite (35) thus

$$P_n(m) = \frac{n(n-1)\ldots(n-m+1)}{m!} \left(\frac{\lambda}{n}\right)^m \left(1-\frac{\lambda}{n}\right)^n \left(1-\frac{\lambda}{n}\right)^{-m}$$

$$= \frac{\lambda^m}{m!}\left(1-\frac{1}{n}\right)\ldots\left(1-\frac{m-1}{n}\right)\left(1-\frac{\lambda}{n}\right)^n\left(1-\frac{\lambda}{n}\right)^{-m}. \tag{2.48}$$

Since $\lambda/n = p$ is assumed to be very small, and m is not large compared with n, Eqn. (48) can be rewritten as

$$P_n(m) \cong \lambda^m (1 - \lambda/n)^n / m!. \tag{2.49}$$

Further

$$(1 - \lambda/n)^n = (1 - \lambda/n)^{\lambda n/\lambda} \cong e^{-\lambda}. \tag{2.50}$$

Substituting (50) into (49), we finally get

$$P(m) = e^{-\lambda}\lambda^m/m!. \tag{2.51}$$

The probability distribution determined by the right-hand side of (51) is called Poisson's law. Recall that this law is an approximation to the binomial law when $p = \lambda/n$ is a very small number and n is not very large.

For the normal law, from (47) follows the relation

$$\int_{-\infty}^{\infty} \rho(v)\, dv = (2\pi)^{-1/2} \int_{-\infty}^{\infty} e^{-v^2/2}\, dv = 1, \tag{2.52}$$

and for the Poisson's law (51) leads to the relation

$$\sum_{m=0}^{\infty} P(m) = e^{-\lambda} \sum_{m=0}^{\infty} \lambda^m/m! = 1. \tag{2.53}$$

Both relations correspond to Eqn. (36) for the binomial law, as should be the case.

Problems

2.1. What are the elementary events in problem 1.10?

2.2. There are 100 families, of which 25 consist of (a) 2 male and 2 female members, 50 have (b) 2 male and 4 female members, and 25 consist of (c) 3 male and 5 female members. One family and then one of its members are chosen at random. What is the probability that the member selected will prove to be female?

2.3. A 'good' coin is tossed at random 10,000 times. What is the probability of the most probable number (k) of the appearances of the 'head'?

2.4. What is the probability, P, that the number of the appearances of the 'head' in 10,000 random tosses will be in the interval between 4950 and 5050?

2.5. How many random tosses of a coin should be made in order that with a probability of 0.9 the frequency of appearance of 'head' would differ from 1/2 less than by 0.005?

Suggested References

1. Buffon, G.: 1777, *Essai d'arithmetique morale* (supplement à l'Histoire Naturelle, Vol. 4). In this book, one can find the original exposition of Buffon's problem together with its theoretical and experimental solution.
2. Fry, T. C.: 1928 (1st edn.) and 1965 (2nd edn.), *Probability and Its Engineering Uses*, D. van Nostrand, New York. This book contains a simple derivation of the normal and Poisson's laws from the binomial one.

Chapter 3

Random Variables

3.1. Definition

We will note at once that the definition of the random variables given here differs from that used in the ordinary probability theory. This should always be borne in mind to avoid misunderstanding.

The term 'unidimensional random variable' will imply any property $t_l \in T$ of concrete objects – elements of a set A – whose values $f_l \in \theta_l$ are real numbers.

Recall that each concrete object $a \in A$ possesses one, and only one, value $f_l \in \theta_l$ of each of its properties $t_l \in T$ (see 1.1 and 1.2). Therefore, mathematically speaking, a unidimensional random variable is simply a real variable determined on a set A of concrete objects, for each of which it has a single definite value. The 'randomness' of a random variable manifests itself only in random tests.

This makes it possible, by using techniques of different branches of mathematics, to introduce various functions of unidimensional random variables into consideration. In particular, the combined property $t_{lm\ldots p} = (t_l, t_m, \ldots, t_p)$, already known to us and formed from k unidimensional random variables $t_l, t_m, \ldots, t_p$, can be considered a k-dimensional random variable (k-dimensional random vector).

Any operations defined for functions of real variables are directly applicable to the same functions of the corresponding random variables. Any relationships between functions of real variables have the same form for the same functions of random variables. If, for instance, the components of a random vector

$$y(t) = (y_1^{(t)}, y_2^{(t)}, \ldots, y_k^{(t)}) \tag{3.1}$$

are determined by the relations

$$y_j^{(t)} = \partial\varphi(t_1, t_2, \ldots, t_k)/\partial t_j, \tag{3.2}$$

where $\varphi(t_1, t_2, \ldots, t_k)$ is a scalar function of the random variables t_j ($j = 1, 2, \ldots, k$), then the following equations hold

$$\partial y_j^{(t)}/\partial t_l = \partial y_l^{(t)}/\partial t_j. \tag{3.3}$$

Further on we will make use of the above considerations as required.

Our definition of the random variable is fully in line with the usual wording. Thus, for instance, the angle of deviation of the thread of a mathematical pendulum from the vertical is a unidimensional random variable, and the coordinates of center of masses of a molecule of a gas in a given vessel is a three-dimensional random variable. But the term 'random variable' is usually also applied to values associated with random events, for instance to the frequency of appearance of some event in a series of independent tests. Can we say then that our definition of the random variable is applicable to such cases?

The answer is yes, because each event occurring in a single test performed over some set A of concrete objects is a concrete event and, hence, a concrete object – an element of the corresponding set $A_{(a)}$ of all concrete events which may take place as a result of these tests. The only value $f_{0(a)}$ of property $t_{0(a)}$ which determines that a concrete event belongs to set $A_{(a)}$ is the fact that this event corresponds to set A under test. The abstract object $f_{0(a)}$ corresponding to set $A_{(a)}$ is a random event – the image of all the concrete events which may occur in performing random tests over set A.

To concrete objects – elements of set $A_{(a)}$ – there can be attached various properties whose values are real numbers, that is, various random variables. If, in particular, the tests are carried out in accordance with Bernoulli's scheme, then to concrete events – sequences of n concrete results of single tests – there can, indeed, be attached the above-mentioned random variable, namely, the 'frequency of appearance of the corresponding event'.

A random variable whose set of values is discrete, that is, is finite or denumerable, will be called a discrete random variable. A random variable possessing a continuous set of values will be termed a continuous random variable. We may, however, also come across mixed-type random variables, whose set of values is discrete on some intervals of the number line, while on the others it is continuous. Random variables of a more complicated type will not be considered here at all.

To each subset $A'(f_0;\, \theta'_{lm\ldots p}) \subset A$ of concrete objects there correspond definite subsets $\theta'_j \subseteq \theta_j$ of values of random variables t_j ($j = 1, 2, \ldots, n$). In particular, to each class $A(f_j)$ of concrete objects there corresponds one definite value $f_j \in \theta_j$ of the random variable t_j ($j = 1, 2, \ldots, n$). A random variable which has a single definite value will be called a random constant. A random constant will be denoted by its value.

3.2. Probability Distribution

We assume that the set θ_l of values of a unidimensional random variable t_l ($l = 1, 2, \ldots, n$) is included in the corresponding unidimensional space $R_1^{(l)}$: $\theta_l \subseteq R_1^{(l)}$. Then the set $\theta_{lm\ldots p}$ of values of the k-dimensional random variable $t_{lm\ldots p}$ is included in the k-dimensional space $R_k^{(lm\ldots p)}$: $\theta_{lm\ldots p} \subseteq R_k^{(lm\ldots p)}$, where $R_k^{(lm\ldots p)} = R_1^{(l)} \times R_1^{(m)} \times \cdots \times R_1^{(p)}$.

It is easy to see that Eqn. (1.11) holds for any subset of points $\theta'_{lm\ldots p} =$

$\theta''_{lm\ldots p} \subseteq R_k^{(lm\ldots p)}$. Indeed, if for some $\theta''_{lm\ldots p}$ the relation $\theta''_{lm\ldots p} \cap \theta_{lm\ldots p} = 0$ holds true, it means that there exist no concrete objects $a \in A$ for which $f_{lm\ldots p} \in \theta''_{lm\ldots p}$, and therefore there exist no corresponding concrete objects in sets $A^{(a)}$ adequate to A, either, and hence $mA'(f_{0ij\ldots k}; \theta''_{lm\ldots p}) = 0$ and $P(f_{0ij\ldots k}; \theta''_{lm\ldots p}) = 0$, which should, indeed, be the case. Rewriting (1.11) as

$$P(f_{0ij\ldots k}; \theta''_{lm\ldots p}) = mA'(f_{0ij\ldots k}; \theta''_{lm\ldots p} \cap \theta_{lm\ldots p})/mA(f_{ij\ldots k}) \tag{3.4}$$

we obtain, among other things,

$$P(f_{0ij\ldots k}; 0) = 0 \tag{3.5}$$

and, by virtue of (1.4) and (1.7),

$$P(f_{0ij\ldots k}; R_k^{(lm\ldots p)}) = mA'(f_{0ij\ldots k}; \theta_{lm\ldots p})/mA(f_{ij\ldots k}) = 1. \tag{3.6}$$

The left-hand side of equality (4) records the probability that for the abstract object $f_{0ij\ldots k}$ the value $f_{lm\ldots p}$ of the random variable $t_{lm\ldots p}$ belongs to the subset $\theta''_{lm\ldots p} \subseteq R_k^{(lm\ldots p)}$. Such form of recording gives the general mode for representing the probability distribution of the random variable $t_{lm\ldots p}$ for the abstract object $f_{0ij\ldots k}$. Important particular modes are representations of the probability distribution of a random variable with the aid: (1) of the probability distribution function; (2) of the probability density.

By the distribution function of a random variable $t_{lm\ldots p}$ for the abstract object $f_{0ij\ldots k}$ we mean the function

$$F(f_{0ij\ldots k}; x_l, x_m, \ldots, x_p) = P(f_{0ij\ldots k}; f_l \leqslant x_l, f_m \leqslant x_m, \ldots, f_p \leqslant x_p). \tag{3.7}$$

This function is obviously obtained from the left-hand side of (4) for $\theta''_{lm\ldots p} = \{f_l \leqslant x_l, f_m \leqslant x_m, \ldots, f_p \leqslant x_p\}$. Therefore, (7), (5), and (6) yield, respectively

$$F(f_{0ij\ldots k}; -\infty, -\infty, \ldots, -\infty) = 0; \qquad F(f_{0ij\ldots k}; \infty, \infty, \ldots, \infty) = 1. \tag{3.8}$$

If the distribution function is known throughout the entire space $R_k^{(lm\ldots p)}$, one can use it to find the probabilities corresponding to any $\theta''_{lm\ldots p} \subset R_k^{(lm\ldots p)}$.

The probability density can be determined as follows: if there exists a function $\rho(f_{0ij\ldots k}; x_l, x_m, \ldots, x_p)$ such that for any set $\theta''_{lm\ldots p} \subseteq R_k^{(lm\ldots p)}$ the following equality holds:

$$\int_{\theta''_{lm\ldots p}} \rho(f_{0ij\ldots k}; x_l, x_m, \ldots, x_p)\, \mathrm{d}\tau = P(f_{0ij\ldots k}; \theta''_{lm\ldots p}), \tag{3.9}$$

where $\mathrm{d}\tau = \mathrm{d}x_l\, \mathrm{d}x_m \ldots \mathrm{d}x_p$, we call this function the probability density of the random variable $t_{lm\ldots p}$ for the abstract object $f_{0ij\ldots k}$. Since the right-hand side of (9) is nonnegative, the probability density cannot be negative either. From (6) follows

$$\int_{R_k^{(lm\ldots p)}} \rho(f_{0ij\ldots k}; x_l, x_m, \ldots, x_p)\,\mathrm{d}\tau = 1. \tag{3.10}$$

The distribution function and the probability density are obviously related by the expression

$$\mathrm{d}F = F(x_l + \mathrm{d}x_l, \ldots, x_p + \mathrm{d}x_p) - F(x_l, \ldots, x_p) = \rho(x_l, \ldots, x_p)\,\mathrm{d}\tau \tag{3.11}$$

where $\mathrm{d}\tau$ has the former meaning (the designation of the abstract object is omitted for simplicity). From (11) we obtain for the case of a unidimensional random variable t_l

$$\rho(x_l) = \mathrm{d}F(x_l)/\mathrm{d}x_l. \tag{3.12}$$

It would seem that the representation of the probability distribution of random variables with the aid of the distribution function is more general than one obtained with the aid of the probability density, since for discrete random variables the equality (9), which determines the probability density, cannot be satisfied with the aid of ordinary functions assigned point by point. This is also indicated by (12), because for discrete random variables $F(x_l)$ changes jumpwise at the points $x_l = f_l^{(j)}$, where $f_l^{(j)} \in \theta_l$ is the value of the random variable t_l, and hence $F(x_l)$ has no derivatives in the usual sense at these points. But the use of the probability density is more convenient in some respects than that of the distribution function. Therefore it is desirable to obviate the indicated difficulty in some way or another. This can be done either by means of so-called generalized functions (in particular, Dirac's-function denoted by $\delta(x)$) or by replacing the initial exact probability distribution with an arbitrarily close to it approximate one.

The use of the generalized function $\delta(x)$ is based on its well-known property that for an arbitrary function $f(x)$ (belonging to a very broad class of functions, which also includes all the functions expandable in a power series) the following equality is true

$$\int_{-\eta_1}^{\eta_2} \delta(x) f(x)\,\mathrm{d}x = f(0), \tag{3.13}$$

where η_1 and η_2 are arbitrary positive numbers. From (13) follows

$$\delta(x) = \begin{cases} 0 & (x \neq 0); \\ \infty & (x = 0); \end{cases} \qquad \int_{-\eta_1}^{\eta_2} \delta(x)\,\mathrm{d}x = 1. \tag{3.14}$$

The probability density $\rho(x)$ of the discrete random variable t_l which has values $f_l^{(j)}$ associated with the probabilities p_j, can now be written

$$\rho(x_l) = \sum_j p_j \, \delta(x_l - f_l^{(j)}). \tag{3.15}$$

Indeed, the increment of $F(x_l)$ has the form

$$\Delta F(x_l) = \int_{x_l}^{x_l + \Delta x_l} \rho(x_l) \, \mathrm{d}x = \sum_j p_j \int_{x_l}^{x_l + \Delta x_l} \delta(x_l - f_l^{(j)}) \, \mathrm{d}x_l. \tag{3.16}$$

It differs from zero and is equal to some one of p_j only when the interval Δx_l contains one corresponding value $x_l = f_l^{(j)}$. It is easy to see that the probability density of type (15) satisfies the equality (10), since $\Sigma_j \, p_j = 1$.

An approximate probability distribution arbitrarily close to (15) can be obtained by replacing the generalized function in (15), $\delta(x_l - f_l^{(j)})$, with the function $\delta_\epsilon(x_l - f_l^{(j)})$, the function $\delta_\epsilon(x)$ being defined as follows

$$\delta_\epsilon(x) = \begin{cases} 0 & (|x| > \epsilon/2), \\ \epsilon^{-1} & (|x| < \epsilon/2), \end{cases} \tag{3.17}$$

where ϵ is a preassignated positive number. For any function $f(x)$ expandable in a power series the following equality is clearly valid:

$$\int_{-\eta_1}^{\eta_2} \delta_\epsilon(x) f(x) \, \mathrm{d}x = f(0) + O(\epsilon^2) \tag{3.18}$$

for any positive η_1 and η_2 exceeding $\epsilon/2$. As regards the equality

$$\int_{-\eta_1}^{\eta_2} \delta_\epsilon(x) \, \mathrm{d}x = 1 \tag{3.19}$$

it follows immediately from definition (17) of the function $\delta_\epsilon(x)$. It is obvious that $\delta_\epsilon(x) \to \delta(x)$ as $\epsilon \to 0$.

No matter what the least distance Δf_l between two adjacent values of $f_l^{(j)}$ of the discrete random variable t_l, one can, by putting $\epsilon < \Delta f_l$, write, in place of (15)

$$\rho(x_l) \cong \sum_j p_j \, \delta_\epsilon(x_l - f_l^{(j)}). \tag{3.20}$$

Equation (20) makes it possible to obtain an approximation of the probability distribution of a discrete random variable to its true distribution, which is determined by Eqn. (15), with any preassigned accuracy; the less the ϵ value, the higher this accuracy is.

The use of (15) and (20) also makes it possible to construct the probability density (true and approximate, respectively) for multidimensional random variables. One can thus assert that the probability distribution can be represented with the aid of the probability density (accurately, or infinitely accurately) for any random variables, or, in other words, that the probability density, as determined by equality (9), exists for any random variables.

The probability density $\rho(x_l, x_m, \ldots, x_p)$ (the designation of the abstract object is omitted again) is evidently the joint probability density of the set of random variables $t_l, t_m, \ldots, t_p$. If it is known, then, by using (9), one can obtain, by integration, the joint distribution function for the entire manifold of these random variables or for some one of their submanifolds, in particular, for each of these random variables separately. From the joint probability density for the entire manifold of random variables it is also possible to obtain by integration the joint probability density for any one of its submanifolds, in particular, for each random variable separately. Thus it is evidently possible to obtain

$$\rho(x_l) = \int_{-\infty}^{\infty} \cdots \int_{-\infty}^{\infty} \rho(x_l, x_m, \ldots, x_p)\, \mathrm{d}x_m \ldots \mathrm{d}x_p. \tag{3.21}$$

We call the probability distributions of the random variables $t_l, t_m, \ldots, t_p$ mutually independent if the density of the joint probability distribution for these random variables has the form

$$\rho(x_l, x_m, \ldots, x_p) = \rho(x_l)\rho(x_m) \ldots \rho(x_p). \tag{3.22}$$

Here, as follows from (11), the following equality also holds:

$$F(x_l, x_m, \ldots, x_p) = F(x_l)F(x_m) \ldots F(x_p). \tag{3.23}$$

To avoid misunderstanding it should be emphasized at once that, in distinction to the ordinary probability theory, we speak of independence of the probability distributions of random variables, rather than of independence of random variables, since random variables have been defined by us irrespective of the probability distributions for them. To illustrate this, we consider two sets of points on a plane: (1) a set of points in a square with a center at the origin of the Cartesian system of coordinates and with a side $2R$ parallel to the coordinate axes; (2) a set of points on a circumference of radius R inscribed in the square. The random variables t_1 = 'the abscissa' and t_2 = 'the ordinate' of a point on the plane with sets of values $\theta_1 = (-R, R)$ and $\theta_2 = (-R, R)$, respectively, are the same for both sets of points. But abstract objects are naturally different for these two cases: 'the point in the square' for set (1), and 'the point on the circumference' for set (2). In the first case the joint probability density has the form: $\rho(x, y) = \rho_1(x)\rho_2(y)$, where $\rho_1(x) = (2R)^{-1}$ and $\rho_2(y) = (2R)^{-1}$. In the second case the joint probability density must have the form

$$\rho(x, y) = c\, \delta(x^2 + y^2 - R^2) \tag{3.24}$$

where c is a constant, which is determined from the condition

$$\int_{-\infty}^{\infty}\int_{-\infty}^{\infty} \rho(x, y)\, dx\, dy = 1, \tag{3.25}$$

since this probability density must be equal to zero everywhere, except at the points of the plane satisfying the equation of the given circumference: $x^2 + y^2 = R^2$.

The probability density (24) evidently cannot be represented as product of a function of x by a function of y. Thus, in the case at hand the probability distributions of the same random variables t_1 and t_2 are mutually independent only for the first set.

Note. Knowing the joint probability density of the random variables $t_l, t_m, \ldots, t_p$, it is possible, as indicated above, to find the probability density for each of these random variables separately. The knowledge of the probability densities of the random variables $t_l, t_m, \ldots, t_p$ separately does not yet, however, guarantee, in the general case, finding the joint probability density for them. This can be done if, and only if the probability distributions of these random variables are mutually independent; then the joint probability density is determined by Eqn. (22).

3.3. Joint Probability Distributions for Functions of Random Variables

Let $\rho(x)$ be the probability density of the random variable t and $\rho_1(y_1)$, the probability density of the random variable $f_1(t)$, which is a single-valued function of t. It is clear that the probability density of $f_1(t)$ must be related to the probability density of t in some way or other. A simple way to find this relationship follows from the consideration of the joint probability density of $f_1(t)$ and t, which has the following form, as can easily be ascertained:

$$\rho(y_1, x) = \delta[y_1 - f_1(x)]\, \rho(x). \tag{3.26}$$

Indeed, such a form of $\rho(y_1, x)$ ensures the vanishing of the joint probability density for values of y_1 and x failing to satisfy the equality

$$y_1 = f_1(x) \tag{3.27}$$

which determines the random variable $f_1(t)$; at the same time the requirement

$$\int \rho(y_1, x)\, dy_1 = \rho(x) \tag{3.28}$$

is evidently fulfilled. Further on we will not, as a rule, indicate the range of integration if it is performed over all the values of the variables or over the entire space (in the latter case we will use only one integral symbol).

The probability density of $f_1(t)$ is now obtained by integrating (26) with respect to x:

$$\rho_1(y_1) = \int \rho(y_1, x)\, dx. \tag{3.29}$$

In integrating we use the well-known equality

$$\int \delta[\omega(x)] f(x)\, dx = \sum_j |\omega'(\alpha_j)^{-1}| \int \delta(x - \alpha_j) f(x)\, dx, \tag{3.30}$$

where the α_j's are the roots of the equation $\omega(x) = 0$, which holds good for any single-valued function $\omega(x)$ expandable in a power series. Recall that this equality, which is equivalent to

$$\delta[\omega(x)] = \sum_j |\omega'(\alpha_j)^{-1}|\, \delta(x - \alpha_j), \tag{3.31}$$

results from the fact that $\omega(x) = 0$ only for $x = \alpha_j$, and in the vicinity of α_j, $\omega(x) = \omega'(\alpha_j)(x - \alpha_j)$. The final result for $\rho_1(y_1)$, expressed as a function of y_1, has a form well-known from the conventional courses of probability theory, namely

$$\rho_1(y_1) = \sum_j |dx_j/dy_1|\, \rho[\varphi_j(y_1)], \tag{3.32}$$

where $\varphi_j(y_1) = x_j$ are the roots of the equation $y_1 - f_1(x) = 0$.

In future we will have to use, more than once, the density of joint probability distribution for functions of random variables, for different purposes. Therefore we will consider a few other, more general cases. From reasoning similar to what has led us to Eqn. (26) we obtain the following expressions for the joint probability density of any number of random variables $f_j(t)$ and the random variable t:

$$\rho(y_1, \ldots, y_l, x) = \prod_{j=1}^{l} \delta[y_j - f_j(x)]\, \rho(x), \tag{3.33}$$

where y_j are the values of the random variables $f_j(t)$, which are supposed to be single-valued functions of t.

If $\rho(x_1, x_2, \ldots, x_k)$ is the probability density of the random variable $t_{12\ldots k} = (t_1, t_2, \ldots, t_k)$, the joint probability density of any number of random variables $f_j(t_1, t_2, \ldots, t_k)$ $(j = 1, 2, \ldots, l)$ and $t_{12\ldots k}$ has the form

$$\rho(y_1, y_2, \ldots, y_l, x_1, x_2, \ldots, x_k) = \prod_{j=1}^{l} \delta[(y_j - f_j(x_1, x_2, \ldots, x_k)] \times$$
$$\times\, \rho(x_1, x_2, \ldots, x_k), \tag{3.34}$$

where y_j are this time the values of the random variables $f_j(t_1, \ldots, t_k)$, supposed to be single-valued functions of $t_1, \ldots, t_k$.

Assume now that the random variables t_j with values x_j ($j = 1, \ldots, k$) are related by the expression

$$\varphi(t_1, \ldots, t_k) = 0. \tag{3.35}$$

To obtain the joint probability density for these random variables we can follow one of two paths. If it is possible to obtain from (35), say, t_k as a function of the other t_j ($j = 1, \ldots, k-1$) in the form of $t_k = f(t_1, \ldots, t_{k-1})$, then the sought-for joint probability density can be written as follows, using (34):

$$\rho(x_k, x_1, \ldots, x_{k-1}) = \delta\,[x_k - f(x_1, \ldots, x_{k-1})]\,\rho(x_1, \ldots, x_k), \tag{3.36}$$

where $\rho(x_1, \ldots, x_{k-1})$ is the joint probability density of the random variables $t_1, \ldots, t_{k-1}$ or, which is the same, the probability density of the random variable $t_{12\ldots(k-1)}$.

We can also do it differently. Let $\rho_0(x_1, \ldots, x_k)$ be the probability density of the random variable $t_{12\ldots k}$ in the absence of any relationship between its components. Then the joint probability density of the random variables in question must evidently have the following form, provided relation (35) is satisfied:

$$\rho(x_1, \ldots, x_k) = c\,\delta\,[\varphi(x_1, \ldots, x_k)]\,\rho_0(x_1, \ldots, x_k), \tag{3.37}$$

where c is a constant determined from the condition

$$\underbrace{\int \cdots \int}_{k} \rho(x_1, \ldots, x_k)\,\mathrm{d}x_1 \ldots \mathrm{d}x_k = 1. \tag{3.38}$$

The expression (37) makes it possible, in principle, to find joint probability densities for any functions of the random variables $t_1, \ldots, t_k$, on condition that (35) is observed.

We will illustrate the application of (36) and (37) on the example on the random variables t_1 = 'the abscissa' and t_2 = 'the ordinate' of a point on a plane, which was considered in the preceding section. For the set of the points on the circumference of radius R these random variables are related by an expression of type (35):

$$t_1^2 + t_2^2 - R^2 = 0 \tag{3.39}$$

whence

$$t_2 = \pm(R^2 - t_1^2)^{1/2}. \tag{3.40}$$

To the positive values of t_2 there corresponds the '+' sign on the right-hand side of (40). Taking advantage of Eqn. (36), we obtain for these values, allowing for (30),

$$\rho(y, x) = \delta\,[y - (R^2 - x^2)^{1/2}]\,\rho_1(x), \tag{3.41}$$

whence

$$\rho_2(y) = y(R^2 - y^2)^{-1/2}\rho_1[(R^2 - y^2)^{1/2}] \tag{3.42}$$

(it is taken into account here that at negative values of x the function $\delta(y - x)$ is everywhere equal to zero).

Equality (42) can be rewritten

$$\rho_2(y)(R^2 - y^2)^{1/2} = \rho_1(x)(R^2 - x^2)^{1/2}. \tag{3.43}$$

If this relation is considered valid for any x and y, the functions $\rho_1(x)$ and $\rho_2(y)$ have one and the same form

$$\rho_1(x) = c(R^2 - x^2)^{-1/2}; \qquad \rho_2(y) = c(R^2 - y^2)^{-1/2}. \tag{3.44}$$

Determining c from the 'normalization condition' $\int \rho_1(x)\,dx = \int \rho_2(y)\,dy = 1$, we finally obtain

$$\rho_1(x) = \pi^{-1}(R^2 - x^2)^{-1/2}; \qquad \rho_2(y) = \pi^{-1}(R^2 - y^2)^{-1/2}. \tag{3.45}$$

Equations (45) are in agreement with the relation (39). Indeed, this relation does not change either on replacement of t_1 by t_2 and t_2 by t_1, nor on reversal of the signs of t_1 and t_2. Derivation of these equations with the aid of (36), however, required the assumption of the validity of (43) for any values of x and y. In the general case Eqn. (36) can be used practically only under the assumption that $\rho(x_1, \ldots, x_{k-1})$ is known. Equation (37), on the other hand, can be used directly in any case without any additional assumption, provided, of course, that $\rho_0(x_1, \ldots, x_k)$ is known.

In the example under consideration $\rho_0(x, y)$ = const, and from (37) and (39) immediately follows (24). Considering $x^2 + y^2 - R^2$ a function of x and using (30), we find

$$\rho_2(y) = c\int_{-\infty}^{\infty} (2|x|)^{-1}\{\delta[x - (R^2 - y^2)^{1/2}] + \delta[x + (R^2 - y^2)^{1/2}]\}\,dx, \tag{3.46}$$

whence we immediately obtain (44) again, and then (45).

Note. If random variables are related by a functional dependence, the joint probability density for them has the form (36) or (37) and hence unfailingly contains the δ-function of some argument as a factor. Therefore one can assert that if the joint probability density of some random variables does not contain a δ-function as a factor, these random variables are not related by a functional dependence.

3.4. Mathematical Expectations. Moments

The mathematical expectation of a random variable t_l, which we denote by $M(f_{0ij\ldots k})t_l$ ($f_{0ij\ldots k}$ indicates to which abstract object the mathematical expectation corresponds; hereafter we will omit the designation of the abstract object except when this may cause misunderstanding), will be defined by the following equality:

$$M(f_{0ij\ldots k})t_l = \int_{R_k^{(lm\ldots p)}} x_l \rho(f_{0ij\ldots k}; x_l, x_m, \ldots, x_p)\, \mathrm{d}\tau. \tag{3.47}$$

This definition must be supplemented by the requirement that there exists an integral of the indicated type with respect to $|x_l|$ which, in turn, stems from the requirement for the existence of an integral corresponding to the positive x_l and one corresponding to the negative x_l.

The meaning of the mathematical expectation of a random variable is clear. Integrating the right-hand side of (47) over all the variables, except x_l, we obtain

$$Mt_l = \int x_l \rho(x_l)\, \mathrm{d}x_l. \tag{3.48}$$

When carrying out a statistical experiment the frequency of revealing the value of a random variable near x_l in a sufficiently small interval Δx_l is approximately equal to $\rho(x_l)\,\Delta x_l$. Therefore the right-hand side of (48), and hence the mathematical expectation of the random variable, are approximately equal to its average statistical value. Thus, mathematical expectation is the theoretically expected average value, which must be approximately equal to the experimental (statistical) average value obtained on a sufficiently large number of random tests.

For any random variable $f(t_l, t_m, \ldots, t_p)$ the mathematical expectation can be obtained by the equation

$$Mf(t_l, t_m, \ldots, t_p) = \int f(x_l, x_m, \ldots, x_p)\rho(x_l, x_m, \ldots, x_p)\, \mathrm{d}\tau. \tag{3.49}$$

Indeed, by definition

$$Mf(t_l, t_m, \ldots, t_p) = \int y\rho(y)\, \mathrm{d}y, \tag{3.50}$$

where y is the value of the random variable $f(t_l, t_m, \ldots, t_p)$, and $\rho(y)$ is its probability density. But according to (34)

$$\rho(y) = \int \delta\,[y - f(x_l, x_m, \ldots, x_p)]\,\rho(x_l, x_m, \ldots, x_p)\, \mathrm{d}\tau. \tag{3.51}$$

Substituting (51) into (50) and integrating with respect to y, we obtain Eqn. (49), with an allowance for (13). This equation is clearly in line with the meaning of the mathematical expectation as a theoretically expected average value of the random variable. Incidentally, (49) is a particular case of a more general equation obtained in the same way

$$\int \varphi(y)\rho(y)\,\mathrm{d}y = \int \varphi[f(x_l, x_m, \ldots, x_p)]\,\rho(x_l, x_m, \ldots, x_p)\,\mathrm{d}\tau, \tag{3.49a}$$

where φ is an arbitrary (maybe even a complex) function.

From (49) follow a number of important corollaries. If $f(t_l, t_m, \ldots, t_p)$ has the form

$$f(t_l, t_m, \ldots, t_p) = \sum_j f_j(t_l, t_m, \ldots t_p), \tag{3.52}$$

then from (49) immediately follows

$$M \sum_j f_j(t_l, t_m, \ldots, t_p) = \sum_j M f_j(t_l, t_m, \ldots, t_p) \tag{3.53}$$

(if the right-hand side of (52) is an infinite sum, the equality (53) is valid only if we assume the admissibility of term-by-term integration of this sum). Equation (53) demonstrates the validity of

THEOREM 1. *The mathematical expectation of a sum of random variables is equal to the sum of their mathematical expectations.*

If $f(t_l, t_m, \ldots, t_p) = C$, where C is a constant, then (49) yields

$$MC = C. \tag{3.54}$$

Hence follows the validity of

THEOREM 2. *The mathematical expectation of a random constant is equal to this constant.*

Let a random variable have the form $\Pi_{j=1}^n f_j(t_l^{(j)}, t_m^{(j)}, \ldots, t_p^{(j)})$, and the corresponding joint probability density, $\Pi_{j=1}^n \rho_j(x_l^{(j)}, x_m^{(j)}, \ldots, x_p^{(j)})$. Then from (49) we get

$$M \prod_{j=1}^n f_j(t_l^{(j)}, t_m^{(j)}, \ldots, t_p^{(j)}) = \int \prod_{j=1}^n f_j(x_l^{(j)}, x_m^{(j)}, \ldots x_p^{(j)}) \times \\ \times \prod_{i=1}^n \rho_i(x_l^{(i)}, x_m^{(i)}, \ldots, x_p^{(i)}) \times \prod_{s=1}^n \mathrm{d}\tau_s, \tag{3.55}$$

where $\mathrm{d}\tau_s = \mathrm{d}x_l^{(s)}\,\mathrm{d}x_m^{(s)} \ldots \mathrm{d}x_p^{(s)}$ (the number k_j of the random variables $t_l^{(j)}, t_m^{(j)}, \ldots, t_p^{(j)}$ in their jth manifold may, of course, be different; integration in (55) is performed over the entire space $R_{\Sigma k_j} = R_{k_1}^{(l_1 m_1 \ldots p_1)} \times R_{k_2}^{(l_2 m_2 \ldots p_2)} \times \ldots \times R_{k_j}^{(l_j m_j \ldots p_j)} \times \ldots \times R_{k_n}^{(l_n m_n \ldots p_n)}$). The right-hand side of (55) can be represented as

$$\prod_{j=1}^{n} f_j(x_l^{(j)}, \ldots, x_p^{(j)}) \prod_{j=1}^{n} \rho_j(x_l^{(j)}, \ldots, x_p^{(j)})\, \mathrm{d}\tau_j = \prod_{j=1}^{n} Mf_j(t_l^{(j)}, \ldots, t_p^{(j)}). \quad (3.56)$$

Therefore (55) finally yields

$$M \prod_{j=1}^{n} f_j(t_l^{(j)}, t_m^{(j)}, \ldots, t_p^{(j)}) = \prod_{j=1}^{n} Mf_j(t_l^{(j)}, t_m^{(j)}, \ldots, t_p^{(j)}). \quad (3.57)$$

A particular case of (57) is the equality

$$M[f_1(t_l) f_2(t_m) \ldots f_k(t_p)] = Mf_1(t_l) Mf_2(t_m) \ldots Mf_k(t_p) \quad (3.58)$$

which demonstrates the validity of

THEOREM 3. *The mathematical expectation of a product of functions of random variables whose probability distributions are mutually independent is equal to the product of the mathematical expectations of these functions.*

This theorem, with due regard for our definition of random variables and their definition in the ordinary probability theory, corresponds to the theorem on the mathematical expectation of a product of independent random variables in the latter.

The mathematical expectation of the random variable $(t_l - a_l)^k$ is called the kth moment (or the moment of the kth order, k being a whole number). If $a_l = 0$, the moments are called initial and if $a_l = Mt_l$, they are called central. In a similar way one introduces moments of the type $Mt_l^{k_l} t_m^{k_m} \ldots t_p^{k_p}$, where k_j ($j = l, m, \ldots, p$) are whole numbers; the order of such moments is defined as $k = k_l + k_m + \cdots + k_p$. The probability distribution of a random variable evidently completely determines all the corresponding moments of different orders. If the probability density depends on several parameters, then moments of all the orders are functions of these parameters and, hence, also of the corresponding number of first moments. Apart from the initial first-order moment, which evidently coincides with the mathematical expectation of the random variable, of great interest is the central second-order moment, which is termed the variance of the random variable. The variance of the random variable t will be denoted by Dt. Then, obviously

$$Df(t_l, t_m, \ldots, t_p) = M[f(t_l, t_m, \ldots, t_p) - Mf(t_l, t_m, \ldots, t_p)]^2. \quad (3.59)$$

If $f(t_l, \ldots, t_p) = C$, then from (59), with due regard for (54), follows

$$DC = M(C - MC)^2 = M0 = 0 \quad (3.60)$$

which is formulated as

THEOREM 4. *The variance of a random constant is equal to zero.*

If a random variable has the form of $cf(t_l, \ldots, t_p)$, then from (59), with an allowance for the equality

$$Mcf(t_l, \ldots, t_p) = cMf(t_l, \ldots, t_p) \tag{3.61}$$

which stems from (49), it follows that

$$Dcf(t_l, \ldots, t_p) = c^2\, Df(t_l, \ldots, t_p). \tag{3.62}$$

If a random variable has the form $\Sigma_j f_j(t_j)$, then from (59) we obtain, with an allowance for (53),

$$D \sum_j f_j(t_j) = M\left[\sum_j f_j(t_j) - M \sum_j f_j(t_j)\right]^2 = M\left[\sum_j \{f_j(t_j) - Mf_j(t_j)\}\right]^2. \tag{3.63}$$

By calculating the right-hand side of (63) we can, with due regard for (59), write it down as

$$\sum_j Df_j(t_j) + \sum_{i,j\,(i \neq j)} M\{[f_j(t_j) - Mf_j(t_j)]\,[f_i(t_i) - Mf_i(t_i)]\}. \tag{3.64}$$

If the joint probability density has the form $\Pi_j\, \rho_j(x_j)$, then by virtue of (58), (49), and (54)

$$M\{[f_j(t_j) - Mf_j(t_j)]\,[f_i(t_i) - Mf_i(t_i)]\}$$
$$= M[f_j(t_j) - Mf_j(t_j)]M[f_i(t_i) - Mf_i(t_i)] = 0 \quad (i \neq j). \tag{3.65}$$

Now, from (63), (64), and (65) we finally obtain

$$D \sum_j f_j(t_j) = \sum_j Df_j(t_j) \tag{3.66}$$

which can be formulated as

THEOREM 5. *The variance of a sum of functions of different random variables, whose probability distributions are mutually independent, is equal to the sum of the variances of these functions.*

Calculating the right-hand side of (59) and performing the necessary rearrangements, we obtain the expression for the variance of a random variable in a form more convenient for practical purposes, namely

$$Df(t_l, \ldots, t_p) = M[f(t_l, \ldots, t_p)]^2 - [Mf(t_l, \ldots, t_p)]^2. \tag{3.59a}$$

Let us consider some examples. For a discrete random variable, whose probability density is given by (15) or (20) (accurately or infinitely accurately, respectively), from (49) and (59), with due regard for (13) or (18), respectively, we get

$$Mt_l = \sum_j p_j f_l^{(j)}, \tag{3.67}$$

$$Dt_l = \sum_j p_j f_l^{(j)2} - \Big(\sum_j p_j f_l^{(j)}\Big)^2. \tag{3.68}$$

We apply Eqns. (67) and (68) to Bernoulli's scheme (see 2.4). The number of appearances of event Q in the lth test is a random variable $t_{(l)}$ whose values, as well as the probability distributions, are independent of l. In each test the event Q may or may not appear. Therefore $t_{(l)}$ has only two values: $f_{(l)}^{(1)} = 1; f_{(l)}^{(2)} = 0$. The corresponding probabilities are: $p_1 = p; p_2 = 1 - p = q$. This yields $Mt_{(l)} = p \cdot 1 + q \cdot 0 = p$. Using Eqn. (59a), we find $Dt_{(l)} = p - p^2 = p(1-p) = pq$.

The number of appearances of the event Q in the sequence (or manifold) of n independent tests is also a random variable t related to the random variables $t_{(l)}$ by the obvious expression: $t = \Sigma_{l=1}^n t_{(l)}$. From (53) we obtain $Mt = \Sigma_{l=1}^n p = np$. Since the probability distributions for $t_{(l)}$ are mutually independent by virtue of the independence of the tests, Dt can be calculated from (66). We get: $Dt = \Sigma_{l=1}^n pq = npq$. For the random variable t/n = 'frequency of appearance of event Q in a series of n independent tests' we find: $Mtn^{-1} = p$, using (54), and $Dtn^{-1} = pqn^{-1}$ (by virtue of (62)). All these results could, of course, be obtained directly from (2.35) but in a somewhat more complicated manner.

In considering one of the limiting cases of the binomial law of probability distribution in Section 2.4, we arrived at (2.47). By appropriate replacement of the variable v we obtain the most general form of the normal law, namely

$$\rho(x) = \sigma^{-1}(2\pi)^{-1/2} \exp\left[-(x-a)^2/2\sigma^2\right]. \tag{3.69}$$

For any random variable t, whose probability density has the form (69), $Mt = a$ and $Dt = \sigma^2$, as can easily be ascertained by trivial calculations. In the general case a and σ^2 are not related. If however, Eqn. (69) corresponds to the limiting case of the binomial law, then σ^2 is a function of a, which follows from the above. In the general case all the higher-order moments are functions of a and σ^2, and in a particular case corresponding to the binomial law they are functions of a.

If the random variables t_1 and t_2 are related by a functional dependence, then moments of the type $Mt_1^l t_2^m$ can be calculated either with the aid of the joint probability density of these random variables, or with the aid of the probability density of one of them, because the product $t_1^l t_2^m$ can be represented as a function of one of these random variables. We will explain this with reference to random variables t_1 and t_2, whose joint probability density has the form (24), where $c = \pi^{-1}$, as has been established before. Any moments must obviously be functions of R. From the form of the right-hand side of (24) it follows that if l or m is an odd number, the moment is zero. Only moments of the form $Mt_1^{2l}t_2^{2m}$ can be actually nonzero. But $t_2^2 = R^2 - t_1^2$. Therefore $Mt_1^{2l}t_2^{2m} = Mt_1^{2l}(R^2 - t_1^2)^m$. As regards the right-hand side of the last equality, it can be calculated with the aid of the probability density $\rho_1(x)$, which has the form (45). And still, even for practical calculation of Mt_1^{2l}, it is more convenient to use (24).

We have

$$Mt_1^{2l} = \pi^{-1} \int_{-\infty}^{\infty} \int_{-\infty}^{\infty} x^{2l}\, \delta(x^2 + y^2 - R^2)\, dx\, dy. \tag{3.70}$$

By using polar coordinates, we get

$$Mt_1^{2l} = \pi^{-1} \int_0^{\infty} \delta(r^2 - R^2) r^{2l+1}\, dr \int_0^{2\pi} \cos^{2l}\varphi\, d\varphi, \tag{3.71}$$

whence using (30) and the tabulated data, we finally find

$$Mt_1^{2l} = Mt_2^{2l} = R^{2l}(2l-1)!!/2l!!\,. \tag{3.72}$$

Since $Mt_1 = Mt_2 = 0$, from (72) it follows, in particular, that

$$Dt_1 = Dt_2 = R^2/2. \tag{3.73}$$

3.5. Characteristic Function

The characteristic function for the joint probability distribution of the random variables $t_l, t_m, \ldots, t_p$ will be defined as follows:

$$\Phi(\epsilon_l, \epsilon_m, \ldots, \epsilon_p) = \int \exp\,[i(\epsilon_l x_l + \epsilon_m x_m + \cdots + \epsilon_p x_p)\rho(x_l, x_m, \ldots, x_p)\, d\tau, \tag{3.74}$$

where $\epsilon_l, \epsilon_m, \ldots, \epsilon_p$ are real variables, which take on values in the interval $(-\infty, \infty)$. Thus, the characteristic function is a Fourier transform of the probability density. The usefulness of the characteristic function is due to its following main properties:

(1) The probability density $\rho(x_l, x_m, \ldots, x_p)$ and the characteristic function $\Phi(\epsilon_l, \epsilon_m, \ldots, \epsilon_p)$ are in one-to-one correspondence. This statement is proved with full rigor in Fourier integral theory and can be substantiated by the following considerations. The functions

$$\exp\,[i(\epsilon_l x_l + \cdots + \epsilon_p x_p)] = \exp\,(i\epsilon_l x_l) \ldots \exp\,(i\epsilon_p x_p) \tag{3.75}$$

form a totality of orthogonal functions, and expansion of some function, in particular $\rho(x_l, \ldots, x_p)$, in them is unique. Further, functions of the type $\exp(i\epsilon x)$ satisfy the relation

$$\int_{-\infty}^{\infty} \exp(i\epsilon x)\exp(-i\epsilon x')\, d\epsilon = 2\pi\,\delta(x - x'). \tag{3.76}$$

Therefore, from (74) it follows, with due regard for (76) and (75), that

$$(2\pi)^{-1}\int_{-\infty}^{\infty}\cdots\int_{-\infty}^{\infty} \Phi(\epsilon_l, \ldots, \epsilon_p)\exp[-i(\epsilon_l x_l + \cdots +$$

$$+ \epsilon_p x_p)\, d\epsilon_l \ldots d\epsilon_p = \rho(x_l, \ldots, x_p). \tag{3.77}$$

By virtue of the above, Eqn. (77), which is the inversion formula for Fourier transforms, determines unambiguously the probability density $\rho(x_l, \ldots, x_p)$ which corresponds to the characteristic function $\Phi(\epsilon_l, \ldots, \epsilon_p)$.

(2) If the probability distributions of the random variables $t_l, \ldots, t_p$ are mutually independent, it follows from (74), with an allowance for (75) and (49a), that the characteristic function corresponding to the probability distribution of the random variable $t = t_l + \cdots + t_p$ has the form

$$\Phi(\epsilon) = \Phi_l(\epsilon) \ldots \Phi_p(\epsilon), \tag{3.78}$$

where $\Phi_j(\epsilon)$ is the characteristic function corresponding to the probability distribution of the random variable t_j.

(3) Calculating $[\partial^{k_l + \cdots + k_p}\Phi(\epsilon_l, \ldots, \epsilon_p)]/[\partial\epsilon_l^{k_l} \ldots \partial\epsilon_p^{k_p}]$ at zero values of $\epsilon_l, \ldots, \epsilon_p$, we obtain from (74)

$$\left[\frac{\partial^{k_l + \cdots + k_p}\Phi(\epsilon_l, \ldots, \epsilon_p)}{\partial\epsilon_l^{k_l} \ldots \partial\epsilon_p^{k_p}}\right]_{(0, \ldots, 0)} = Mt_l^{k_l} \ldots t_p^{k_p}. \tag{3.79}$$

Equation (79) enables one to calculate the required moments directly from the characteristic function obtained by means of (78) without using the corresponding probability density. Various examples of utilization of the characteristic function can be found in conventional courses of probability theory and mathematical statistics.

Important remark. The characteristic function is an auxiliary mathematical quantity with the aid of which one can determine unambiguously the corresponding probability distribution, but which has no probability meaning in its own right. Therefore even its purely formal conventional definition as the mathematical expectation of the corresponding complex random variable is inexpedient, because it creates a false impression of the existence of a probabilistic meaning in a mathematical quantity which does not have it.

We have previously familiarized ourselves with various methods for representing the probability distribution of a random variable, in particular, with the aid of the distribution function and of the probability density, which possess a direct probabilistic meaning. The characteristic function was the first example of representation of the probability distribution with the aid of an auxiliary mathematical quantity which has no such meaning. But it is not the only quantity of this kind. Any other integral transformation possible for the probability density of some random variable leads to the corresponding function, which also possesses the property (1) of the characteristic function and therefore also represents unambiguously the probability distribution for this random variable without having a direct probabilistic meaning. In Chapter 5 we will introduce the reader to other possible methods for representing the probability distribution of a random variable with the aid of auxiliary mathematical quantities.

Problems

3.1. What are the random variables related to a die?

3.2. What are the random variables related to a point in space?

3.3. *Schrödinger's Cat.* Schrödinger [*Naturwiss.* **23**, 807 (1935)] suggested the following thought experiment. A cat and a flask of poison are enclosed together in a hermetic container. When the flask is broken, the cat is killed by the poison. The breaking occurs due to the discharge of a Geiger counter placed behind one of two holes made in a screen whose front side is bombarded at random by electrons; the source of them being very weak, so that only one electron is emitted during the whole experiment. Thus, there is a certain probability that the cat will die as a result of the experiment.

What are the random variables related to this experiment, and what role is played by the cat in it?

3.4. What are the random variables related to a free particle?

3.5. Which of the random variables in the preceding problem are related by a functional dependence? What are the probability distributions for the random variables of that problem?

3.6. Write down the probability density of the x-coordinate of a free particle by means of the δ_ϵ-function.

3.7. Calculate the different order initial moments and variances of p and x for a free particle moving in the positive direction of the x-axis.

3.8. A two-dimensional random variable $t_{12} = (t_1, t_2)$ has the probability density $\rho(x, y)$, where x and y are the values of t_1 and t_2, respectively. What is the probability density of the random variable $t = t_1 + t_2$?

3.9. What is the probability density of the random variable $t = t_1/t_2$ for t_1 and t_2 the same as in the preceding problem?

Suggested References

1. Cramer, H.: 1946, *Mathematical Methods of Statistics*, Princeton U.P., Princeton. This book supplies abundant material on mathematical methods for solving various problems dealing with probability distributions, in particular on the theory and utilization of characteristic functions.
2. Mayants, L. S.: 1973, *Found. Phys.* **3**, 413. This paper contains the initial formulation of the definition of the random variable given here and also references to the author's previous work, which substantiates the expedience of discarding the conventional definition.
3. Mayants, L. S.: 1977, *Found. Phys.* 7, 3. Here the reader will find the initial consideration of the joint probability density of functions of random variables.
4. van der Waerden, B. I.: 1957 (1st edn.) and 1971 (3rd edn.), *Mathematische Statistik*, Springer, Berlin. This book includes various examples of the utilization of the characteristic function for calculation purposes.

Chapter 4

Some Aspects of Statistics

4.1. Preliminary Considerations

The term statistics derives, as is well known, from the Latin 'Status' (state) and was originally used in connection with the description of the state of a country by means of a manifold of factual data characterizing this state. The characteristic of the state of a country, say, in relation to its population, is given by so-called statistical data (statistical material), which are obtained, for instance, by interviewing concrete people inhabiting the country (population census). These statistical data are then classified according to different features (properties) of the country's inhabitants (sex, age, education, etc.) and are presented in the form of a summary indicating the distribution of the population according to the values of these properties, that is, the absolute or relative number of people (the absolute or relative measure of a set of people) belonging to each category accounted for (to each subset accounted for) of the country's population (of the entire set of concrete people in the country). The initial statistical material is obtained by experimentation (interviewing the population), and the experimenters deal with each concrete person (each concrete object) separately. The classification of this experimental material leads to abstraction of the individual peculiarities of concrete persons and to the description of the state of the country's population (or, in other words, the state of an abstract inhabitant of the country) with the aid of the measure of the set of concrete people belonging to different categories of the population. But by definition (see Sect. 1.5) the relative measure of a subset of concrete objects is the probability that the respective abstract object has the corresponding subset of values of the properties of these concrete objects. Therefore the classification of the statistical material leads, in this case, to the description of the state of an abstract inhabitant of the country by indicating the probability distribution for him to have different values of the properties of its inhabitants.

At present the term 'statistics' is associated with various kinds of phenomena of a random character, irrespective of their nature, and hence with the term 'probability'. The present-day terminology in this field, which does not account for the necessity of making distinction between concrete and abstract objects (see Sects. 1.1–1.3), and hence between the experimental and theoretical aspects of probabilistics, often results in misunderstandings (this will be treated in more detail in the methodological part of the book). To avoid such misunderstandings we will adhere strictly to the meaning of the term 'statistics' as referring to experimental investigation of the distribution of a set of concrete objects according to the values of the properties of these objects, that is, as relating to experimental measurement of the probabilities of different values of these properties. To establish distinctly what we are dealing with – the experimental aspect of probabilistics (statistics) or its theoretical aspect (probability theory), we will make use of the following rigorous criterion: If concrete objects are studied, we are dealing with statistics; and, conversely, if abstract objects (of any degree of abstraction) are studied, we are dealing with the application of probability theory. Using this criterion, we can assert, among other things, that so-called statistical physics, which studies abstract physical systems of definite types, is a particular branch of probabilistic physics, which deals with probabilistic consideration of any physical systems. Similarly, mathematical statistics, which deals with probabilistic consideration of the results of any experiments, is the application of probability theory to analysis of any experimental data.

We will consequently assume that the most general task of statistics, which defines its essence, is experimental finding of the measures (relative ones, as a rule) of various subsets of any set of concrete objects, irrespective of how the experiment is conducted.

4.2. Statistical Experiment

Exact determination of the measures of subsets of concrete objects is possible only in relatively few cases, and even then on condition that these measures remain unchanged in the course of time. Even in the above example it is impossible to actually obtain exact data on the population of a country, because within the time required for obtaining and classifying these data the population composition undergoes some changes (see also 1.5). In the vast majority of cases statistical experiment thus yields only approximate values of the sought-for measures. The specific method for carrying out statistical experiment – the statistical method (see 1.6) – indeed, always yields (see Axiom 1) only approximate results (that is, approximate values of probabilities), no matter how large the number n of random tests.

Let us consider some peculiarities of the statistical method in more detail (see also 1.6–1.9). We recall, first of all, that the frequency of appearance of an accidental result is approximately equal to the probability of this result, provided the number of single random tests is large enough. Therefore only the entire set of a sufficiently large number of single random tests should be regarded as a single measurement of probabilities, whereas separate random tests do not enable one to determine the probabilities of any results whatsoever. Separate random tests cannot, consequently, confirm the correctness of the choice of an adequate set, either. However, they can disprove this choice. If several random tests yield results which are impossible (or hardly possible at all) with the adopted choice of the adequate set, this choice is erroneous (it contradicts the facts) and must be discarded.

A natural question arises: What number of random tests can be considered sufficiently large? The answer depends on the intended accuracy of measurement of the probability. In solving some practical problems or other this accuracy, in turn, depends on the nature of the problems themselves. One usually strives to obtain the necessary results with the desired accuracy by running the least possible number of random tests. On occasions, however, the number of random tests is extremely large because of the specific nature of the experiment itself. Such cases include, in particular, all experiments associated with particle diffraction.

Despite the great diversity of methods for carrying out diffraction experiments the device used for this purpose consists of the following three basic parts: (1) Particle source; (2) diffraction grating (artificial or natural), and (3) display of the diffraction pattern (we will further on assume it to be a flat screen on which the 'point' of impingement of a particle is recorded in any arbitrary way). The diffraction pattern appearing on the screen naturally depends on the experimental conditions. It is significant, however, that under all experimental conditions there exists a corresponding probability desnity $\rho(x, y)$ of particle impingement on a point of the screen with coordinates (x, y). Diffraction experiment makes it possible, in principle, to determine $\rho(x, y)$ approximately, because the number of particles hitting a sufficiently small area $\Delta x \, \Delta y$ in the vicinity of the point (x, y) is approximately proportional to $\rho(x, y) \, \Delta x \, \Delta y$. The number of independent random tests carried out in a diffraction experiment is enormous as a rule, especially when particle collisions with the screen are recorded photographically (in such cases the experimenter could not restrict himself to a small number of tests, even if he wanted to). Independence of random tests in diffraction experiment has been proved experimentally [1].

In the experiment at hand an electron beam of such a low intensity was used (~4000 e/sec) that the average time-interval between the entry of separate electrons into the device was about 10^5 times as large as the average time of passage of an electron through it. Therefore in that experiment the electrons practically passed through the device singly. The diffraction pattern thus obtained turned out to be the same as in a similar experiment using an electron beam 10^7 times more intensive, which proves the independence of random tests, each of which consists in passing one electron through the diffraction device and recording the point on which it impinges on the screen.

When using the statistical method one must, of course, provide such conditions for carrying out random tests which would ensure, in the first place, that the tests are performed over one and the same set of concrete objects and, secondly, that they are actually random tests (see 1.6). Only in this case can we be sure that the random tests will, at least approximately, be able to reproduce the actual probability distribution for the state of the abstract object which is of interest to us. In cases where the abstract object corresponds to quite a definite set of concrete objects, which is completely in the hands of the experimenter (as, for instance, with the set of balls in an urn), there will hardly be any doubt as to what set is subjected to random tests. But in even slightly more complicated cases, for instance, in studying the quality of the products of some enterprise, a potential possibility of confusion already exists.

Assume that the flow of finished products passes a quality-control device, which rejects substandard products. For one and the same abstract object 'product' the probability distribution of the property 'deviation from the average standard value' (the word standard is understood here in its ordinary technical meaning, of course) is different before the quality-control device and after it, since the corresponding sets of concrete products for this abstract object are different. If random tests are performed over products after the quality-control device, but the device is now working and now idle, confusion is quite possible, because it may not be known to which of the two sets of concrete objects each random test belongs.

The situation is still more complicated with experiments staged for studying or using some natural phenomena. In all such experiments, no matter what their objective, the investigator deals with random tests, irrespective of his own wish (and he may not even realize it). Wishing to obtain reproducible results, the experimenter tries to conduct his experiments under identical conditions. If he succeeds in it, he will thereby ensure work with a definite set of concrete objects, although he may not know (and this, as a rule, is the case) what kind of set it is. We will explain this for the case of a diffraction experiment with particles.

It is well known that at present particle diffraction is widely used for various purposes, in particular for studying the structure of matter, since all things being equal, the statistical distribution of the particles on the display screen depends on the parameters of the diffraction grating of the device, which is simulated by a specimen of the substance under investigation. The diffraction pattern helps determine (with a greater or lesser accuracy, of course) the interatomic distances in the substance at hand by using the appropriate theoretical concepts. If it is assumed theoretically that the experiment is carried out with the aid of a parallel beam of particles possessing a definite momentum, then the researcher does his best to ensure that all the concrete particles in the beam have identical and constant values of the momentum vector. In other words, he strives to obtain such a set of concrete particles, for which the corresponding abstract object (a 'particle') would have a quite definite value of the three-dimensional random variable 'momentum'. Such an ideal beam cannot be obtained in principle, and the experimenter has to use a set of concrete particles with a more or less narrow distribution of the values of the momentum vector.

If the device has no diffracting grating and the particles impinge directly on the display screen, their statistical distribution on the screen (experimental probability distribution) would be definite, corresponding to these experimental conditions. The investigator does not know precisely what kind of a set of concrete particles he actually uses (that is, he does not know what momentum each concrete particle possesses, still less at each concrete instant), but if the conditions are the same in different experiments, he can practically be sure that the particle distribution in the beam according to the values of the three-dimensional momentum is the same in all the experiments (to put it differently, an abstract particle corresponding to the set of concrete particles in the beam in these experiments is in one and the same state).

In the presence of a diffraction grating the statistical distribution of the particles on the screen changes. This means that the diffraction grating alters the original distribution of the particles in the beam according to the values of the three-dimensional momentum (that is, it changes the initial state of the abstract particle), and each diffraction grating does this in its own way. The researcher does not know what kind of a set of concrete particles he deals with after the diffraction grating either, that is, what is the new distribution with respect to the momenta (and this does not interest him at all), but he knows for sure that the statistical distribution of the particles on the screen (and that is all he can investigate experimentally), and hence the probability distribution of particle impingement on the screen, is in one-to-one correspondence (all other things being equal) with the parameters of this lattice, whose determination is the purpose of the experiment.

Incidentally, in principle, if we abstract ourselves from the difference in the aims and ways of utilization of the statistical method, the diffraction experiment is similar to the above example with the study of the flow of finished products. A diffraction grating corresponds to the quality-control device in that example; the particle beam before the diffraction grating corresponds to the flow of finished products before the quality-control device, and the set of concrete particles after the diffraction grating, to the set of concrete products after the quality-control device. An essential difference between these two examples, however, lies in the fact that in the example with finished products the set of concrete objects is accessible to direct study both before and after the quality-control device, while in the example with particle diffraction the set of concrete objects is inaccessible to direct study either before or after the diffraction grating – in this example direct study is possible only with respect to the statistical distribution of particles on the display screen (the probability distribution of particle impingement on the screen), whose variation is determined by the variation in the set of particles moving towards the screen (the variation in the state of the abstract particle).

Experimental determination of probability distribution is in the general case a practically unsolvable problem. It is easier to carry out experimental verification of the correctness of the hypothetical probability distribution following from some kind of theoretical considerations (or, what is the same, the correctness of the choice of an adequate set corresponding to that of the concrete objects under study). In solving such a problem, if it refers to the probability distribution of random variables, one can use either the statistical method (without a very large number of random tests) or direct measurement of some kind of characteristics of the distribution, provided this is possible. A more or less detailed consideration of the solution of such problems, actually on the basis of the use of the statistical method, is given in conventional courses of mathematical statistics. Here, we will dwell on just one interesting example of solving a problem of this type by direct measurement of certain characteristics of probability distribution.

Each separate gas molecule, contained in a set of molecules in a concrete vessel, has, at each concrete instant, definite values of all its intrinsic properties, in particular, the position of the center of mass in space, of the velocity vector of translational motion, of the momentum vector of translational motion, of energy, etc. All these properties are clearly random variables, and one can pose the question of probability distribution for them at some preassigned conditions. For the idealized case (the so-called ideal gas) the problem has been completely solved theoretically. The resultant probability distribution of energy (Boltzman's law) is found to be dependent on a single parameter, which is the mathematical expectation of the energy per each degree of freedom of the translational motion. This parameter differs from the absolute temperature by the universal factor (Boltzman's constant). Using this probability distribution, it is possible to calculate the probability distribution for the velocity of the translational motion of a molecule and then obtain the mathematical expectation of the force acting from the side of the gas molecules on a unit surface area of the gas-containing vessel. This mathematical expectation, which is well known to be nothing more than the gas pressure, also depends on the absolute temperature in a definite way. Thus probabilistic consideration of the ideal gas leads to the establishment of the theoretical relationship between its temperature and pressure, which are the mathematical expectations of the corresponding random variables. But both these quantities can be measured directly in experiment, which will verify the correctness of the probabilistic consideration (the correctness of the choice of set $A^{(a)}$ adequate to set A of the concrete gas molecules being considered). As is well known, experimental data confirm the correctness of this theoretical consideration. Moreover, from the theoretical probability distribution for energy follows quite a number of thermodynamical relations, which can also be checked experimentally and generally prove valid (we are not now interested in the refinements which would have to be made in the probabilistic consideration for bringing theory in line with experiment).

This example is of fundamental interest, since many associated experimental results were obtained long before their theoretical explanation was supplied; the theoretical approach was also insufficiently consistent, which found its reflection in the name of the discipline concerned with such problems, 'statistical physics'. Later on, in particular in Chapter 9, we will revert to this question and will ascertain the possibility of a purely probabilistic approach to the solution of such problems. We have adduced this example here in order to emphasize the statistical nature of an experiment in which the temperature, pressure, or other thermodynamical quantities are measured. The results of this experiment may be used to check the correctness of the corresponding probabilistic consideration, since the quantities measured are mathematical expectations of the relevant random variables.

4.3. Numerical Statistical Experiment

For experimental solution of various problems relating to the measures of subsets of some set A of concrete objects, it is by no means mandatory (as explained in 1.8) to perform random tests over precisely this very set A. Instead, one can use any set $A^{(a)}$ adequate to A and apply the statistical method to it. Further, as mentioned in (1.9), for any set of concrete objects, which may be subjected to random tests, one of the sets adequate to it is some definite set of points in the corresponding Euclidean space. Thus, precisely this set can be subjected to random tests in order to solve experimentally the problem stated for the initial set of concrete objects. The question may arise: Why is this random experiment necessary at all? Indeed, if a set of points is given, say, in an n-dimensional Euclidean space with the aid of some surfaces bounding this set, then the determination of its measure and those of its various subsets is simply a problem of integral calculus, which can be solved by conventional calculation methods without any experiment. But in actuality the calculation of certain integrals, and at $n > 1$ to boot, is always far from a simple task. At the same time, experimental 'calculation' of these integrals, based on the use of the statistical method, although giving only approximate results, is always practicable in principle.

How do we then perform random tests over a set of points in an n-dimensional Euclidean space? Since each concrete point of this space is defined by a totality of n real numbers, which are the Cartesian coordinates of this point, the set of points we are interested in is, in turn, adequate to a set of manifolds of n real numbers, each of which is a coordinate of the corresponding point at R_n. Therefore it is possible to subject a set of points in an n-dimensional Euclidean space to random tests by performing random tests over the set of points in the space R_1. The manifold of the results of n independent random tests over points of R_1 is the result of a single random test over points in R_n. On the other hand, each random test performed over a set of points in R_1 yields some real number. The numbers obtained as a result of such random tests are called random numbers. Since any number is recorded with the aid of numbers from the set $\{0, 1, 2, 3, 4, 5, 6, 7, 8, 9\}$ (in the decimal system), then in order to obtain a random number containing k digits it will suffice to perform k independent random tests over this initial set of numbers. The random tests themselves can be performed in different ways (for instance, with the aid of roulette or by withdrawing, one ball after another from an urn containing 10 identical balls, marked with numbers from 0 to 9, and then placing them back, etc.). But in any method of performing random tests the requirement for randomness of these tests is mandatory (see also 1.6). If we multiply a manifold of k-digit random numbers by 10^{-k}, the obtained manifold of new random numbers will fall into the interval (0, 1). It must evidently correspond to a random variable t, whose probability density is unity in the interval (0, 1) and 0 outside it. For this random variable, from (3.48) and (3.59a) follow, respectively, $Mt = 1/2$ and $Dt = 1/12$. Therefore, if randomness of the tests is actually ensured, our manifold of random numbers must have an average value close to 1/2 and an experimental dispersion close to 1/12, which is the first criterion for the fulfillment of the requirement for the randomness of the tests which have lead to the given set of random numbers.

A set of random numbers once obtained can be used as many times as may be required for solving any problems which can be solved with their aid. This is precisely the advantage of the numerical experiment (that is, of the use of random numbers) as compared with the other methods for conducting a statistical experiment. But far from all probabilistic problems can be solved by the method of number experimentation. From the above substantiation of this method it follows that it is applicable only to those problems for which one can construct, from strict theoretical considerations, at least one set $A^{(a)}$ adequate to the set of concrete objects studied. In all other cases this method is inapplicable. Thus, for instance, it cannot be applied for solving the problem on the probability of opening up the 'head' in a single tossing of an irregular coin or of six points coming up on a single throw of a loaded die, etc. In such cases it is impossible to skip a direct statistical experiment over the set of concrete objects itself.

We will now illustrate the numerical experimentation on a simple, but important example, namely on the Bertrand problem. It is very difficult to solve this problem by direct experiment. Let us assume that we draw a circle of unit radius on a horizontal plane and drop a sufficiently long needle on it in a random manner, noting the number of cases when the needle intersects the circle at two points, and the number of cases among them when the length of the chord made by the needle exceeds $3^{1/2}$. How can we ensure the randomness of such tests? If we do not try to drop the needle on the circle, the number of misses will be enormous. Conversely, if we do try to hit the circle, the randomness is sure to be disturbed. Hence, it is necessary to construct a device, which will perform such 'drops', automatically ensuring their randomness. This will hardly be a simple matter, however. Let us see how this problem can be solved with the aid of random numbers.

In 1.8–1.9 we constructed two sets $A^{(a)}$, adequate to the set of chords in the circle, which can be used to obtain a solution to the Bertrand problem, always the same into the bargain. We will now try to find an explicit expression for the probability density of the random variable 'the chord length' by using equations of Section 3.3. Since $l = 2(1 - x^2 - y^2)^{1/2}$, where x and y are the Cartesian coordinates of the midpoint of a chord of length l in a reference system with the origin at the center of the circle, then

$$\rho(l, x, y) = c\,\delta\,[l - 2(1 - x^2 - y^2)^{1/2}], \tag{4.1}$$

where c is a constant. Therefore

$$\rho(l) = c \int_{-\infty}^{\infty} \int_{-\infty}^{\infty} \delta\,[l - 2(1 - x^2 - y^2)^{1/2}]\,\mathrm{d}x\,\mathrm{d}y. \tag{4.2}$$

Integrating in polar coordinates, we obtain

$$\rho(l) = c' \int_{0}^{\infty} \delta\,[l - 2(1 - r^2)^{1/2}]\,r\,\mathrm{d}r, \tag{4.3}$$

whence, using (3.31), we find

$$\rho(l) = c''l \tag{4.4}$$

which is an expression of the type (1.10). We thus obtain the solution (again the same) to the Bertand problem in one more way.

To be able to apply random numbers for solving the Bertrand problem we formulate it in an equivalent form, more amenable to calculation. What is the probability that the square of the length of a chord of a unit radius circle exceeds 3? This problem, in turn, is equivalent to the problem: What is the probability that the sum of the squares of the Cartesian coordinates of the midpoint of a chord of a unit radius circle is less than 1/4? Before we proceed to solve this problem with the aid of random numbers, let us consider it analytically.

Denoting $x^2 + y^2 = z$, we obtain from (3.34)

$$\rho(z, x, y) = \delta(z - x^2 - y^2)\rho_1(x)\rho_2(y). \tag{4.5}$$

It is most convenient to use random numbers from the range (0, 1). The probability densities $\rho_1(x)$ and $\rho_2(y)$ corresponding to these random numbers, with an allowance for the type of function $\delta_\epsilon(x)$ (3.17), can be written as

$$\rho_1(x) = \delta_1(x - \tfrac{1}{2}); \qquad \rho_2(y) = \delta_1(y - \tfrac{1}{2}) \tag{4.6}$$

(since $\epsilon = 1$). Then from (5), by virtue of (6) follows

$$\rho(z) = \int_{-\infty}^{\infty}\int_{-\infty}^{\infty} \delta(z - x^2 - y^2)\,\delta_1(x - \tfrac{1}{2})\,\delta_1(y - \tfrac{1}{2})\,dx\,dy. \tag{4.7}$$

Integrating with respect to y, we obtain

$$\rho(z) = \int_{-\infty}^{\infty} 2^{-1}(z - x^2)^{-1/2}\,\delta_1(x - \tfrac{1}{2})\,\delta_1(z - x^2 - \tfrac{1}{2})\,dx. \tag{4.8}$$

The integrand is different from zero and equal to $\frac{1}{2}(z - x^2)^{-1/2}$ only at those values of the arguments in the functions δ_1 which lie simultaneously in the range (0, 1). Therefore

$$\begin{aligned} &\rho(z) = 0 \quad (z < 0; z > 2),\\ &\rho(z) = \int_0^{z^{1/2}} 2^{-1}(z - x^2)^{-1/2}\,dx = \pi/4 \quad (0 < z < 1),\\ &\rho(z) = \int_{z^{1/2}-1}^{1} 2^{-1}(z - x^2)^{-1/2}\,dx = 0.5 \arcsin(2z^{-1} - 1) \quad (1 < z < 2), \end{aligned} \tag{4.9}$$

which can be abridged to

$$\rho(z) = 4^{-1}\pi\,\delta_1(z - \tfrac{1}{2}) + 2^{-1}\int_1^2 \arcsin(2z^{-1} - 1)\,\delta(z - z')\,dz'. \tag{4.10}$$

We can now calculate the sought-for probability. Since we are concerned with chords, the condition $z < 1$ must be fulfilled. Hence, the sought-for probability, which will be denoted by $P(l > 3^{1/2})$, is equal to

$$P(l > 3^{1/2}) = \int_0^{1/4} \delta_1(z - \tfrac{1}{2})\,dz \Big/ \int_0^1 \delta_1(z - \tfrac{1}{2})\,dz = \tfrac{1}{4}. \tag{4.11}$$

We have again obtained the familiar answer (see 1.9).

With the aid of random numbers this problem is solved as follows. In a sequence of random numbers we take the first pair of numbers and calculate the sum of their squares. We obtain the first random value of z. Then we take the second pair of numbers and, by the same operations, obtain the second random value of z. This computational process is continued until the necessary (or possible for this sequence of random numbers) number of random values of z is obtained. Assume that N random values of z are obtained. Suppose, further, that among them n values do not exceed 1, and m values do not exceed 1/4. Then m/n gives the approximate value of $P(l > 3^{1/2})$. Incidentally, the side result obtained is also of interest, namely, the relation $(N - n)/N$ must be approximately equal to $P(z > 1)$.

Using the 1250 two-digit random numbers listed in Table IV from Ref. [4], one can obtain 250 pairs of five-digit decimal fractions of the type 0. – – – – –, placing the first digit at the first decimal place in the first number of the first pair and the second digit of the first number of the first row of the table in the second number of the same pair, the corresponding digits of the second number of the same row at the second places, and so on, until all the five decimal places are filled; the following pairs of numbers are constructed in the same way from the following two-digit numbers. Thus, using all the random numbers of this table, we obtain $N = 250$. It is easy to ascertain with the aid of this table that $n = 196$ and $m = 51$, whence $P(l > 3^{1/2}) = 0.26$. This experimental value coincides remarkably with the theoretical value obtained above in different ways (namely, 1/4). Besides, the experimental value $P(z > 1) = 54/250 = 0.216$ also coincides remarkably with the theoretical value $(1 - \pi/4 = 0.215)$, which confirms the correctness of Eqn. (10).

4.4. Concluding Remarks

In the conventional courses of statistics the initial concepts are 'population' and 'sample' (random). The population actually coincides with the (tested) set of concrete objects in probabilistics, when this set is finite or denumerable. If statistics also considered populations containing nonenumerable sets of elements, then in any case population would differ from a set of concrete objects only in name. A sample is a finite set of elements randomly sampled from among a population. Therefore, sample is evidently just another name for a set of concrete objects revealed upon a finite number of random tests over the entire corresponding set of concrete objects.

One of the main tasks of statistics is to estimate the probabilistic characteristics of a population by some characteristics of samples from it. Since a sample corresponds to a sequence of random tests, and hence of random events, this problem appears to be purely probabilistic, which would seem to predetermine the possibility of only probabilistic judgements of the estimates obtained. For instance, if we wished to estimate the probability p of some event by the frequency m/n of its appearance in a sample consisting of a large number n of results of the corresponding random tests, we could only assert, by virtue of (2.45), that with a given n the probability that the relation $|m/n - p| < \epsilon$ will hold is the higher, the larger is ϵ, but for a finite ϵ it never reaches unity.

The lack of complete confidence in the correctness of the estimate obtained appears inacceptable, however, from the practical standpoint, and therefore the concept of practical 'confidence' is introduced in statistics. An estimate is considered practically confidential if its probability is 'practically' equal to unity. The probability which can be considered practically equal to unity (cf. the term 'degree of confidence' in 1.7) depends primarily on the practical task itself. With a given probability (regarded as practical confidence) the range containing the estimate of interest to a practical man, so-called 'confidence interval', depends on the problem to be tackled, on the volume of the sample (the number of elements in it), and on the method for solving the problem. But with any correct method for solution the confidence interval for each problem is the wider, the higher is the probability chosen as certainty, and it is the narrower, the larger is the sample volume.

To illustrate, we revert to the above example of solution of Bertrand problem with the aid of random numbers. The theoretical values of p and $q = 1 - p$ for this problem are equal to: $p = 1/4$; $q = 3/4$. Since n was found to be equal to 196 in the experiment, the conditions under which Eqn. (2.45) is approximately correct are fulfilled, and the equation can be used for finding the confidence interval corresponding to the given probability (practical confidence). Let us choose the probability of 0.95. Then, using the table containing the values of the integral on the right-hand side of (2.45) as a function of z, we find $\epsilon = 0.08$. And this is precisely the sought-for confidence interval. The experimental value $|m/n - p| = 0.01$ evidently actually lies within this interval. Taking the probability of 0.997, we obtain the confidence interval: $\epsilon = 0.12$. If we wished to reduce the confidence interval by a factor of 10, we would have to increase the number of tests (n) hundred-fold, that is, take 25,000 pairs of random numbers.

We will note in conclusion that the introduction of practical confidence in statistics is an artificial mode. Probabilistics has no need for such artificial devices, since from the very outset it proceeds from the experimentally established fact that with a large number of tests the frequency of appearance of an event is approximately equal to its probability, which is defined as a relative measure of the corresponding subset of the set of concrete objects under test (this set being replaced by a set adequate to it, if necessary).

Suggested References

1. Biberman, L., N. Sushkin, and V. Fabricant: 1949, *DAN SSSR* 66, 185.
2. Cramer, H.: 1946, *Mathematical Methods of Statistics*, Princeton U.P., Princeton.
3. Hudson, D.: 1963, *Lectures on Elementary Statistics and Probability*, CERN, Geneva.
4. Storm, R.: 1967 (2nd edn.) and 1969 (3rd edn.), *Wahrscheinlichkeitsrechnung, Mathematische Statistik und Statistische Qualitatskontrolle*, Fachbuchverlag, Leipzig.

Chapter 5

States of Abstract Objects

5.1. Introductory Remarks

The term 'state' is wide-spread both in various fields of human activity and common life. One may speak, for instance, of the state of some affairs, of the economic state of a country, as well as of the state of a physical system, and so on. Although this term can hardly be defined exactly by reference to some simpler notions, its meaning is, as a rule, clear enough and understandable. As the above examples show, when speaking of a state we always mean some qualitative or quantitative set of characteristics of an object. One may, for instance, say that the state of health of Mr. N is good (bad or satisfactory). The same qualitative estimates may be used when one characterizes crops, economics, etc.

It is possible, however, to give also a quantitative characteristic of Mr. N's health, having indicated numbers by means of which different indices of health are usually estimated in medicine – number of heart beats per minute (pulse), blood pressure, content of haemoglobin in the blood, etc. For crops and other objects, for which a qualitative estimation of states is possible, a quantitative one is possible too. But for physical systems only a quantitative characterization of a state makes sense. For instance, an equilibrium state of a gas is characterized by its temperature, pressure, and specific volume. A state of a mechanical system consisting of N particles is characterized by their coordinates and momenta.

As has been indicated in Section 4.1, in the last analysis, according to the very origin of this word, statistics deals with a description of states. It might seem that the term 'state' relates here to the whole set of concrete objects under examination. But, first, as has been noted before, the set is not always given wholly, and even the measure (L) cannot always be introduced in it. Secondly, characterization of a state of a set of concrete objects by means of the relative measures of its subsets corresponding to different values of the properties of the concrete objects coincides with the indication of the distribution of the probability for the abstract object, which is the image of this set, to have the respective values of these properties. Therefore, in probabilistics we will regard the term 'state' as referring to abstract objects and suppose that a state of an abstract object (sometimes we will say: "a state in which an abstract object is") is characterized by probability distribution for it to have different values of the properties of the concrete objects corresponding to it. This makes sense for any sets of concrete objects, which are capable of being subjected to random tests (see 1.8–1.9), even in the cases when it is impossible to introduce measure (L) into them (because of which the term 'state' cannot be referred directly to these sets).

Let us recall (Chap. 1) that to each abstract object $f_{0ij\ldots k} \in \theta$, which is the image of class $A(f_{ij\ldots k})$ of the concrete objects of A, there corresponds a definite distribution of probabilities over the values of all the remaining properties of these concrete objects, in particular, over the values $f_{lm\ldots p} \in \theta_{lm\ldots p}$ of any combined property $t_{lm\ldots p} = (t_l, t_m, \ldots, t_p)$. Therefore, a definite state corresponds to each abstract object $f_{0ij\ldots k}$, namely, the state determined by the value $f_{ij\ldots k}$ of the property $t_{ij\ldots k}$. This fact allows one to say that the abstract object $f_{0ij\ldots k}$ is the abstract object f_0 being in the state $f_{ij\ldots k}$ (henceforth, we will denote a state by the value of the corresponding property of the concrete objects, which determines it). It is also possible to say that the abstract object $f_{0ij\ldots k}$ is the abstract object f_{0i} being in the state $f_{j\ldots k}$, and so on. Generally, any abstract object (except f_0, which is the most abstract one for the given set of concrete objects) can be considered any one of the pertinent more abstract objects being in the corresponding definite state.

So, for instance, a pendulum oscillating with a frequency ω and an amplitude l is the abstract object 'pendulum' being in the state for which the property 'frequency' has the value ω and the property 'amplitude', the value l. At the same time, one can also regard it both as the abstract object 'pendulum of frequency ω' being in the state for which the property 'amplitude' has the value l and as the abstract object 'pendulum of amplitude l' being in the state for which the property 'frequency' has the value ω. The state of such pendulum, at any indicated interpretation, is uniquely characterized by the probability distribution of the property 'deviation from the equilibrium position'. For a harmonic pendulum this probability distribution can be easily obtained theoretically. It does not depend on time and has the form as given in (3.43), where R is replaced by l and x (or y) is the small deviation of the pendulum from its equilibrium position.

In view of the note at the end of Section 1.1, we can always regard states of any abstract objects as being characterized by probability distributions of some appropriate random variables. This fact makes it possible to approach the description of states of any abstract objects from a unique point of view.

5.2. Description of States

We will say that a mathematical quantity κ describes the state $f_{ij\ldots k}$ of the abstract object f_0 (in short, the state $f_{0ij\ldots k}$) with respect to the random variable $t_{lm\ldots p}$, if κ uniquely determines the probability distribution of this random variable in the given state. It is obvious that if κ_1 describes a state and κ_2 uniquely determines κ_1, then κ_2 also describes the state.

In Section 3.2 we became acquainted with two natural ways of representing the probability distribution of a random variable, namely, with the aid of the distribution function and the probability density. It follows from the above definition of the notion 'description of a state' that these mathematical quantities may both serve for the description of any state $f_{0ij\ldots k}$ with respect to each random variable $t_{lm\ldots p}$ attached to the set of concrete objects under consideration. The distribution function $F(f_{0ij\ldots k}; x_l, x_m, \ldots, x_p)$ defined by Eqn. (3.7) always gives the exact description of a state. As regards the probability density defined by Eqn. (3.9) it is necessary to stress that if the random variable has a discrete set (or subset) of values, then the exact description of a state is only possible when the Dirac δ-function is used. Let us consider this problem in more detail, using the example of an one-dimensional random variable t_l.

We will confine ourselves to the case when the whole set of the values f_l of this random variable contains a finite or denumerable subset of discrete values $f_l^{(j)}$ in the intérval $(-\infty, \alpha_1)$, while the rest of its values is continuous in the interval (α_1, α_2). The probability density $\rho(x_l)$ then has the form (the designation of the state being omitted)

$$\rho(x_l) = \begin{cases} \sum_j p_j\, \delta(x_l - f_l^{(j)}) & (-\infty < x_l < \alpha_1) \\ \rho_1(x_l) & (\alpha_1 < x_l < \alpha_2) \\ 0 & (\alpha_2 < x_l < \infty), \end{cases} \tag{5.1}$$

where p_j is the probability of the value $f_l^{(j)}$ and $\rho_1(x_l)$ is the probability density of t_l in its continuity interval.

This probability density can also be written in the equivalent form

$$\rho(x_l) = \sum_j p_j\, \delta(x_l - f_l^{(j)}) + \int_{\alpha_1}^{\alpha_2} \rho_1(x_l')\, \delta(x_l - x_l')\, \mathrm{d}x_l'. \tag{5.2}$$

Equation (2) gives the exact description of a state with respect to the random variable t_l, since by using it we can get the distribution function $F(x_l)$ for this state, which has the form

$$F(x_l) = \int_{-\infty}^{x_l+\eta} \rho(x_l)\, \mathrm{d}x_l = \begin{cases} 0 & (-\infty < x_l < f_l^{(1)}) \\ \sum_{j=1}^{k} p_j & (f_l^{(k)} < x_l < f_l^{(k+1)}) \\ \sum_j p_j + \int_{\alpha_1}^{x_l} \rho_1(x_l)\, \mathrm{d}x_l & (\alpha_1 < x_l \leqslant \alpha_2) \\ 1 & (\alpha_2 < x_l < \infty), \end{cases} \tag{5.3}$$

where η is an arbitrary small positive number.

Having replaced δ-function by δ_ϵ-function defined by Eqn. (3.17), in the first term on the right-hand side of (2), we will obtain

$$\rho(x_l) = \sum_j p_j \, \delta_\epsilon(x_l - f_l^{(j)}) + \int_{\alpha_1}^{\alpha_2} \rho_1(x_l') \, \delta(x_l - x_l') \, \mathrm{d}x_l'. \tag{5.4}$$

Equation (4) gives only an approximate description of a state, that is the more exact, the less ϵ. But for some reasons, which will be elucidated later on, in a number of cases the approximate description of a state with the aid of (4) proves to be more convenient than its exact description by means of (2). Equation (4) can be obtained directly from the expression (3) for the distribution function $F(x_l)$ by using the following simple considerations.

Let us introduce the operation of 'ϵ-differentiation', having defined the 'ϵ-derivative' of a function, $\mathrm{d}_\epsilon/\mathrm{d}x$, by the following relation

$$\frac{\mathrm{d}_\epsilon f(x)}{\mathrm{d}x} = \frac{f(x + \epsilon/2) - f(x - \epsilon/2)}{\epsilon}, \tag{5.5}$$

where ϵ is a given positive number. This definition will be used frequently in what follows. If there exists $\lim_{\epsilon \to 0} \mathrm{d}_\epsilon f(x)/\mathrm{d}x$, then it coincides with the derivative $\mathrm{d}f(x)/\mathrm{d}x$. By applying this operation to $\mathrm{d}_\epsilon f(x)/\mathrm{d}x$, we get $\mathrm{d}_\epsilon^2 f(x)/\mathrm{d}x^2$. Upon applying it to $\mathrm{d}_\epsilon^2 f(x)/\mathrm{d}x^2$, we obtain $\mathrm{d}_\epsilon^3 f(x)/\mathrm{d}x^3$, and so on. Any function $f(x)$ proves thus to be unlimitedly ϵ-differentiable.

The sequence of the ϵ-derivatives of the Heviside function $H(x)$ defined as

$$H(x) = \begin{cases} 0 & (-\infty < x < 0) \\ 1 & (0 < x < \infty) \end{cases} \tag{5.6}$$

is of special interest. It is easily seen, in particular, that in view of (3.17)

$$\frac{\mathrm{d}_\epsilon H(x)}{\mathrm{d}x} = \delta_\epsilon(x). \tag{5.7}$$

One can consider the Heviside function the distribution function describing a state of an abstract object with respect to a random variable t. Then the right-hand side of (7), which is approximately equal to the probability density of t, gives the description of the same state, but only an approximate one. By integrating $\delta_\epsilon(x)$ between $-\infty$ and x, we obtain the approximate distribution function corresponding to it, instead of Heviside function, namely

$$F(x) = \begin{cases} 0 & (-\infty < x < -\epsilon/2) \\ \tfrac{1}{2} + \epsilon^{-1} x & (-\epsilon/2 < x < \epsilon/2) \\ 1 & (\epsilon/2 < x < \infty). \end{cases} \tag{5.8}$$

It follows from (8), by the way, that

$$\int_{-\eta_1}^{0} \delta_e(x)\,dx = \int_{0}^{\eta_2} \delta_\epsilon(x)\,dx = \tfrac{1}{2} \tag{5.9}$$

for any η_1 and η_2, greater than $\epsilon/2$, at any ϵ. Therefore Eqn. (9) is also valid for the Dirac δ-function with any positive η_1 and η_2.

The above statement shows that Eqn. (4) can be obtained from (3) as a result of the ϵ-differentiation of $F(x_l)$ in the interval, where its form is 'stepped', and the ordinary differentiation in the remaining interval. If the set of discrete values $f_l^{(j)}$ is finite, it is possible to choose ϵ to be smaller than the least distance between two adjacent values, $f_l^{(j)}$ and $f_l^{(j+1)}$ (although nothing prevents one from choosing an own ϵ_j for each $f_l^{(j)}$). The case of a denumerable set of discrete $f_l^{(j)}$-values does not present any specific difficulties for its consideration either, but for simplicity we omit it here.

In a number of important cases δ_ϵ-functions also give exact descriptions of states. In particular, this is the case for all states, where the probability density of the corresponding random variable t is constant in some interval (a, b) and equals zero outside it. Such states are described by the probability density $\rho(x)$, which has the form

$$\rho(x) = \delta_{b-a}\left(x - \frac{a+b}{2}\right). \tag{5.10}$$

The direct calculation by Eqns. (3.48) and (3.59a), respectively, gives for these states the following exact values of Mt and Dt:

$$Mt = (a+b)/2; \qquad Dt = (b-a)^2/12. \tag{5.11}$$

In Section 4.3 we have already dealt with these kind of states described by the probability density of the form $\delta_1(x - \frac{1}{2})$, for which $a = 0$ and $b = 1$. In case $a = -b$ Eqn. (10) becomes

$$\rho(x) = \delta_{2b}(x) \tag{5.12}$$

and Eqn. (11) gives

$$Mt = 0; \qquad Dt = b^2/3. \tag{5.13}$$

In Section 2.5 we have acquainted ourselves with one more way for the representation of the probability distribution of random variables, namely, with the aid of the characteristic function $\Phi(f_{0ij\ldots k}; \epsilon_l, \epsilon_m, \ldots, \epsilon_p)$ defined by Eqn. (3.74) (in that equation the designation of the state is omitted). It has been noted there that in distinction to the distribution function and probability density, which have an immediate probabilistic meaning, the characteristic function does not have such a meaning; and is an auxiliary mathematical quantity only. Since the characteristic function uniquely determines the distribution function, also that may serve for the description of states. It is easy to make sure by direct calculation with the aid of Eqn. (3.74) that, in particular, a state, to which there corresponds the probability density of the form (12) is described by the characteristic function

$$\Phi(\epsilon_l) = \frac{\sin \epsilon_l b}{\epsilon_l}. \tag{5.14}$$

It is possible, of course, to invent many other auxiliary mathematical quantities describing states. The most interesting are, however, two of them, namely, the state vector ψ and the density operator $\hat{W}$, whose introduction has lead to extremely significant results. But before proceeding to a consideration of these quantities let us recall some mathematics and obtain additional results, which will be necessary for the further exposition.

5.3. Necessary Mathematics

Let two unique operations, 'addition' and 'multiplication' by a number (generally complex), be defined on a set R of arbitrary elements $x, y, \ldots, z, \ldots$ The operation of addition will be denoted by '+' and the operation of multiplication, by '$\cdot$' (the latter symbol will be, as a rule, omitted). Further, let for any elements x, y, and z of R, and any numbers α and β, the following conditions be satisfied:

(1) $x + y = y + x$;
(2) $(x + y) + z = x + (y + z)$;
(3) $0 \cdot x = 0$, where 0 at the right is the only element of R obeying this equality for any $x \in R$ (the 0 at the left is the number zero);
(4) $1 \cdot x = x$;
(5) $\alpha(\beta x) = (\alpha\beta)x$;
(6) $(\alpha + \beta)x = \alpha x + \beta x$;
(7) $\alpha(x + y) = \alpha x + \alpha y$.

Then set R is said to be a vector space, and its elements are called vectors.

Vectors $\mathbf{x}, \mathbf{y}, \ldots, \mathbf{u}$ of R are said to be linearly independent, if the equation

$$\alpha x + \beta y + \cdots + \delta u = 0 \tag{5.15}$$

may be satisfied only when all the numbers $\alpha, \beta, \ldots, \delta$ are simultaneously equal to zero. Otherwise, if this equation is satisfied without all its coefficients being zero, these vectors are called linearly dependent. If in the space R there exist n linearly independent vectors, whereas any $n + 1$ its vectors are linearly dependent, then the space is called n-dimensional. But if in a space there exist a denumerable or nonenumerable set of linearly independent vectors, then the space is said to be infinite-dimensional.

Any system of ordered linearly independent vectors $\mathbf{e}_1, \mathbf{e}_2, \ldots$ is called a basis of the space (space basis), if any vector x of this space is linearly dependent with this system, that is, if the vectors $x, e_1, e_2, \ldots$ are linearly dependent. If that is the case, then an arbitrary vector x may be represented in the form

$$x = x_1 e_1 + x_2 e_2 + \cdots \tag{5.16}$$

Such representation of a vector is also called an expansion of a vector in terms of the basis vectors. The totality of the expansion coefficients $x_1, x_2, \ldots$ may be written, in the same sequence, in the form of either a column or a row. In the first case a column matrix is obtained, which we will write as $(x_1, x_2, \ldots)$, and in the second case, a row matrix, which will be written as $\| x_1 x_2 \ldots \|$. The same form of record for column and row matrices will be kept in what follows.

In the calculus of matrices, a sum of matrices of equal dimensions is defined as a matrix, whose elements are sums of the corresponding elements of the matrices added up; and a product of a matrix by a number, as a matrix whose elements are the products of the corresponding elements of the initial matrix by that number. The conditions (1)–(7) defining a vector space obviously hold valid for both column and row matrices. Therefore, in particular, the column matrices of the coefficients of expansion of vectors of the space R in terms of the basis vectors chosen are vectors themselves of the corresponding vector space, which will be called the e-representation space (for the initial vector space R). In this space the 0-vector (zero vector) is the column matrix, whose elements all are zeros. The basis vector e_j is represented in this space by the jth column of the unit matrix, whose elements are zeros, except the jth one, which is 1. Marking the vectors in the e-representation space by the inferior parenthetic index '(e)', we may write

$$e_{j(e)} = (0, 0, \ldots, 0, {}^{j}1, 0, \ldots). \tag{5.17}$$

Instead of column matrices, one might use row matrices for the representation of the vectors of the space R. This would not, however, essentially change anything. The transition from a column matrix to the row matrix with the same elements ordered in the same sequence, as well as the inverse transition, is called transposition of matrices. In what follows, the operation of transposition will be designated by the prime '′' with the letter denoting a matrix. A double transposition leads to the initial matrix. We may thus write

$$x_{(e)} = (x_1, x_2, \ldots); \qquad x'_{(e)} = \| x_1 x_2 \ldots \|; \qquad x''_{(e)} = x_{(e)}. \tag{5.18}$$

A column or a row matrix is a certain kind of recording of an ordered set of numbers – its elements – each of which is uniquely determined by its ordinal number in the set. Each element x_k of the matrix x (or x') may be thus regarded as the value at 'point' $j = k$ of some function $x(j) = x_j$ of the integer j. From this standpoint nothing will change if a discrete set of integers is replaced by a continuous set of real numbers forming some interval (a, b). Therefore, the function

$f(\alpha)$ uniquely assigning the corresponding value $f(\alpha')$ of the function to each value $\alpha' \in (a, b)$ of the argument may be regarded, at one's wish, as both a column and row matrix. The set $R_{(f)}$ of one-valued functions $f(\alpha)$ determined on the set (a, b) of the values of the argument α obeys the above conditions (1)–(7), which conforms to the possibility of regarding them as matrices. Therefore, $R_{(f)}$ is a vector space, namely, the so-called function space. The set of the basis vectors of this space is nonenumerable. Under certain conditions the record (16) for the expansion of a vector in terms of the basis vectors is replaced here by

$$f(\alpha) = \int_a^b f(\alpha')e(\alpha')\,d\alpha' = \int_a^b f(\alpha')\,\delta(\alpha - \alpha')\,d\alpha'. \tag{5.16a}$$

The basis vector $e(\alpha')$ is the generalized function (Dirac delta function) $\delta(\alpha - \alpha')$ analogous to the basis vector $e_{j(e)}$ in (17) whose elements are δ_{jk} (Kroneker dalta).

If in the space R an operation may be introduced which transforms each vector x of the space into the vector x' of the same or some other space (R'), then it is said that in the space R an operator is defined performing the transformation mentioned. We will write such transformation as

$$x' = Ax, \tag{5.19}$$

where A designates the operator. An operator A is called linear if it satisfies the conditions

$$A(x + y) = Ax + Ay, \tag{5.20}$$

where x and y are any vectors, and

$$A(\alpha x) = \alpha Ax, \tag{5.21}$$

where x is any vector and α is an arbitrary number. Henceforth we will deal with linear operators only.

Two operators A and B are considered equal if the equality

$$Ax = Bx \tag{5.22}$$

is satisfied for any vector x. By the sum of two operators A and B defined in one and the same space is meant such an operator $C = A + B$ which satisfies the equation

$$Cx = Ax + Bx \tag{5.23}$$

for any vector x. By the product of an operator A by a number α such an operator $C = \alpha A$ is meant for which any vector x satisfies the equality

$$Cx = \alpha Ax. \tag{5.24}$$

An operator may generally be defined only on a certain set of vectors of the space R, not in the entire space. The set of vectors on which the operator is defined is called the domain of definition of the operator. The set of vectors into which the operator transforms the vectors of its domain of definition is termed the domain of values of the operator. If the domain of definition of the operator B intersects with the domain of values of the operator A, then the sequence of operations performed first by operator A and then by operator B makes sense. By the product of operators B and A one means such an operator $C = BA$ for which each corresponding vector x of the domain of definition of A satisfies the equation

$$Cx = BAx. \tag{5.25}$$

It follows from the above that the domain of definition of the operator $C = BA$ does not generally coincide with that of operator A. Let $R_{d(A)}$ and $R_{v(A)}$ be the domains of definition and, respectively, of values of operator A, and let $R_{d(B)}$ be the domain of definition of operator B. Then the domain of definition of the operator C should satisfy the relation

$$R_{d(C)} \subseteq R_{d(A)} \tag{5.26}$$

the equality sign corresponding to the case when $R_{v(A)} = R_{d(B)}$. If $R_{v(A)} \cap R_{d(B)} = 0$, then the domain of definition of the operator C is empty, that is, this operator cannot be introduced for vectors of $R_{d(A)}$. The operator C makes sense only in the case when $R_{v(A)} \cap R_{d(B)} \neq 0$. If $R_{v(C)}$ is the domain of values of the operator C, then the relation

$$R_{v(C)} \subseteq R_{v(B)} \tag{5.27}$$

is obviously valid, where $R_{v(B)}$ is the domain of values of operator B.

In the general case

$$BA \neq AB. \tag{5.28}$$

Moreover, it may happen that operator BA makes sense, whereas operator AB does not, that is, such an operator cannot be introduced for the set of vectors under consideration at all.

If the domain of values of an operator A coincides with the domain of its definition, that is, if $R_{d(A)} = R_{v(A)}$, then there exist the operators $AA = A^2$, $AA^2 = A^3, \ldots, AA^n = A^{n+1}$ for any positive integer n. The operator E transforming any vector x into itself, according to the equation

$$Ex = x \tag{5.29}$$

is called the unit operator (or the operator of identical transformation). It follows from (29) that $AE = EA = A$. Therefore, we may also write: $E = A^0$. If the domains of definition and of values of the operators $A, B, \ldots, D$ coincide, then there exist the products of these operators: $D \ldots BA, AB \ldots D, A \ldots BD, \ldots$, as well as any integer powers of these operators and of any of their products, including, of course, $E = A^0 = B^0 = \cdots = D^0$.

For the operation of multiplication of operators by operators, as is easily seen, the following relations are fulfilled:

$$A(B + C) = AB + AC; \qquad (A + B)C = AC + BC; \qquad (\alpha A)B = A(\alpha B) = \alpha AB. \quad (5.30)$$

The operator B obeying the equation

$$BA = E \tag{5.31}$$

is called the inverse operator for A and is designated by A^{-1}. Not for any operator there exists an inverse one, but if for each of the operators $A, B, \ldots, D$ there exist inverse ones, then it follows from (31) that

$$(D \ldots BA)^{-1} = A^{-1}B^{-1} \ldots D^{-1}. \tag{5.32}$$

The inverse operator for A^n is A^{-n}. Therefore, for any l and m integer (both positive and negative), including zero, the relation $A^l A^m = A^{l+m}$ is valid.

Let the operator A transform each vector of the space R into the corresponding vector of the space R'. Further, let the vectors e_j ($j = 1, 2, \ldots$) be the basis in R and vectors e'_k ($k = 1, 2, \ldots$), the basis in R'. Then operator A transforms each basis vector e_j into a definite vector of the space R', which can be expanded in terms of e'_k. We may write this as

$$Ae_j = \sum_k a_{kj} e'_k. \tag{5.33}$$

From (16) and (33), in view of the linearity of the operator A, it follows that to the transformation (19) from the R-space vectors to the R'-space vectors, performed with the aid of operator A, there corresponds the transformation

$$x'_{(e')} = Ax_{(e)} \tag{5.34}$$

from the vectors of the e-representation space for R to those of the e'-representation space for R', performed by the matrix $A = \| a_{ij} \|$ (the first index numbers the rows, the second one the columns of the matrix). The right-hand side of (34) is the product of the column matrix $x_{(e)}$ by the matrix A, which is calculated according to the rules of calculus of matrices, namely: If $A = \| a_{ij} \|$, $B = \| b_{ij} \|$, $C = \| c_{ij} \|$, and $C = BA$, then

$$c_{ij} = \sum_k b_{ik} a_{kj}. \tag{5.35}$$

If the spaces R and R' do not coincide, the matrix A does not obviously represent operator A either in the representation space for R or in that for R'. But if R and R' do, then at the given choice of the basis e_j ($j = 1, 2, \ldots$) the matrix $A_{(e)}$ is an operator in the representation space and, therefore, represents operator A in this space. In this case, each operator defined in the space R is represented by the corresponding matrix in the e-representation space. In particular, the unit operator E is represented by the unit matrix $E = \| \delta_{ij} \|$, the inverse operator A^{-1}, if it exists, by the matrix $A_{(e)}^{-1}$. Hence it follows, by the way, that the operator A^{-1} exists if $\operatorname{Det} A_{(e)} \neq 0$, since only in this case $A_{(e)}^{-1}$ does.

When the basis of the R-space changes, so do both the column matrices representing vectors and the square ones representing operators, that is, the representation space corresponding to the space R changes wholly. If the new basis vectors g_j are related to the vectors e_j by

$$e_j = \sum_k b_{kj} g_k, \tag{5.36}$$

then the column matrix $x_{(g)}$ representing the vector x in this basis is connected with the column matrix $x_{(e)}$ representing the same vector in the old one by the relation

$$x_{(g)} = Bx_{(e)} \tag{5.37}$$

where $B = \| b_{ij} \|$. The matrix $A_{(g)}$ representing operator A defined in the R-space in the new basis is connected with the matrix $A_{(e)}$ representing this operator in the old one by

$$A_{(g)} = BA_{(e)}B^{-1}. \tag{5.38}$$

Equation (38) defines the so-called similarity of the matrices $A_{(e)}$ and $A_{(g)}$; its right-hand side is known to be the similarity transformation. The analogous operator relations are also called similarity transformations, and operators connected by such relations are said to by similar operators.

If in a space R there exists a vector v such that the operator A transforms it according to the equation

$$Av = \lambda v \tag{5.39}$$

where λ is a number, then this vector is called the eigenvector, and λ the eigenvalue of the operator A. As can be easily proven, similar operators have equal eigenvalues. Since Eqn. (39) holds valid upon multiplying v by a non-zero number, the eigenvectors of an operator can be determined with an accuracy to a numerical factor only. The eigenvectors corresponding to different eigenvalues are linearly independent. If to an eigenvalue there correspond a few eigenvectors, then their linear combination also is an eigenvector relating to the same (so-called multiple, or degenerate) eigenvalue. Among these linear combinations it is always possible to choose linearly independent ones. Therefore, the totality of eigenvectors of a linear operator may always be regarded as a system of linearly independent vectors. If any vector of the space under consideration can be expanded in terms of the vectors of this system, then the operator is said to have a total system of eigenvectors. The totality of ordered eigenvectors of such operator may, hence, be chosen as the basis of the space.

On the ground of (33), the matrix representing the operator A in the basis of its eigenvectors v_j has the form

$$A_{(v)} = \Lambda = \| \lambda_j \, \delta_{jK} \| . \tag{5.40}$$

The totality of the equations

$$A_{(e)} v_{j(e)} = \lambda_j v_{j(e)} \tag{5.41}$$

which the eigenvectors and eigenvalues of the operator A obey in the e-representation, may be now rewritten in the form

$$A_{(e)} L_{(e)} = L_{(e)} \Lambda, \tag{5.42}$$

where $L_{(e)}$ is the matrix whose jth column is $v_{j(e)}$. From (42) it follows that

$$A_{(e)} = L_{(e)} \Lambda L_{(e)}^{-1}. \tag{5.43}$$

So, the matrix corresponding to the operator A in any representation space is connected with the diagonal matrix Λ of its eigenvalues by the corresponding similarity transformation.

Let us designate the kth row of the matrix $L_{(e)}^{-1}$ by $w'_{k(e)}$. Then (43) may be rewritten in the form

$$A_{(e)} = \sum_j \lambda_j Z_{j(e)}, \tag{5.44}$$

where

$$Z_{j(e)} = v_{j(e)} w'_{j(e)}. \tag{5.45}$$

The matrices $Z_{j(e)}$ are said to be the components of the matrix $A_{(e)}$, and Eqn. (44) is the expansion of a matrix in terms of its components. From $L_{(e)} L_{(e)}^{-1} = E$ it follows that

$$w'_{j(e)} v_{k(e)} = \delta_{jK}. \tag{5.46}$$

Therefore,

$$Z_{j(e)} v_{k(e)} = \delta_{jk} v_{k(e)}; \qquad Z_{j(e)} Z_{k(e)} = \delta_{jk} Z_{k(e)}; \qquad Z_{j(e)}^2 = Z_{j(e)}. \tag{5.47}$$

Any vector of the representation space can be expanded in terms of the eigenvectors of the operator A in this representation. According to the first equation in (47), the matrix $Z_{j(e)}$ annuls all $v_{k(e)}$, except for $v_{j(e)}$ which remains unchanged. Hence, as a result of the action of $Z_{j(e)}$ on any vector in the representation space, only its component does not vanish, which corresponds to the eigenvector $v_{j(e)}$. In the initial space R to every matrix $Z_{j(e)}$ there corresponds the operator $A_{j(z)}$ whose action on any vector in R leads to the detention of only that component which corresponds to the eigenvector v_j of the operator A, whereas all its other components are deleted. Therefore, the operators $A_{j(z)}$ in terms of which the operator A can be expanded, according to the equation

$$A = \sum_j \lambda_j A_{j(z)} \tag{5.48}$$

corresponding to Eqn. (44), are termed the projection operators. For the projection operators there are obviously valid the relations analogous to those in (47), namely:

$$A_{j(z)}v_k = \delta_{jk}v_k; \qquad A_{j(z)}A_{k(z)} = \delta_{jk}A_{k(z)}; \qquad A^2_{j(z)} = A_{j(z)}. \tag{5.49}$$

A function $f(A)$ of the operator A is defined as such an operator whose eigenvectors coincide with those of A and the corresponding eigenvalues are equal to $f(\lambda_j)$. The expansion of $f(A)$ in terms of the projection operators obviously has the form

$$f(A) = \sum_j f(\lambda_j)A_{j(z)}. \tag{5.50}$$

For $f(A) = A^0 = E$, Eqn. (50) gives

$$E = \sum_j A_{j(z)}. \tag{5.51}$$

Equation (51) is known to be the expansion of unity. Such an expansion exists for any operator possessing a total system of eigenvectors, which may, hence, be taken as a basis of the space.

In any matrix representation, to Eqns. (50) and (51) there correspond the respective matrix equations:

$$f(A_{(e)}) = \sum_j f(\lambda_j)Z_{j(e)} \tag{5.52}$$

and

$$E = \sum_j Z_{j(e)}. \tag{5.53}$$

Equation (52) may be rewritten in the equivalent form

$$f(A_{(e)}) = L_{(e)} f(\Lambda) L_{(e)}^{-1} \tag{5.54}$$

analogous to (43), where $f(\Lambda)$ is the diagonal matrix $\|f(\lambda_j)\,\delta_{jk}\|$. From (54) it follows, in turn, that

$$f(\Lambda) = L_{(e)}^{-1} f(A_{(e)}) L_{(e)} \tag{5.55}$$

whence, in view of the above notation

$$w'_{j(e)} f(A_{(e)}) v_{k(e)} = \delta_{jk} f(\lambda_k). \tag{5.56}$$

The particular cases of (56) are

$$w'_{j(e)} A_{(e)} v_{k(e)} = \delta_{jk} \lambda_k \tag{5.57}$$

and (46).

Transposition of a matrix transforms each of its columns into the corresponding row with the same index (and, hence, also each row into the corresponding column with the same index). It follows from (35) that transposition of the product of matrices leads to the product of transposed ones inversely ordered, that is, if $C = AB$, then, denoting the transposed matrix by a prime,

$$C' = B'A'. \tag{5.58}$$

Transposing (54) and multiplying the equation obtained by $L'^{-1}_{(e)}$ from the right, we find

$$f(A'_{(e)}) L'^{-1}_{(e)} = L'^{-1}_{(e)} f(\Lambda), \tag{5.59}$$

where, as follows from (52),

$$f(A'_{(e)}) = \sum_j f(\lambda_j) Z'_{j(e)}. \tag{5.60}$$

From (59) it is seen that the column matrix $w_{j(e)}$, which is the jth column of the matrix $L'^{-1}_{(e)}$, relates to the jth eigenvector w_j of a certain operator represented by the matrix $f(A'_{(e)})$; this eigenvector corresponds to the eigenvalue $f(\lambda_j)$ of this operator, which is coinciding with the respective eigenvalue of the operator $f(A)$. Operators having the same sets of eigenvalues and represented in one and the same basis by transposed matrices will be termed transposed operators. The transposed operator will also be designated by the prime '′' with it. Then the connection between the projection operators for the operators A and A' may be indicated as follows

$$A'_{j(z)} = A_{j(z')}. \tag{5.61}$$

The trace of a matrix $A = \|a_{ij}\|$, recorded $\operatorname{Tr} A$, is defined as the sum of its diagonal elements

$$\operatorname{Tr} A = \sum_j a_{jj}. \tag{5.62}$$

It is easy to make sure that for any matrices A and B, for which the matrix AB exists and is a square one

$$\operatorname{Tr} AB = \operatorname{Tr} BA \tag{5.63}$$

(for matrices of infinite dimensions the possibility of changing the order of summation on the left-hand side of (63) must be valid too). Since transposition of a square matrix does not change its diagonal elements, we obtain, in view of (58)

$$\operatorname{Tr} AB = \operatorname{Tr} B'A' = \operatorname{Tr} A'B'. \tag{5.64}$$

It follows from (64), (45), and (46) that

$$\operatorname{Tr} Z_{j(e)} = \operatorname{Tr} Z'_{j(e)} = 1. \tag{5.65}$$

From (52) and (60), in view of (65), we get

$$\operatorname{Tr} f(A_{(e)}) = \operatorname{Tr} f(A'_{(e)}) = \sum_j f(\lambda_j). \tag{5.66}$$

In particular

$$\operatorname{Tr} A_{(e)} = \operatorname{Tr} A'_{(e)} = \sum_j \lambda_j. \tag{5.67}$$

The trace of an operator is defined as the trace of the matrix representing it, at any choice of the basis of the space. It follows from the above that the trace of an operator is equal to the sum of its eigenvalues, therefore similar operators have equal traces.

All results related to operators, which do not depend on the dimension of the space R, hold valid for function spaces too. But those in which space dimension figures explicitly must be appropriately altered. Formally that comes to the following: the discrete indices are replaced by continuous variables, the summation by the integration; the matrices representing operators are replaced by functions of two variables (arguments), called the kernels of operators (the first variable corresponds to the row number, and the second one to the column number); transposition of the matrix representing an operator is replaced by the transposition of the arguments in its kernel.

Equation (33) thus goes over into equation

$$Ae(\alpha) = \int a(\beta, \alpha)e(\beta)\,\mathrm{d}\beta. \tag{5.68}$$

Instead of Eqn. (35) we have

$$C(\alpha, \gamma) = \int b(\alpha, \beta)a(\beta, \gamma)\,\mathrm{d}\beta. \tag{5.69}$$

Equation (39) determining the eigenvectors and eigenvalues of an operator A goes over into the equations:

$$\text{(a)}\quad \int a(\beta, \alpha)\varphi_\lambda(\alpha)\,\mathrm{d}\alpha = \lambda\varphi_\lambda(\beta) \tag{5.70a}$$

with a set of solutions which is nonenumerable, or

$$\text{(b)}\quad \int a(\beta, \alpha)\varphi_j(\alpha)\,\mathrm{d}\alpha = \lambda_j\varphi_j(\beta) \tag{5.70b}$$

whose set of solutions is denumerable.

If (a) the system of functions $\varphi_\lambda(\alpha)$, or (b) that of functions $\varphi_j(\alpha)$ is complete, then it may be taken as a basis of the corresponding function space. An operator A is represented on this basis (a) by the function ('matrix') $\Lambda = \lambda\,\delta(\lambda - \lambda')$, (b) by the diagonal matrix of the form given in (40). Equation (42) thereby takes the form

$$\text{(a)}\quad \int a(\beta, \alpha)\varphi_\lambda(\alpha)\,\mathrm{d}\alpha = \int \lambda'\,\delta(\lambda - \lambda')\varphi_{\lambda'}(\beta)\,\mathrm{d}\lambda', \tag{5.71a}$$

$$\text{(b)}\quad \int a(\beta, \alpha)\varphi_j(\alpha)\,\mathrm{d}\alpha = \sum_k \lambda_k\,\delta_{jk}\varphi_k(\beta). \tag{5.71b}$$

To the matrix L^{-1} there corresponds then such a function (a) $\eta_\lambda(\alpha)$, (b) $\eta_j(\alpha)$, which obey the equations

$$\text{(a)}\quad \left\{\begin{aligned} &\int \varphi_{\lambda'}(\alpha)\eta_\lambda(\alpha)\,\mathrm{d}\alpha = \delta(\lambda - \lambda') && (5.72\text{a})\\ &\int \varphi_\lambda(\alpha)\eta_\lambda(\alpha')\,\mathrm{d}\lambda = \delta(\alpha - \alpha'), && (5.73\text{a})\end{aligned}\right.$$

$$\text{(b)}\quad \left\{\begin{aligned} &\int \varphi_j(\alpha)\eta_k(\alpha)\,\mathrm{d}\alpha = \delta_{jk} && (5.72\text{b})\\ &\sum_j \varphi_j(\alpha)\eta_j(\alpha') = \delta(\alpha - \alpha'). && (5.73\text{b})\end{aligned}\right.$$

The kernels of the corresponding projection operators are:

(a) $a_{\lambda(z)}(\beta, \alpha) = \varphi_\lambda(\beta)\eta_\lambda(\alpha)$, (5.74a)

(b) $a_{j(z)}(\beta, \alpha) = \varphi_j(\beta)\eta_j(\alpha)$. (5.74b)

It is possible, hence, to treat Eqns. (73a) and (73b) also as the corresponding expansions of unity analogous to (51). To Eqn. (50), which is the so-called spectral expansion of the operator $f(A)$, there correspond the equations;

(a) $a_f(\beta, \alpha) = \int f(\lambda)\varphi_\lambda(\beta)\eta_\lambda(\alpha)\, d\lambda$, (5.75a)

(b) $a_f(\beta, \alpha) = \sum_j f(\lambda_j)\varphi_j(\beta)\eta_j(\alpha)$, (5.75b)

where $a_f(\beta, \alpha)$ is the kernel of the operator $f(A)$. The kernel of the transposed operator has the form $a'_f(\beta, \alpha) = a_f(\alpha, \beta)$. Therefore

(a) $a'_f(\beta, \alpha) = \int f(\lambda)\varphi_\lambda(\alpha)\eta_\lambda(\beta)\, d\lambda$, (5.76a)

(b) $a'_f(\beta, \alpha) = \sum_j f(\lambda_j)\varphi_j(\alpha)\eta_j(\beta)$. (5.76b)

The functions (a) $\eta_\lambda(\alpha)$, (b) $\eta_j(\alpha)$, are the eigenfunctions of the operator A' obeying equations analogous to Eqns. (70a), respectively, (70b).

By the scalar product (x, y) of a pair of vectors x and y given in a definite order is meant a complex number satisfying the following conditions: (1) $(y, x) = \overline{(x, y)}$; (2) $(\alpha x, y) = \alpha(x, y)$; $(x + y, z)$ $(x, z) + (y, z)$ (the line '—' over a letter denotes its complex conjugate; α is a number). From these conditions the following result: (4) $(x, \alpha y) = \bar{\alpha}(x, y)$; (5) $(x, y + z) = (x, y) + (x, z)$.

It follows from condition (1) that $(x, x) \geqslant 0$. It is usually agreed that the equality $(x, x) = 0$ may be valid if, and only if, $x = 0$. An important result is that for any vectors x and y the Schwarz's inequality

$$|(x, y)| \leqslant (x, x)^{1/2}(y, y)^{1/2} \tag{5.77}$$

is true.

Let for the vectors e_i and e_j of a basis of the space R $(e_i, e_j) = g_{ij}$. Then $g_{ji} = \bar{g}_{ij}$. A matrix $A = \|a_{ij}\|$ satisfying the condition $a_{ji} = \bar{a}_{ij}$ is called a Hermitian matrix. Thus, the matrix $G = \|g_{ij}\|$ is a Hermitian matrix. In view of (16), the scalar product (x, y) looks in the e-representation as

$$(x, y) = x'_{(e)} G \bar{y}_{(e)}. \tag{5.78}$$

Vectors x and y are called orthogonal if $(x, y) = 0$. A vector x is said to be a unit vector if $(x, x) = 1$. A basis is termed orthonormal if $g_{ij} = \delta_{ij}$ for it, that is, if $G = E$. In an orthonormal basis Eqn. (78) goes over into

$$(x, y) = x'_{(e)}\bar{y}_{(e)}. \tag{5.78a}$$

In a function space the scalar product of the vectors $\psi_{(1)}$ and $\psi_{(2)}$ has, by analogy with (78), the form

$$(\psi_{(1)}, \psi_{(2)}) = \iint \psi_{(1)}(\alpha)G(\alpha, \beta)\bar{\psi}_{(2)}(\beta)\, \mathrm{d}\alpha\, \mathrm{d}\beta, \tag{5.79}$$

where $G(\alpha, \beta) = \bar{G}(\beta, \alpha)$ is the function corresponding to the scalar products of the basis vectors of this space. If the basis is orthogonal, then $G(\alpha, \beta) = F(\alpha)\,\delta(\alpha - \beta)$, where $F(\alpha)$ is a real positive function which is sometimes called the weight function. If the basis is orthonormal, then $G(\alpha, \beta) = \delta(\alpha - \beta)$. In the first case the scalar product has the form

$$(\psi_{(1)}, \psi_{(2)}) = \int \psi_{(1)}(\alpha)\bar{\psi}_{(2)}(\alpha)F(\alpha)\, \mathrm{d}\alpha, \tag{5.79a}$$

and in the second one

$$(\psi_{(1)}, \psi_{(2)}) = \int \psi_{(1)}(\alpha)\bar{\psi}_{(2)}(\alpha)\, \mathrm{d}\alpha. \tag{5.79b}$$

The Schwarz's inequality takes on the form

$$|(\psi_{(1)}, \psi_{(2)})| \leqslant \left[\int |\psi_{(1)}(\alpha)|^2\, \mathrm{d}\alpha\right]^{1/2} \left[\int |\psi_{(2)}(\alpha)|^2\, \mathrm{d}\alpha\right]^{1/2}. \tag{5.77a}$$

In view of (78a), the scalar product (Ax, y), where x and y are arbitrary vectors and A is an operator, has in the e-representation, provided an orthonormal basis is chosen, the following form

$$(Ax, y) = x'_{(e)}A'_{(e)}\bar{y}_{(e)}. \tag{5.80}$$

The operator A^* is said to be the adjoint of the operator A if for arbitrary vectors x and y

$$(Ax, y) = (x, A^*y). \tag{5.81}$$

From the consideration of the e-representation of Eqn. (81) it is seen that $\overline{A^*_{(e)}} = A'_{(e)}$.

An operator A is called self-adjoint (or Hermitian) if $A^* = A$. Although there may be a difference in the strict mathematical definitions of self-adjoint and of Hermitian operators (cf., for instance, Ref. [2]), there is no need to distinguish between them in what follows. In any orthonormal basis to a Hermitian operator there obviously corresponds a Hermitian matrix.

An operator U satisfying for any vectors x and y the equation

$$(Ux, Uy) = (x, y), \tag{5.82}$$

that is, such an operator that keeps unchanged the scalar product of any vectors, is termed the unitary operator. For a unitary operator the relation $U^*U = UU^* = E$ is obviously valid. The matrix representing a unitary operator in an orthonormal basis obeys the analogous equation: $U^*_{(e)}U_{(e)} = U_{(e)}U^*_{(e)} = E$, and is, therefore, a unitary matrix. The similarity transformation performed by a unitary operator is termed a unitary transformation.

Since for a Hermitian operator A

$$(Ax, Ay) = (A^2x, y) = (x, A^2y) \tag{5.83}$$

for arbitrary x and y, any whole power of a Hermitian operator is also a Hermitian operator. Moreover, any real function of a Hermitian operator is a Hermitian operator too. But the product of two Hermitian operators A and B is a Hermitian operator if, and only if they commute. The sum of Hermitian operators is also a Hermitian operator.

The product of any number of unitary operators is a unitary operator too. But the sum of unitary operators is not a unitary operator.

The eigenvalues of operator A^* are obviously the complex conjugate of those of operator A. Therefore, the eigenvalues of a Hermitian operator are real. Since for a unitary operator $U^* = U^{-1}$, the absolute magnitude of any of its eigenvalues is unity. For both a Hermitian and a unitary operator the system of the eigenvectors of the operator itself coincides with that of its adjoint. But the eigenvectors of operator A^* are complex conjugate with those of A'. Therefore, both Hermitian and unitary operators have, as follows from (46) and (78a), respectively, orthonormal systems of eigenvectors, which prove to be total. Hence, for these operators the matrix $L_{(e)}$ proves unitary, Eqn. (45) takes on the form

$$Z_{j(e)} = \bar{v}_{j(e)}v'_{j(e)} \tag{5.84}$$

and Eqn. (52), the form

$$f(A_e) = \sum_j f(\lambda_j)\bar{v}_{j(e)}v'_{j(e)} \tag{5.85}$$

(a function of a unitary operator is not, generally, a unitary operator).

It follows from (49) and (50) that two operators commute if, and only if they have a common total system of eigenvectors. This statement is valid, of course, for Hermitian operators too. A system of eigenvectors of any Hermitian operator may be taken as an orthonormal basis of the space. If Hermitian operators commute, then the matrices representing them can be simultaneously reduced to the diagonal form with the aid of one and the same unitary matrix, which corresponds to the choice of the common complete orthonormal system of eigenvectors of these operators as the basis of the space. It is said, sometimes, that commuting Hermitian operators are simultaneously reducible to the diagonal form, in their common complete system of eigenvectors, or that they are simultaneously reducible to such form by a unitary transformation.

Let A be a Hermitian operator and x be a unit vector. Then in the orthonormal basis of eigenvectors of A, that is, in the A-representation, in view of (80), we will obtain

$$(f(A)x, x) = \sum_j f(\lambda_{j(A)})|x_{j(A)}|^2, \tag{5.86}$$

where $\lambda_{j(A)}$ are eigenvalues of A (the inferior index 'A' designates the A-representation). The relation

$$(x, x) = \sum_j |x_{j(A)}|^2 = 1 \tag{5.87}$$

is obviously also valid. The equation

$$(A_{j(z)}x, x) = |x_{j(A)}|^2, \tag{5.88}$$

where $A_{j(z)}$ is the jth projection operator for the Hermitian operator A, is valid in view of (49). Let us note, by the way, that, in view of the properties of the scalar product of vectors and the relation $x = \Sigma_j x_{j(A)} v_{j(A)}$, Eqn. (86) results immediately from (88) and (50), and Eqn. (87) from (88) and (57).

It follows from (80), in view of (63) and (64), that

$$(Ax, y) = \operatorname{Tr} A W_{(x,y)}, \tag{5.89}$$

where $W_{(x,y)}$ is the operator assigned to the vectors x and y, which is represented in any orthonormal basis by the corresponding matrix

$$W_{(x,y)} = x\bar{y}'. \tag{5.90}$$

From (88) and (89) we now obtain

$$|x_{j(A)}|^2 = \operatorname{Tr} A_{j(z)} W_{(x,x)} \tag{5.91}$$

whence follows, in view of (50)

$$\sum_j f(\lambda_j)|x_{j(A)}|^2 = \operatorname{Tr} f(A) W_{(x,\,x)}. \tag{5.92}$$

It is necessary to stress that the operator $W_{(x,\,x)}$ is determined by only the vector x and, if this is a unit vector, then

$$\operatorname{Tr} W_{(x,\,x)} = 1 \tag{5.93}$$

which follows, in particular, from (91) and (51). For $x = v_{k(A)}$, Eqns. (86) and (92), in view of $x_{j(A)} = \delta_{jk}$, yield

$$(f(A)v_{k(A)}, v_{k(A)}) = \operatorname{Tr} f(A) W_{(v_k,\, v_k)} = f(\lambda_k). \tag{5.94}$$

In a function space Eqn. (80) goes over, if (79b) is used, into

$$(Ax, y) = \iint \psi_{(1)}(\alpha) a'(\alpha, \beta) \overline{\psi}_{(2)}(\beta)\, \mathrm{d}\alpha\, \mathrm{d}\beta, \tag{5.95}$$

where $a(\alpha, \beta)$ is the kernel of the operator A'. The kernel of the operator A^* then has the form $\overline{a}(\alpha, \beta)$. To the Hermitian operator A there corresponds a kernel satisfying the equation $a(\beta, \alpha) = \overline{a}(\alpha, \beta)$. A kernel of a unitary operator U should satisfy the condition

$$\int u(\alpha, \beta)\overline{u}(\gamma, \beta)\, \mathrm{d}\beta = \int u(\eta, \alpha)\overline{u}(\eta, \gamma)\, \mathrm{d}\eta = \delta(\alpha - \gamma). \tag{5.96}$$

As follows from (74a) and $\eta_\lambda(\alpha) = \overline{\varphi}_\lambda(\alpha)$, and respectively from (74b) and $\eta_j(\alpha) = \overline{\varphi}_j(\alpha)$, the kernels of the projection operators for a Hermitian operator A have the forms

(a) $$a_{\lambda(z)}(\beta, \alpha) = \varphi_\lambda(\beta)\overline{\varphi}_\lambda(\alpha), \tag{5.97a}$$

(b) $$a_{j(z)}(\beta, \alpha) = \varphi_j(\beta)\overline{\varphi}_j(\alpha). \tag{5.97b}$$

The expansion of a unit vector ψ in terms of eigenfunctions of a Hermitian operator A is determined by the equations:

(a) $$\psi(\alpha) = \int b_{\lambda(A)}\varphi_{\lambda(A)}(\alpha)\, \mathrm{d}\lambda_{(A)}, \tag{5.98a}$$

(b) $$\psi(\alpha) = \sum_j b_{j(A)}\varphi_{j(A)}(\alpha), \tag{5.98b}$$

provided the respective conditions

$$\text{(a)} \quad \int |b_{\lambda(A)}|^2 \, d\lambda_{(A)} = 1, \tag{5.99a}$$

$$\text{(b)} \quad \sum_j |b_{j(A)}|^2 = 1 \tag{5.99b}$$

analogous to (87) are fulfilled. Equation (88) goes over into

$$\text{(a)} \quad \iint a_{\lambda(z)}(\beta, \alpha)\psi(\alpha)\overline{\psi}(\beta) \, d\alpha \, d\beta = |b_{\lambda(A)}|^2, \tag{5.100a}$$

$$\text{(b)} \quad \iint a_{j(z)}(\beta, \alpha)\psi(\alpha)\overline{\psi}(\beta) \, d\alpha \, d\beta = |b_{j(A)}|^2. \tag{5.100b}$$

In accordance with the definition, the trace of the operator A with the kernel $a(\beta, \alpha)$ has, if existing, the form

$$\operatorname{Tr} A = \int a(\alpha, \alpha) \, d\alpha. \tag{5.101}$$

Since the kernel of the operator $W_{(\psi_{(1)}, \psi_{(2)})}$ is $\psi_{(1)}(\beta)\overline{\psi}_{(2)}(\alpha)$, the equation

$$\operatorname{Tr} AW_{(\psi_{(1)}, \psi_{(2)})} = \iint a(\beta, \alpha)\psi_{(1)}(\alpha)\overline{\psi}_{(2)}(\beta) \, d\alpha \, d\beta \tag{5.102}$$

is valid, which, as should be the case, coincides with (95). Equations (100a) and (100b) may also be rewritten in the forms:

$$\text{(a)} \quad \operatorname{Tr} A_{\lambda(z)} W_{(\psi, \psi)} = |b_{\lambda(A)}|^2, \tag{5.103a}$$

$$\text{(b)} \quad \operatorname{Tr} A_{j(z)} W_{(\psi, \psi)} = |b_{j(A)}|^2 \tag{5.103b}$$

respectively, where it is taken into account that the kernel of the operator $W_{(\psi, \psi)}$ is $\psi(\alpha)\overline{\psi}(\beta)$. Since

$$(\psi, \psi) = \operatorname{Tr} W_{(\psi, \psi)}, \tag{5.104}$$

the equation of the form (93) is valid, where x is replaced by ψ, provided $(\psi, \psi) = 1$. The equality $W^2_{(\psi, \psi)} = W_{(\psi, \psi)}$ is also valid. For a unit vector ψ from Eqns. (100) and (98) follow the respective equations analogous to Eqns. (86) and (92):

$$\text{(a)} \quad (f(A)\psi, \psi) = \operatorname{Tr} f(A) W_{(\psi, \psi)} = \int f(\lambda)|b_{\lambda(A)}|^2 \, d\lambda \tag{5.105a}$$

$$\text{(b)} \quad (f(A)\psi, \psi) = \operatorname{Tr} f(A) W_{(\psi, \psi)} = \sum_j f(\lambda_j)|b_{j(A)}|^2. \tag{5.105b}$$

If ψ is an eigenfunction of the operator A corresponding to its eigenvalue (a) λ, or (b) λ_j, then the right-hand side of Eqns. (105a), and (105b), yields $f(\lambda)$, respectively $f(\lambda_j)$ (it is supposed that the set of eigenfunctions of A: (a) satisfies the condition $\int \psi_{\lambda(A)}(\alpha)\overline{\psi}_{\lambda'(A)}(\alpha)\,d\alpha = \delta(\lambda - \lambda')$; (b) is orthonormal, that is, satisfies the condition $\int \psi_{j(A)}(\alpha)\overline{\psi}_{k(A)}(\alpha)\,d\alpha = \delta_{jk}$).

The above results related to one-dimensional vectors (and, to functions of one variable) may easily be extended to the case of multi-dimensional vectors (respectively, to functions of many variables). In what follows, we will make use of this fact, without going into details connected with such extension. Additional items will be considered, as need arises.

5.4. Two Specific Modes of Description of a State

Let $\rho(x_l, \ldots, x_p)$ be the probability density of the random variable $t_{l\ldots p}$ for a certain state $f_{0ij\ldots k}$. Consider the function

$$\psi(x_l, \ldots, x_p) = \rho^{1/2}(x_l, \ldots, x_p)\exp\,[i\eta(x_l, \ldots, x_p)]\,, \tag{5.106}$$

where η is an arbitrary real function. This function uniquely determines $\rho(x_l, \ldots, x_p)$, since $|\psi(x_l, \ldots, x_p)|^2 = \rho(x_l, \ldots, x_p)$, and hence also describes the state with respect to the random variable $t_{l\ldots p}$. The function $\psi(x_l, \ldots, x_p)$ may be interpreted as the representation of a definite k-dimensional vector $\psi_{(l\ldots p)}$, when a particular choice of the basis in the vector space is made, namely, as the x-representation of this vector. Since the vector $\psi = \psi_{(l\ldots p)}$ uniquely determines $\psi(x_l, \ldots, x_p)$ and, therefore, $\rho(x_l, \ldots, x_p)$, too, it also describes the state with respect to the random variable $t_{l\ldots p}$.

To the vector ψ there can be assigned the operator $W = W_{(l\ldots p)}$, whose kernel in the x-representation has the form

$$W(x_l, \ldots, x_p; x'_l, \ldots, x'_p) = \psi(x_l, \ldots, x_p)\overline{\psi}(x'_l, \ldots, x'_p). \tag{5.107}$$

Since operator W uniquely determines $\rho(x_l, \ldots, x_p)$, because in the x-representation $W(x_l, \ldots, x_p; x_l, \ldots, x_p) = \rho(x_l, \ldots, x_p)$, it describes, too, the state with respect to the random variable $t_{l\ldots p}$. In accordance with the common terminology of quantum physics, we will call the vector ψ the state vector and the operator W the density operator, respectively.

Like the characteristic function, neither the state vector nor the density operator have a direct probabilistic meaning, and they are auxiliary mathematical quantities only. The characteristic function has, however, a number of useful peculiarities, which makes it possible to use it for examination and solution of various probabilistic problems, in particular, for computation of mathematical expectations of random variables. Does the state vector or the density operator have any analogous possibilities? It is easy to make sure that the question is answered in the affirmative.

Making use of the x-representation of the state vector, it is possible, for instance, to write the mathematical expectation of the random variable $\varphi(t_l)$ as

$$M\varphi(t_l) = \iint \varphi(x_l')\,\delta(\tau - \tau')\,\psi(x_l, \ldots, x_p)\,\overline{\psi}(x_l', \ldots, x_p')\,\mathrm{d}\tau\,\mathrm{d}\tau', \tag{5.108}$$

where $\delta(\tau - \tau') = \delta(x_l - x_l') \ldots \delta(x_p - x_p')$, $\mathrm{d}\tau = \mathrm{d}x_l \ldots \mathrm{d}x_p$, and $\mathrm{d}\tau' = \mathrm{d}x_l' \ldots \mathrm{d}x_p'$. Utilizing the mathematical formalism of the preceding section, it is possible to rewrite Eqn. (108) in the two following equivalent forms:

$$M\varphi(t_l) = (\varphi(\hat{t}_l)\psi, \psi), \tag{5.109}$$

$$M\varphi(t_l) = \mathrm{Tr}\,\varphi(\hat{t}_l)W, \tag{5.110}$$

where $\hat{t}_l$ is an operator corresponding to the random variable t_l, ψ the state vector, and W is the density operator. It follows from the connection of ψ and W with the probability density of the random variable t_l that

$$(\psi, \psi) = \mathrm{Tr}\,W = 1 \tag{5.111}$$

and, therefore, ψ is a unit vector. The totality of the eigenvalues (spectrum) of the operator $\hat{t}_l$ should obviously coincide with the set θ_l of the values f_l of the random variable t_l. But according to our definition of a random variable (see 3.1), these values are real numbers. Therefore, the operator $\hat{t}_l$ is Hermitian, since it is precisely the Hermitian operators which necessarily have real eigenvalues, while to have all the eigenvalues real proves sufficient for an operator to be Hermitian. On the basis formed of the orthonormal eigenvectors of $\hat{t}_l, \ldots$, and $\hat{t}_p$ the state vector and the density operator are represented by (106) and (107), respectively.

Thus, the introduction of a state vector or a density operator for description of a state leads simultaneously to the introduction of a Hermitian operator $\hat{t}_l$ for each random variable t_l. It follows from (109) that the eigenvector of the operator $\hat{t}_l$ corresponding to its eigenvalue $f_l \in \theta_l$ describes the state for which the random variable t_l has a definite value, namely, f_l. The expansion of the operator $\hat{t}_l$ in terms of its projection operators has the form (for simplicity the index 'l' is omitted)

$$\hat{t} = \int f\hat{t}_{f(z)}\,\mathrm{d}f, \tag{5.112}$$

where $\hat{t}_{f(z)}$ is the projection operator corresponding to the eigenvalue f of the operator $\hat{t}$. The matrices representing the projection operators in the orthonormal basis (normalized to the δ-function) formed of the eigenvectors of the operator $\hat{t}$ look as

$$t_{f(z)}(x, x') = \delta(f - x)\,\delta(f - x'). \tag{5.113}$$

It is easy to make sure, using the mateiral of the preceding section, in particular Eqn. (100a), that

$$(\hat{t}_{f(z)}\psi, \psi) = |b_{f(t)}|^2 = \rho(f), \tag{5.114}$$

where $b_{f(t)}$ is the coefficient in the expansion of the state vector in terms of the eigenvectors of $\hat{t}$ corresponding to the eigenvector v_f of this operator, which conforms to its eigenvalue f; $\rho(f)$ is the probability density of the random variable t at the point $x = f$ (see Eqn. (99a)); in the calculation it is taken into account that an eigenfunction of t in the t-representation is $\delta(x - f)$, that the operator $\hat{t}_l$ acts only on the variable x_l, and that

$$\int \cdots \int |\psi(x_l, x_m, \ldots, x_p)|^2 \, dx_m \ldots dx_p = \rho(x_l). \tag{5.115}$$

Equation (114) shows that the probability density of a random variable, in a state described by a state vector ψ normalized to unity, is equal to the magnitude squared of the coefficient of the expansion of this vector in terms of the eigenvectors of the Hermitian operator corresponding to this random variable. The same conclusion follows from the equivalent to (114), equation

$$\mathrm{Tr}\, \hat{t}_{f(z)} W = |b_{f(t)}|^2 = \rho(f) \tag{5.116}$$

which, in turn, results from (103a) and (106).

In the case when a random variable possesses a discrete spectrum, partly or completely, Eqns. (109)–(111) retain their form, but the integration in Eqns. (112) and (114–116) is replaced by the summation in the intervals corresponding to the discrete values of the random variable, the Kroneker's δ_{jk} being used in (113), instead of the δ-function.

It follows from the definition of a random variable, and the connection of (108) with (109) and (110), that the operator corresponding to a sum of random variables is equal to the sum of the operators corresponding to the summands (recall that a sum of Hermitian operators also is Hermitian). Considering this statement the definition of the operator for a sum of random variables in the general case, and making use of the properties of the scalar product of vectors (or, respectively, of those of the trace of an operator), we will obtain directly from (109), as well as (110), Theorem 1 concerning the mathematical expectation of the sum of random variables (Sect. 3.4). Theorem 2 on the mathematical expectation of a random constant follows directly from (109) or (110), since multiplication of a vector by a constant c corresponds to the transformation of the vector by the scalar operator cE. Not more difficult is the proof of Theorem 3 concerning the mathematical expectation of the product of random variables in a state where the probability distributions for them are mutually independent, but it seems appropriate now to discuss a related problem of principle, before proceeding any further.

The product of random variables should have, by the definition of a random variable, a set of values which are equal to products of the values of the random variables multiplied. If Eqns. (109) and (110) are considered, the different records of Eqn. (108) only, that is, if it is actually believed that all the operators $\hat{t}_l, \ldots, \hat{t}_p$, as well as the operators corresponding to other possible random variables, have the same common system of eigenvectors, then for any function of these random variables, of course, including their products, it is possible to write an equation analogous to (108). In this case the above definition of the product of random variables proves to be valid. But if one believes that Eqns. (109) and (110) have a more general, independent meaning, whereas the equations of the form (108) are valid only in the corresponding particular cases, then the definition of an operator conforming to the product of two random variables, say t_l and t_m, encounters difficulties of principle. Indeed, to each of these random variables there corresponds its own Hermitian operator, $\hat{t}_l$ and $\hat{t}_m$, respectively. But the product of two Hermitian operators also is a Hermitian operator only in the case when these operators commute, that is, when they have a common system of eigenvectors. In the general case, however, the Hermitian operators for the random variables t_l and t_m do not need to commute, and therefore, neither $\hat{t}_l\hat{t}_m$, nor $\hat{t}_m\hat{t}_l$ are Hermitian operators and they cannot, hence, be considered the operators corresponding to the product of the random variables t_l and t_m.

This fact might seem to impede the proof of Theorem 3. However, the theorem has nothing to do with the general case, since it only relates to states in which the probability distributions for the random variables multiplied are mutually independent. But in this case from $\rho(x_l, x_m) = \rho_1(x_l)\rho_2(x_m)$ follows $\psi(x_l, x_m) = \psi_1(x_l)\psi_2(x_m)$, provided the phase factor is chosen in the form $\eta(x_l, x_m) = \eta_1(x_l) + \eta_2(x_m)$. This means that the operators $\hat{t}_l$ and $\hat{t}_m$ have a common system of eigenvectors and thus necessarily commute. Now we may write

$$\begin{aligned} Mt_l t_m &= (\hat{t}_l\hat{t}_m \psi_l \times \psi_m, \psi_l \times \psi_m) = (\hat{t}_l\psi_l \times \hat{t}_m\psi_m, \psi_l \times \psi_m) \\ &= (\hat{t}_l\psi_l, \psi_l)(\hat{t}_m\psi_m, \psi_m) = Mt_l Mt_m \end{aligned} \tag{5.117}$$

which proves Theorem 3 (ψ_l and ψ_m are the vectors represented by $\psi_1(x_l)$ and $\psi_2(x_m)$, respectively).

The calculations above have been based on the following considerations. Let the vector space R be the direct product of the vector spaces $R_{(l)}$ and $R_{(m)}$: $R = R_{(l)} \times R_{(m)}$. This means that each vector x in R is the direct product of a vector $x_{(l)}$ in $R_{(l)}$ and a vector $x_{(m)}$ in $R_{(m)}$: $x = x_{(l)} \times x_{(m)}$. If the scalar product of vectors is defined in both $R_{(l)}$ and $R_{(m)}$, then it does in R, and

$$(x, y) = (x_{(l)} \times x_{(m)}, y_{(l)} \times y_{(m)}) = (x_{(l)}, y_{(l)})(x_{(m)}, y_{(m)}). \tag{5.118}$$

Since Eqns. (109) and (110) are equivalent, Theorem 3 may be demonstrated by using the latter too.

Now let us revert to the problem of finding a Hermitian operator for the product of random variables in the general case. The difficulty indicated above may in some cases be overcome, in particular, by choosing the 'symmetrized' product of the operators, $(\hat{t}_l\hat{t}_m + \hat{t}_m\hat{t}_l)/2$, which obviously is a Hermitian operator, as the operator for the product of the random variables t_l and t_m. Such choice might be confirmed by the fact that in the particular case when the operators commute it leads to correct results. The correctness of this choice in certain cases is corroborated, too, by other considerations, which will be given in Chapter 6. It is essential, however, that like the case of the choice of an adequate set, the decisive criterion is the confirmation of the theoretical results by the statistical experiment (for the choice of an operator both for a random variable, in general, and for the product of random variables, in particular). It is necessary to emphasize at once that the assignment of a Hermitian operator to every random variable is not unique. First, if to a random variable there corresponds a certain Hermitian operator, then any other one similar to it, that is connected with it by the unitary transformation, corresponds to the random variable too. Secondly, since a Hermitian operator assigned to a random variable is merely an auxiliary mathematical quantity, there is no reason to believe that a unique one corresponds to every random variable. As noted above, the definitive criterion for the correctness of the choice of an operator among many Hermitian operators seeming appropriate for the given random variable is the confirmation of the theory by the statistical experiment. If this proves to be compatible with the choice of several different operators for one and the same random variable, then they are all equally conforming to it.

It directly follows from the definition of variance as stated above that Theorems 4 and 5 concerning variance (see 3.4) can also be obtained with the aid of (109) or (110). It is easy to make sure of this.

Now, it is necessary to stress several highly essential circumstances related to the description of a state of an abstract object with the aid of a state vector or a density operator. First, as follows from the Note at the end of Section 3.2, the transition from the description of a state with respect to a k-dimensional random variable $t_{lm\ldots p}$ to the description of the same state with respect to the $k-j$-dimensional one ($j = 1, 2, \ldots, k-1$), which is formally obtained from $t_{lm\ldots p}$ by omitting some its j indices, is unique, but not vice versa. Therefore, if the vector ψ describing a state, say, with respect to the k-dimensional random variable $t_{lm\ldots p}$ is known, it is impossible to immediately get from it the vector describing the same state, for instance, with respect to the $k+1$-dimensional random variable $t_{lm\ldots pu}$.

Incidentally, the transition from the description of a state with respect to a k-dimensional random variable to its description with respect to a $k-j$-dimensional one by means of a state vector is, in the general case, a rather complicated procedure. In order to perform it one has to integrate $|\psi|^2$ over all the j eliminated variables, thereby having thus obtained the probability density of the corresponding $k-j$-dimensional random variable. Then, making use of this new probability density, one can obtain a new function of the form (106), which describes the state with respect to this random variable. With the aid of the new ψ-function it is possible to form also the kernel of the new density operator describing the state with respect to the $k-j$-dimensional random variable under consideration. The new state vector and density operator may then be used in Eqns. (109) and (110), respectively. The procedure described becomes significantly simplified in the case, when the probability distributions of the $k-j$-dimensional random variable and of the eliminated j-dimensional one are mutually independent. In this case the function η in the unessential phase factor in (106) may be represented as the sum of the functions η_1 and η_2 of the corresponding different systems of variables, because of which the integration over the eliminated variables amounts to replacing the function of them by unity. It is possible, however, to avoid the whole procedure also in the general case, if a state is described by the density operator, not by the state vector. Indeed, upon replacing on the right-hand side of (107) the primed eliminated variables by the non-primed ones and integrating over them, the result is the kernel of some other density operator (differing from the above), by means of which it is possible to find the mathematical expectations of the corresponding random variables, using Eqn. (110). The kernel of the new density operator thus obtained cannot be represented, in the general case, in the form (107), which does not, however, prevent one from direct utilization of it in Eqn. (110), without changing the form of this equation. The density operator thus has the advantage compared to the state vector, that when performing the description of a state with respect to a smaller number of random variables it needs no transformation. At the same time, as is easily seen, Eqn. (116) makes it possible to directly calculate the probability density of any random variable with the aid of a density operator describing the state with respect to any greater number of random variables.

Secondly, it follows from the way the state vector was introduced (for the description of a state of an abstract object with respect to random variables) that the representations of different vectors describing one and the same state with respect to the same random variables, in a certain chosen basis, may differ not more than by a nonessential phase factor. Such vectors may be said to be nonessentially different. Then the essentially different vectors, whose representations in a basis chosen differ by the fore-exponent and which correspond, therefore, to the different probability densities of the same random variables, describe different states. Since two nonessentially different vectors describe one and the same state, it is possible to use either and, if necessary, replace one by the other. Two essentially different vectors correspond to two unlike states, and it is therefore impossible to deal with them in this way. The sum of two essentially different vectors is a third vector, which obviously essentially differs from either of them. Therefore, each summand and the sum correspond to three different states, that is, to three different abstract objects (see Sect. 1). It was stressed in a Note at the end of Section 2.2 that the addition theorem for probabilities expressed by Eqn. (2.8) is true only for events related to one and the same abstract object (hence, to one and the same state). The operation of addition in a vector space (see Sect. 3) thus has nothing to do with the theorem (law) of addition of probabilities.

Thirdly, since addition of probabilities makes sense only for one and the same state (for one abstract object), this operation should be considered for one state also. To the two incompatible events Q_1 and Q_2 conforming to one state described by the state vector ψ or the density operator W there correspond the probabilities

$$P(Q_1) = \int_{\Gamma_1} \rho(f_l)\, df_l; \qquad P(Q_2) = \int_{\Gamma_2} \rho(f_l)\, df_l, \tag{5.119}$$

where Γ_1 and Γ_2 are nonintersecting sets of the values $f_l \in \theta_l$. Then, in accord with Eqn. (2.8), to the sum of the events there corresponds the probability

$$P(Q_1 + Q_2) = \int_{\Gamma_1 + \Gamma_2} \rho(f_l)\, df_l = P(Q_1) + P(Q_2). \tag{5.120}$$

Equation (120), in view of (114) and (116), can be represented in the two following equivalent forms:

$$P(Q_1 + Q_2) = \left(\int_{\Gamma_1} \hat{t}_{f(z)}\, df\psi, \psi \right) + \left(\int_{\Gamma_2} \hat{t}_{f(z)}\, df\psi, \psi \right), \tag{5.121}$$

$$P(Q_1 + Q_2) = \mathrm{Tr} \int_{\Gamma_1} \hat{t}_{f(z)}\, dfW + \mathrm{Tr} \int_{\Gamma_2} \hat{t}_{f(z)}\, dfW. \tag{5.122}$$

These are the records of the theorem of the probability of the sum of two incompatible events concerning random variables, when using the state vector and the density operator, respectively.

Finally, the requirement that the operator assigned to a random variable be Hermitian, is an extremely strong condition limiting the set of feasible probability distributions for the random variable and leading to a number of far reaching conclusions, in particular, to the so-called 'quantization' of the values of random variables (see Chap. 6). That limitation is connected with the fact that the definition domain of a Hermitian operator for a random variable may be only a part of the entire vector space under consideration. In view of the special interest of this problem, we will return to it in the next section.

If a state is described with respect to a discrete random variable (see Sect. 3.1), it seems impossible to form function (106) with the aid of the probability density expressed by Eqn. (3.15), since the functions $\delta^{1/2}(x_l - x_l')$ is usually believed not to exist. In this case, for the exact description of a state, it would be possible to use, instead of the $\delta^{1/2}$-function, the corresponding column matrix, whose elements are the square roots of the respective discrete probabilities. One may, however, use the approximate expression of the form (3.20), too, for the formation of the function (106), when the state is described approximately. In order that it would be possible to use here the mathematical methods of the preceding section, it is necessary to replace the operator of differentiation by its approximate equivalent – the operator of ϵ-differentiation, d_ϵ/dx, defined by Eqn. (5). The formalism related will also be considered in the next section.

5.5. Additional Mathematics

The definition of the Hermitian operator has proceeded from the definition of the adjoint of a given operator expressed by Eqn. (81). An operator A is termed Hermitian, if it coincides with its adjoint. When passing to the e-representation (to the matrix representation) by using Eqn. (80), the possibility arises, as might seem, to define the adjoint of an operator A as the operator A^* whose matrix in any matrix representation is the complex conjugate of the transpose (that is, the adjoint) of the matrix of the operator A in the same representation. Then a Hermitian operator might be defined as the operator whose matrix in any representation is the adjoint of itself. It might seem that such a definition of the Hermitian operator, which does not care for the definition domain, would be valid irrespective of the choice of the basis of the space under consideration. We will show by a simple example that, as a matter of fact, this is not the case.

Consider the operator $-i\, d/d\varphi$ in the space of continuous functions of φ defined on the interval $(0, 2\pi)$. Having chosen $\psi_1 = \varphi^n$, $\psi_2 = \varphi^m$ and using Eqn. (79b), we will get

$$\left(-i \frac{d\varphi^n}{d\varphi}, \varphi^m\right) = -in \int_0^{2\pi} \varphi^{n+m-1}\, d\varphi, \tag{5.123}$$

$$\left(\varphi^n, -i \frac{d\varphi^m}{d\varphi}\right) = im \int_0^{2\pi} \varphi^{n+m-1}\, d\varphi. \tag{5.124}$$

Since the right-hand sides of (123) and (124) are not equal to each other, the operator $-i\, d/d\varphi$ is not Hermitian on the set of the functions φ^m for any m. At the same time, this operator is Hermitian on the set of functions $\exp(im\varphi)$ for any whole m (both positive and negative), including zero, and therefore, on the set of any linear combinations of these functions, which contain a finite number of terms (this follows from the properties of the scalar product of vectors). Indeed,

$$\left(-i \frac{d\, e^{in\varphi}}{d\varphi}, e^{im\varphi}\right) = \left(e^{in\varphi}, -i \frac{d\, e^{im\varphi}}{d\varphi}\right) = 2\pi m\, \delta_{nm} \tag{5.125}$$

which proves the above statement. Note, by the way, that the functions $\exp(im\varphi)$ are the eigenfunctions of the Hermitian operator $-i\, d/d\varphi$ and, therefore, the functions $(2\pi)^{-1/2} \exp(im\varphi)$ $(m = 0, \pm 1, \pm 2, \ldots)$ form a total orthonormal system, which may be chosen as a basis of the space. A question now may arise: If so, then the functions φ^m may also be expanded in terms of this basis – and why is the operator under consideration not Hermitian on the set of these functions? The answer to this question will be given in more detail in Chapter 6. Here we will merely take notice of the fact that the functions φ^m can be expanded only in an infinite series in terms of the basis functions given, this expansion reproducing the original function at all the points, except for $\varphi = 0$ and $\varphi = 2\pi$. But it is precisely these points which turn out to be essential for a function to belong to the definition domain of the Hermitian operator $-i\ \mathrm{d}/\mathrm{d}\varphi$.

The example considered illustrates the fact that the definitiòn of the Hermitian operator proceeding from the matrix representation of operators is incorrect in the general case, and an attempt to introduce such a definition may lead to misunderstandings connected, in particular, with the neglect of the existence of the definition domain for the Hermitian operator. The matrix approach makes it possible to give the definition of the Hermitian operator for sure only in the case when we deal with a finite-dimensional space. In the case of an infinite-dimensional space the right-hand side of (80) is ambiguous, since the result of the double summation meant there may depend on the sequence in which it has been performed. Therefore, consideration of Eqn. (81) proceeding from Eqn. (80) is inadmissible in this case. We come thus to the conclusion that the matrix representation of the left-hand side of the equations of the type of Eqn. (80) is not always equivalent to those expressions themselves, which one must not forget. We will finish the consideration of this problem by the demonstration of two propositions.

PROPOSITION 1. *Let $R_{\mathrm{d}(A)}$ and $R_{\mathrm{d}(B)}$ be the definition domains of the Hermitian operators A and B, respectively, these operators satisfying the relation*

$$AB - BA = icE, \tag{5.126}$$

where c is a non-zero real number and E is the unit operator. Let, further, $x_\alpha \in R_{\mathrm{d}(B)}$ be an eigenvector of A. Then Bx_α does not belong to $R_{\mathrm{d}(A)}$.

Proof. Suppose $Bx_\alpha \in R_{\mathrm{d}(A)}$, then $(Bx_\alpha, Ax_\alpha) = (ABx_\alpha, x_\alpha)$ and, in view of (126)

$$\begin{aligned}\mathrm{Im}\,(Bx_\alpha, Ax_\alpha) &= -2^{-1}i[(Bx_\alpha, Ax_\alpha) - (Ax_\alpha, Bx_\alpha)]\\ &= -2^{-1}i([AB - BA]x_\alpha, x_\alpha) = 2^{-1}c(x_\alpha, x_\alpha).\end{aligned} \tag{5.127}$$

But $Ax_\alpha = \lambda_\alpha x_\alpha$, where λ_α is a real number. Therefore, $(Bx_\alpha, Ax_\alpha) = \lambda_\alpha(Bx_\alpha, x_\alpha)$, and this, in virtue of the hermiticity of the operator B, gives: $\mathrm{Im}\,(Bx_\alpha, Ax_\alpha) = 0$, which contradicts Eqn. (127). So, proposition (1) is proven.

PROPOSITION 2. *Equation (126) can only be valid for operators defined in an infinite-dimensional space.*

Proof. Suppose this equation to be valid in a finite-dimensional space. Then

$$\mathrm{Tr}\,(AB - BA) = icN, \tag{5.128}$$

where N is the dimension of the space. But in this case, in view of (63), Tr $(AB - BA) = \text{Tr}\,AB - \text{Tr}\,BA = 0$, which contradicts Eqn. (128). Thereby proposition (2) is proven.

In Chapter 6 we will show that if two operators are related by an equation of the form (126), then on the basis where one of them is the operator of multiplication by its eigenvalues, the other is the operator of differentiation with respect to them.

We will now get several helpful relations connected with the use of the δ_ϵ-functions and the ϵ-derivatives. If F is any algebraic function, then $F[\delta_\epsilon(x)]$ is obtained from (3.17) merely by replacement of ϵ^{-1} by $F(\epsilon^{-1})$ there, since $0^\alpha = 0$ $(\alpha \geqslant 0)$. This can be represented in the form

$$F[\delta_\epsilon(x - x')] = F(\epsilon^{-1})\epsilon\,\delta_\epsilon(x - x'). \tag{5.129}$$

In particular,

$$[\delta_\epsilon(x - x')]^\alpha \equiv \delta_\epsilon^\alpha(x - x') = \epsilon^{-\alpha+1}\,\delta_\epsilon(x - x') \quad (\alpha \geqslant 0). \tag{5.130}$$

From Eqn. (129) we get for the product of $F_1\,[\delta_\epsilon(x - x')] \equiv F_1^0(x')$ by $F_2\,[\delta_\epsilon(x - x'')] \equiv F_2^0(x'')$,

$$F_1^0(x')F_2^0(x'') = F_1(\epsilon^{-1})F_2(\epsilon^{-1})\epsilon^2\,\delta_\epsilon(x - x')\,\delta_\epsilon(x - x''). \tag{5.131}$$

In the particular case, when $x' = x''$, Eqn. (131), in view of (130), yields

$$F_1^0(x')F_2^0(x') = F_1(\epsilon^{-1})F_2(\epsilon^{-1})\epsilon\,\delta_\epsilon(x - x'). \tag{5.132}$$

If $x' \neq x''$, then for any $\epsilon < |x' - x''|$, in view of (3.17)

$$\delta_\epsilon(x - x')\,\delta_\epsilon(x - x'') = 0 \tag{5.133}$$

and, therefore, first, the right-hand side of Eqn. (131) vanishes, and second, the equation

$$(\delta_\epsilon(x - x'), \delta_\epsilon(x - x'')) = \int_{-\infty}^{\infty} \delta_\epsilon(x - x')\,\delta_\epsilon(x - x'')\,\mathrm{d}x = 0 \tag{5.134}$$

is valid. From (132), in view of (3.19), we obtain

$$(F_1^0(x'), F_2^0(x')) = \int_{-\infty}^{\infty} F_1^0(x')F_2^0(x')\,\mathrm{d}x = F_1(\epsilon^{-1})F_2(\epsilon^{-1})\epsilon. \tag{5.135}$$

The right-hand side of Eqn. (135) is equal to 1, independently of what ϵ is, only if $F_1(\epsilon^{-1})F_2(\epsilon^{-1}) = \epsilon^{-1}$. Therefore, in particular

$$(\delta_\epsilon^{1/2}(x - x'), \delta_\epsilon^{1/2}(x - x')) = 1. \tag{5.136}$$

It follows from the definition of the ϵ-differentiation that the operator of the ϵ-derivative defined by Eqn. (5) is linear. Further, it is easy to see that

$$\frac{d_\epsilon x^n}{dx} = nx^{n-1} + O(\epsilon^2)x^{n-3} \tag{5.137}$$

for any integer n. Therefore, in particular,

$$\frac{d_\epsilon C}{dx} = 0; \qquad \frac{d_\epsilon x}{dx} = 1; \qquad \frac{d_\epsilon x^2}{dx} = 2x. \tag{5.138}$$

Then, for any other value of α

$$\frac{d_\epsilon x^\alpha}{dx} = \alpha x^{\alpha-1} + O(\epsilon^2), \tag{5.139}$$

and for any functions $f_1(x)$ and $f_2(x)$

$$\frac{d_\epsilon [f_1(x)f_2(x)]}{dx} = f_1(x)\frac{d_\epsilon f_2(x)}{dx} + \frac{d_\epsilon f_1(x)}{dx} f_2(x) + O(\epsilon^2). \tag{5.140}$$

Applying the operator d_ϵ/dx to an arbitrary function $f(x)$ n times in succession, we will get

$$\frac{d_\epsilon^n f(x)}{dx^n} = \epsilon^{-n} \sum_{j=0}^{n} (-1)^j \binom{n}{j} f\left[x + \left(\frac{n}{2} - j\right)\epsilon\right], \tag{5.141}$$

which can be proven by induction. Indeed, for $n = 1$ (141) coincides with (5). Further, in view of (5) and (141)

$$\begin{aligned}\frac{d_\epsilon^{n+1} f(x)}{dx^{n+1}} &= \epsilon^{-1}\left[\frac{d^n f(x+\epsilon/2)}{dx^n} - \frac{d^n f(x-\epsilon/2)}{dx^n}\right] \\ &= \epsilon^{-(n+1)}\left\{\sum_{j=0}^{n} (-1)^j \binom{n}{j} f\left[x + \left(\frac{n+1}{2} - j\right)\epsilon\right] - \right. \\ &\quad \left. - \sum_{j=0}^{n} (-1)^j \binom{n}{j} f\left[x + \left(\frac{n+1}{2} - j - 1\right)\epsilon\right]\right\}. \end{aligned} \tag{5.142}$$

When $j - 1$ is replaced by $\bar{j}$ in the second sum, the minus sign before the sum turns to plus sign, the summation goes from 1 up to $n + 1$, and in view of

$$\binom{n}{j} + \binom{n}{j-1} = \binom{n+1}{j},$$

(142) goes over into (141) replacing n by $n + 1$, which proves the validity of (141).

Applying (141) to $\delta_\epsilon(x - x')$, we find

$$\frac{\mathrm{d}_\epsilon^n\, \delta_\epsilon(x - x')}{\mathrm{d}x^n} \equiv \delta_\epsilon^{(n)}(x - x')$$

$$= \epsilon^{-n} \sum_{j=0}^{n} (-1)^j \binom{n}{j} \delta_\epsilon \left[x - x' + \left(\frac{n}{2} - j\right) \epsilon \right]. \qquad (5.143)$$

Further, from (143) and (3.18) we obtain

$$(\delta_\epsilon^{(n)}(x - x'), f(x)) = \int_{-\infty}^{\infty} \delta_\epsilon^{(n)}(x - x') f(x)\, \mathrm{d}x$$

$$= \epsilon^{-n} \sum_{j=0}^{n} (-1)^j \binom{n}{j} f\left[x' - \left(\frac{n}{2} - j\right)\epsilon\right] + O(\epsilon^2). \quad (5.144)$$

Introducing $j = n - k$ into the right-hand side of Eqn. (144) and taking into consideration (141), we come after simple rearrangements to the equation

$$(\delta_\epsilon^{(n)}(x - x'), f(x)) = (-1)^n \frac{\mathrm{d}_\epsilon^n f(x)}{\mathrm{d}x^n} + O(\epsilon^2). \qquad (5.145)$$

Replacing j by $n - k$ on the right-hand side of (143), we will get, after simple calculations and with due allowance for $\delta_\epsilon(x) = \delta_\epsilon(-x)$,

$$\delta_\epsilon^{(n)}(x - x') = (-1)^n\, \delta_\epsilon^{(n)}(x' - x). \qquad (5.146)$$

If the ordinary nth-order derivative $\mathrm{d}^n f(x)/\mathrm{d}x^n$ exists, then it differs from the corresponding ϵ-derivative by $O(\epsilon^2)$ too. Therefore, on the right-hand side of (145) it is possible to replace the ϵ-derivative by the ordinary one.

Equations (145) and (146) are also valid for $\epsilon \to 0$, that is, for the derivatives of the Dirac δ-function (the term $O(\epsilon^2)$ vanishing thereby). As follows from (145), expression $(-1)^n\, \delta_\epsilon^{(n)}(x - x')$ is the kernel of the operator of the nth ϵ-derivative in the x-representation. This expression makes it possible to comprehend what is the structure of the kernel of the operator of the ordinary nth derivative in this representation. It follows from the above statement that in some cases it could be helpful to use for the calculations the δ_ϵ-function and then go to the limit $\epsilon \to 0$, intead of the direct use of the Dirac δ-function. Making use of this method, we can approximately describe a state with respect to a discrete random variable t, which has values $f^{(j)}$ with the probabilities p_j, by the function $\psi(x)$ of the form

$$\psi(x) = \sum_j p_j^{1/2}\, \delta_\epsilon^{1/2}(x - f^{(j)})\, \mathrm{e}^{i\eta_j(x)}, \qquad (5.147)$$

where η_j are real functions and ϵ is a positive number satisfying the condition: $\epsilon < \min |f^{(j+1)} - f^{(j)}|$ (we do not consider here the case of a denumerable set of values $f^{(j)}$). Indeed, it follows from (130), (131), and (138) that the function (147) obeys the required condition: $|\psi(x)|^2 = \rho(x) = \Sigma_j p_j \delta_\epsilon(x - f^{(j)})$, as should be the case. For $\epsilon \to 0$ the description of the state becomes exact, which makes it possible to conditionally introduce the function $\delta^{1/2}(x - x') = \lim_{\epsilon\to 0} \delta_\epsilon^{1/2}(x - x')$, whose square is $\delta(x - x')$.

Problems

5.1. In problem 3.3, a thought experiment was outlined involving a cat (Schrödinger's Cat). Can one speak of a state of the cat in that experiment?

5.2. What abstract object is meant and what state of it is examined when casting a die at random?

5.3. Describe the state of the abstract object in the preceding problem in different 'classical' ways, provided the die is regular.

5.4. Describe the same state by making use of the methods of linear algebra.

5.5. Calculate the mathematical expectation of the random variable t_1 in problem 2.

5.6. One usually believes that the eigenfunctions of the operator of the multiplication by a function $f(x)$ are the generalized functions $\delta(x - x')$, since $f(x)\,\delta(x - x') = f(x')\,\delta(x - x')$. What is wrong with it?

5.7. Demonstrate that $\delta^{1/2}(x - x')$ is the eigenfunction of the operator of the multiplication by a function $f(x)$, which corresponds to its eigenvalue $f(x')$, and that it is normalized to unity.

Suggested References

1. Bellman, R.: 1960 (1st edn.), 1970 (2nd edn.), *Introduction to Matrix Analysis*, McGraw-Hill, New York. This excellent book contains much important information on linear algebra and its application.
2. Dunford, N., and J. Schwartz: 1958–1971, *Linear operators*, Parts 1–3, Interscience, New York. This book is a monumental handbook which will give more information on various specific problems concerning linear algebra.
3. Gantmacher, F.: 1959, *Theory of Matrices*, tr. from Russian by K. A. Hirsh, Vols. 1 and 2, Chelsea, New York; reprinted in 1960 and 1977. This perfect book contains a lot of necessary material for studying linear algebra.
4. Mayants, L. S.: 1973, *Found. Phys.* 3, 413. This paper introduces for the first time the two specific modes of describing states.
5. Mayants, L. S.: 1974, *Found. Phys.* 4, 335. This paper contains a discussion of the meaning of functions describing states and of operators corresponding to random variables. The necessity of carefully using the properties of Hermitian operators is also discussed here.

6. Mayants, L. S.: 1977, *Found. Phys.* 7, 3. This paper discusses in more detail the problem of finding operators for random variables, the meaning of these operators and of the functions describing states, as well as the peculiarities resulting from the introduction of Hermitian operators.

Chapter 6

Hamiltonian Random Variables

6.1. Lagrangian Equations and Hamiltonian Equations

We will consider, first, some types of ordinary second- and first-order differential equations, which are known to play an extremely important part in various branches of physics. The results obtained will then be applied to the examination of the corresponding types of random variables.

Let us write a set of real, second-order ordinary differential equations in the matrix form

$$A_0\ddot{x}^{(1)} + A_1\dot{x}^{(1)} + Q = 0. \tag{6.1}$$

Here $x^{(1)}$ is the column matrix of the variables x_j $(j = 1, \ldots, n-1)$; A_0 and A_1 are square matrices, and Q is a column matrix; the elements of all of these matrices depending generally on all x_j, x_n, and $\dot{x}_j$; the dot denotes derivative with respect to the variable x_n.

In this chapter we shall use the following notation: e_j is the jth column of a unit matrix; $\partial\varphi/\partial\tilde{a}$ is the column matrix with elements $\partial\varphi/\partial a_j$, where φ is a scalar, a is the column matrix with elements a_j, and the tilde is the transposition sign; $\tilde{a}$ is thus the corresponding row matrix with the same elements as in a.

Equations (1) are called Lagrangian equations, provided they can be represented as

$$\frac{\partial L}{\partial \tilde{x}^{(1)}} - \frac{\mathrm{d}}{\mathrm{d}x_n}\frac{\partial L}{\partial \dot{\tilde{x}}^{(1)}} = 0, \tag{6.2}$$

where $L = L(x_j, \dot{x}_j)$ $(j = 1, \ldots, n)$ is a function of all the x_j and $\dot{x}_j$. In the calculus of variations (2) is the direct consequence of the variational condition

$$\delta \int_{x_n^{(1)}}^{x_n^{(2)}} L\,\mathrm{d}x_n = 0, \tag{6.3}$$

where $x_n^{(1)}$ and $x_n^{(2)}$ are definite values of x_n.

Now, let us introduce the auxiliary variables

$$x_{n+j} = \partial L/\partial \dot{x}_j, \quad j = 1, \ldots, n-1 \tag{6.4}$$

and form the Hamiltonian function

$$H = \tilde{x}^{(2)}\dot{x}^{(1)} - L, \tag{6.5}$$

where $x^{(2)}$ is the column matrix with elements x_{n+j} ($j = 1, \ldots, n-1$), and all the $\dot{x}_j$ are expressed in terms of the x_k ($k = 1, \ldots, n, n+1, \ldots, 2n-1$); x_{n+j} and x_j are called canonically conjugate variables. From (5), (2), and (4) one obtains, instead of (1), the set of Hamiltonian equations

$$\dot{x}^{(1)} = \partial H/\partial \tilde{x}^{(2)}, \qquad \dot{x}^{(2)} = -\partial H/\partial \tilde{x}^{(1)} \tag{6.6}$$

which are of first order, and the equation

$$\partial H/\partial x_n = -\partial L/\partial x_n. \tag{6.7}$$

Supposing that in (1) the x_j are functions of a certain parameter χ and writing $dx_j/d\chi = x_j'$, we obtain from (3), in view of $dx_n = x_n'\, d\chi$ and $x_j' = \dot{x}_j x_n'$,

$$\delta \int_{\chi_1}^{\chi_2} L(x_j, x_j'/x_n')x_n'\, d\chi = 0. \tag{6.3'}$$

Thereby the Lagrangian function $L(x_j, \dot{x}_j)$ is replaced by the parametric Lagrangian function

$$L'(x_j, x_j') = L(x_j, x_j'/x_n')x_n'. \tag{6.8}$$

From (8), in view of the above, we obtain the auxiliary variables

$$x_{n+j}' = \partial L'/\partial x_j' = x_{n+j}, \quad j = 1, \ldots, n-1 \tag{6.4'}$$

and

$$x_{2n}' = \partial L'/\partial x_n' = L(x_j, \dot{x}_j) - \tilde{x}^{(2)}\dot{x}^{(1)}. \tag{6.4''}$$

The last equation, in view of (5), yields

$$x_{2n} = -H \tag{6.9}$$

irrespective of the choice of χ (the prime on x_{2n} is therefore omitted).

It should be emphasized at once that Eqn. (9) is not an identity. It expresses the connection between the variable x_{2n} canonically conjugate to x_n and the rest of the variables x_k ($k = 1, \ldots, n, n+1, \ldots, 2n-1$), provided the variables x_j ($j = 1, \ldots, n$) satisfy (1). In other words, (9) determines in the $2n$-dimensional space, whose points are sets of x_k ($k = 1, \ldots, n, n+1, \ldots, 2n$) the $(2n-1)$-dimensional hypersurface where the variables x_j ($j = 1, \ldots, n$) obey (1).

The parametric Hamiltonian function, in accordance with (5) is

$$H' = \tilde{x}^{(2)} x^{(1)\prime} + x_{2n} x_n' - L' = (x_{2n} + H) x_n'. \tag{6.5'}$$

In view of (9),

$$H' = 0 \tag{6.10}$$

is valid, also irrespective of the choice of χ. So, one can take $\chi = x_n$, obtaining thereby

$$H' = x_{2n} + H \tag{6.11}$$

as the parametric Hamiltonian function. When making use of H' instead of H, the parametric Hamiltonian equations equivalent to (1) take the form

$$y^{(1)\prime} = \partial H'/\partial \tilde{y}^{(2)}, \qquad y^{(2)\prime} = -\partial H'/\partial \tilde{y}^{(1)}, \tag{6.6'}$$

where $y^{(1)}$ and $y^{(2)}$ are the column matrices with the elements x_j and x_{n+j} ($j = 1, \ldots, n$), respectively; thus $y^{(1)}$ differs from $x^{(1)}$ by one more element x_n, and $y^{(2)}$ does from $x^{(2)}$ by the additional element x_{2n}. In view of (5′), this yields (6), the equation

$$\dot{x}_{2n} = -\partial H/\partial x_n, \tag{6.12}$$

and the identity $x_n' = x_n'$.

If $F = F(x_k)$ is a function of all the x_k ($k = 1, \ldots, n, n+1, \ldots, 2n$), the variables x_j ($j = 1, \ldots, n$) obeying (1), then the equation

$$\dot{F} = \frac{\partial F}{\partial x_n} + \dot{\tilde{x}}^{(1)} \frac{\partial F}{\partial \tilde{x}^{(1)}} + \dot{\tilde{y}}^{(2)} \frac{\partial F}{\partial \tilde{y}^{(2)}} \tag{6.13}$$

in view of (6′) yields

$$\dot{F} = \frac{\partial F}{\partial x_n} + \frac{\partial H}{\partial x^{(2)}} \frac{\partial F}{\partial \tilde{x}^{(1)}} - \frac{\partial H}{\partial x^{(1)}} \frac{\partial F}{\partial \tilde{x}^{(2)}} - \frac{\partial H}{\partial x_n} \frac{\partial F}{\partial x_{2n}}. \tag{6.13'}$$

Making use of the expanded Poisson bracket $[u, v]'$ defined by

$$[u, v]' = \frac{\partial u}{\partial y^{(2)}} \frac{\partial v}{\partial \tilde{y}^{(1)}} - \frac{\partial v}{\partial y^{(2)}} \frac{\partial u}{\partial y^{(1)}} = \frac{\partial u}{\partial x_{2n}} \frac{\partial v}{\partial x_n} - \frac{\partial v}{\partial x_{2n}} \frac{\partial u}{\partial x_n} + [u, v], \tag{6.14}$$

where

$$[u, v] = \frac{\partial u}{\partial x^{(2)}} \frac{\partial v}{\partial \tilde{x}^{(1)}} - \frac{\partial v}{\partial x^{(2)}} \frac{\partial u}{\partial \tilde{x}^{(1)}} \tag{6.15}$$

is the conventional Poisson bracket, we can rewrite (13) as

$$\dot{F} = [H', F]'. \tag{6.16}$$

Let us note, by the way, that the expanded Poisson bracket obviously has the same properties as the conventional one, namely:

$$[u, v]' = -[v, u]'; \qquad [u, C]' = 0; \qquad [u + w, v]' = [u, v]' + [w, v]';$$

$$[Cu, v]' = C[u, v]'; \qquad [uw, v]' = u[w, v]' + [u, v]'w;$$

$$[u, [v, w]']' + [v, [w, u]']' + [w, [u, v]']' = 0 \tag{6.17}$$

(C is a constant). Further, since $\partial x_{n+j}/\partial y^{(2)} = \tilde{e}_j$, $\partial x_l/\partial \tilde{y}^{(1)} = \tilde{e}_l$, and $\partial x_{n+j}/\partial y^{(1)} = \partial x_l/\partial y^{(2)} = 0$, it follows from (14) that

$$[x_{n+j}, x_l]' = \delta_{jl},\ [x_{n+j}, x_{n+l}]' = [x_j, x_l]' = 0, \quad j, l = 1, \ldots, n. \tag{6.18}$$

In view of (11), Eqn. (16) can be rewritten as

$$\dot{F} = [x_{2n}, F]' + [H, F]'. \tag{6.16$'$}$$

On the right-hand side of (16′) the first term brings out the explicit, the second one the implicit, dependence of F on x_n. If the x_k ($k = 1, \ldots, n, n + 1, \ldots, 2n$) are taken as F, then (16′) yields Eqns. (6) and (12) as well as the identity $\dot{x}_n = 1$. Substituting into (16′) H instead of F, we get

$$\dot{H} = \partial H/\partial x_n. \tag{6.19}$$

This follows from the nonzero first term, whereas (12) follows from the nonzero second term on the right-hand side of (16′). Therefore, (19) is the result of the explicit dependence of H on x_n, while (12) is the result of the implicit dependence of x_{2n} on x_n. This fact confirms once more that (9) is not an identity. We also conclude from (16′) that the Hamiltonian function H cannot depend on x_n implicitly, whereas x_{2n} cannot depend on x_n explicitly. Therefore, in view of (12) and (19), if the Hamiltonian function H does not include x_n, neither it nor x_{2n} depends on x_n, explicitly or implicitly. But if H includes x_n, then it depends on x_n explicitly, while x_{2n} does implicitly.

If it is possible to rewrite the relation (10) as

$$G'_{n+k} = \varphi(x_{n+k}) + G_{n+k}(x_1, \ldots, x_n, x_{n+1}, \ldots,$$
$$\ldots, x_{n+k-1}, x_{n+k+1}, \ldots, x_{2n}) = 0 \tag{6.10$'$}$$

or as

$$G'_k = -f(x_k) + G_k(x_1, \ldots, x_{k-1}, x_{k+1}, \ldots, x_n, x_{n+1}, \ldots, x_{2n}) = 0 \tag{6.10$''$}$$

for certain values of the index k, then a peculiar equation for the derivative of F with respect to the variable canonically conjugate to that singled out, similar to (16′), corresponds to each of the relations obtained, namely

$$\frac{dF}{dx_k} = \frac{[G'_{n+k}, F]'}{d\varphi/dx_{n+k}} = \frac{[\varphi, F]'}{d\varphi/dx_{n+k}} + \frac{[G_{n+k}, F]'}{d\varphi/dx_{n+k}}, \tag{6.16''}$$

$$\frac{dF}{dx_{n+k}} = \frac{[G'_k, F]'}{df/dx_k} = -\frac{[f, F]'}{df/dx_k} + \frac{[G_k, F]'}{df/dx_k} \tag{6.16'''}$$

(for details see Appendix 1). This shows that all the x_j are equivalent, and the same is valid for all the x_{n+j}. It is obvious that $G'_{2n} = H'$ and $G_{2n} = H$. For $F = x_l$ ($l = 1, \ldots, n, n+1, \ldots, 2n$), Eqns. (16″) and (16‴) yield, respectively,

$$\frac{dy^{(1)}}{dx_k} = \left(\frac{d\varphi}{dx_{n+k}}\right)^{-1} \frac{\partial G'_{n+k}}{\partial \tilde{y}^{(2)}}; \qquad \frac{dy^{(2)}}{dx_k} = -\left(\frac{d\varphi}{dx_{n+k}}\right)^{-1} \frac{\partial G'_{n+k}}{\partial \tilde{y}^{(1)}}; \tag{6.6''}$$

$$\frac{dy^{(1)}}{dx_{n+k}} = \left(\frac{df}{dx_k}\right)^{-1} \frac{\partial G'_k}{\partial \tilde{y}^{(2)}}; \qquad \frac{dy^{(2)}}{dx_{n+k}} = -\left(\frac{df}{dx_k}\right)^{-1} \frac{\partial G'_k}{\partial \tilde{y}^{(1)}}. \tag{6.6'''}$$

These equations are analogous to (6′) and may be called generalized Hamiltonian equations.

If it is possible to express x_{n+j} ($j = 1, \ldots, n-1$) in terms of x_j and $\dot{x}_j$ ($j = 1, \ldots, n$), utilizing the first matrix equation in (6), one can go from the Hamiltonian equations (6) to the original Lagrangian equations (1). Similarly, if it is possible to express x_j ($j = 1, \ldots, n-1$) in terms of x_n, x_{n+j}, and $\dot{x}_{n+j}$ ($j = 1, \ldots, n-1$), making use of the second matrix equation in (6), we can get another set of second-order equations from (6). Of course, they are also Lagrangian, but include derivatives of the x_{n+j} with respect to x_n instead of those of the x_j. From (6″) and (6‴), respectively, one can generally obtain, in the same way, other Lagrangian equations. But all those Lagrangian equations as well as the corresponding Hamiltonian equations conform to one and the same set of n pairs of canonically conjugate variables x_j and x_{n+j} ($j = 1, \ldots, n$) and determine in the corresponding $2n$-dimensional space one and the same $(2n-1)$-dimensional hypersurface described by any of the relations (10), (10′), or (10″). This confirms once more the equivalence of all the x_j and all the x_{n+j}, respectively ($j = 1, \ldots, n$).

Now, let a set of $2n-2$ real ordinary first-order differential equations be of the matrix form

$$\dot{x} + R = 0, \tag{6.20}$$

where x and R are the column matrices

$$x = (x^{(1)} \vdots x^{(2)}), \qquad R = (R^{(1)} \vdots R^{(2)}) \tag{6.21}$$

with the components $x^{(1)} = (x_1, \ldots, x_{n-1})$; $x^{(2)} = (x_{n+1}, \ldots, x_{2n-1})$; $R^{(1)} = (r_1, \ldots, r_{n-1})$; $R^{(2)} = (r_{n+1}, \ldots, r_{2n-1})$; elements of R are functions of the elements of x and of the variable x_n, with respect to which the x matrix is differentiated.

Equations (20) will be called Hamiltonian equations if they can be represented in the form given in (6). At least one set of Lagrangian equations can be formed for every set of Hamiltonian equations, provided the calculations needed are feasible. The natural questions arise: What are the conditions guaranteeing that Eqns. (1) are Lagrangian, and what are those ensuring that Eqns. (20) are Hamiltonian? The answer to the second question is clear: Equations (20) are Hamiltonian equations if the conditions

$$\frac{\partial R^{(1)}}{\partial x^{(2)}} = \frac{\partial \tilde{R}^{(1)}}{\partial \tilde{x}^{(2)}}; \qquad \frac{\partial R^{(2)}}{\partial x^{(1)}} = \frac{\partial \tilde{R}^{(2)}}{\partial \tilde{x}^{(1)}}; \qquad \frac{\partial R^{(1)}}{\partial x^{(1)}} = -\frac{\partial \tilde{R}^{(2)}}{\partial \tilde{x}^{(2)}} \tag{6.22}$$

are satisfied. Making use of these, one can also find the general conditions ensuring that Eqns. (1) are Lagrangian. But we will confine ourselves to some special cases only.

One important case for which Eqns. (20) are Hamiltonian occurs when

$$R^{(1)} = Bx^{(2)}, \qquad R^{(2)} = \partial\varphi/\partial\tilde{x}^{(1)}, \tag{6.23}$$

where $\varphi = \varphi(x_j)$ $(j = 1, \ldots, n)$ and $B = \tilde{B}$ is a nonsingular constant matrix. In this case

$$H = -2^{-1}\tilde{x}^{(2)}Bx^{(2)} + \varphi, \tag{6.24}$$

$$L = 2^{-1}\dot{\tilde{x}}^{(1)}B^{-1}\dot{x}^{(1)} + \varphi, \tag{6.25}$$

and this, in turn, corresponds to the conditions

$$A_0 = \tilde{A}_0 = B^{-1}, \qquad A_1 = 0, \qquad Q = -\partial\varphi/\partial\tilde{x}^{(1)}, \tag{6.26}$$

ensuring that Eqns. (1) are Lagrangian. Another important instance where Eqns. (1) are Lagrangian occurs for

$$A_0 = \tilde{A}_0 = \frac{\partial^2\varphi_1}{\partial\dot{\tilde{x}}^{(1)}\,\partial\dot{x}^{(1)}}, \qquad A_1 = -\tilde{A}_1 = \frac{\partial\eta}{\partial x^{(1)}} - \frac{\partial\tilde{\eta}}{\partial\tilde{x}^{(1)}}, \qquad Q = \frac{\partial\varphi_2}{\partial\tilde{x}^{(1)}}, \tag{6.27}$$

where $\varphi_1 = \varphi_1(\dot{x}_j)$, $\varphi_2 = \varphi_2(x_j)$ $(j = 1, \ldots, n)$, and η is a column matrix with elements $\eta_k(x_j)$ $(k, j = 1, \ldots, n-1)$. One may here choose

$$L = \varphi_1 - \varphi_2 + \dot{\tilde{x}}^{(1)}\eta \tag{6.28}$$

as a Lagrangian function.

If H does not depend on a variable x_j explicitly, then it follows from (6′) that the corresponding canonically conjugate variable x_{n+j} is a constant. In particular, if Eqns. (1) or (20) do not depend on x_n explicitly, then

$$x_{2n} = -H = \text{const.} \tag{6.29}$$

Equation (29) determines in the $(2n-2)$-dimensional space, whose points are defined by sets of x_k $(k = 1, \ldots, n-1, n+1, \ldots, 2n-1)$, the $(2n-3)$-dimensional hypersurface where the variables x_j $(j = 1, \ldots, n)$ satisfy Eqns. (1).

In case Eqns. (1) or Eqns. (20) separate into subsets, each one of which includes only its own variables, the variable x_n being common to them, then all the results obtained above hold for each subset separately. The Lagrangian function L and the Hamiltonian function H for the whole set of equations are in this case

$$L = \sum_k L_k, \qquad H = \sum_k H_k, \tag{6.30}$$

where L_k and H_k correspond to the kth subset. If all the equations belonging to those subsets do not depend on x_n explicitly, then, in accordance with (29).

$$x_{2n}^{(k)} = -H_k = \text{const} \tag{6.29′}$$

and the constant value of x_{2n} for the whole set of the equations is

$$x_{2n} = \sum_k x_{2n}^{(k)}. \tag{6.31}$$

Equation (1) determines the x_j as functions of x_n and the initial values: $x_j^0 = x_j(0)$ and $\dot{x}_j^0 = \dot{x}_j(0)$ $(j = 1, \ldots, n-1)$. Therefore, the derivatives of the x_j with respect to x_n, as well as any functions of the x_j and these derivatives, are also functions of x_n and certain constants C_k $(k = 1, \ldots, 2n)$, which, in turn, are definite functions of the x_j^0 and $\dot{x}_j^0$. But this means that both the $\dot{x}_j$ and any functions of them and of the x_j can be represented as functions of only the x_j and certain C_k. In order to do this, it is only necessary to eliminate x_n and some of the C_k from the expressions for the x_j and those for the functions involved. In particular, all the x_{n+j} are functions of the x_j and certain C_k.

To illustrate the above, let us consider the equation

$$\ddot{x}_1 + q(x_1) = 0. \tag{6.32}$$

In this case, $n = 2, \dot{x}_1 = \mathrm{d}x_1/\mathrm{d}x_2$, and

$$L = 2^{-1}\dot{x}_1^2 - \varphi(x_1) \tag{6.33}$$

where $\varphi(x_1)$ is related to $q(x_1)$ by $q(x_1) = \mathrm{d}\varphi/\mathrm{d}x_1$. Further, $x_3 = \partial L/\partial \dot{x}_1 = \dot{x}_1$ and

$$H = 2^{-1}x_3^2 + \varphi(x_1), \qquad H' = x_4 + H = 0. \tag{6.34}$$

Since Eqn. (32) does not depend on x_2 explicitly, we have x_4 = const. Should we find x_1 and x_3 as functions of x_2, x_1^0, and $\dot{x}_1^0$, we would obtain x_3 as a function x_1 and a certain constant. But in the case under consideration one can proceed more simply. For every constant value $x_4^0 = -2^{-1}E^0$ one finds, from (34),

$$x_3 = \pm[E^0 - 2\varphi(x_1)]^{1/2}. \tag{6.35}$$

Finally, let us emphasize that all the results obtained are purely mathematical and concern the ordinary second- and first-order differential equations of the forms involved, irrespective of the origin of these equations.

6.2. Hamiltonian Random Variables

If (20) represents Hamiltonian equations and the values x_j and x_{n+j} of the random variables t_j and t_{n+j} ($j = 1, \ldots, n$), respectively, satisfy both (20) and (12), then, in accordance with what has been stated in Section 3.1, one may write

$$t + R(t) = 0, \qquad t_{2n} = -\partial H(t)/\partial t_n, \tag{6.20$'$}$$

where we have replaced in (20) and (12) each variable x_k by the corresponding random variable t_k. The random variables satisfying (20′) will be called Hamiltonian random variables. The first n of them obviously satisfy equations similar to (1).

The formalism developed above can be used to solve certain probabilistic problems concerning objects possessing Hamiltonian random variables, irrespective of the nature of the objects. First, we consider the probability distributions for Hamiltonian random variables. In accordance with the foregoing results, random variables t_{n+j} as well as any functions of t_j and t_{n+j} ($j = 1, \ldots, n$) are functions only of the t_j and certain C_k determining a state of the corresponding abstract object. Let us now apply the equations of Section 3.3 to the Hamiltonian random variables. The relation between them is as given in (10), provided all x_k are replaced by the corresponding t_k. Therefore, in view of (3.37), the differential of the joint distribution function for the Hamiltonian random variables is of the form

$$dF(x_1, \ldots, x_n, x_{n+1}, \ldots, x_{2n}) = c\,\delta(H')\,dv, \tag{6.36}$$

where

$$dv = \rho_0(x_1, \ldots, x_n, x_{n+1}, \ldots, x_{2n})\,dx_1 \ldots dx_n\,dx_{n+1} \ldots dx_{2n}. \tag{6.37}$$

Utilizing (36), one can in principle find probability distributions for any functions of the Hamiltonian random variables. If, for all the random variables t_k the sets θ_k of their values $x_k \in \theta_k$ ($k = 1, \ldots, n, n+1, \ldots, 2n$) are identical, e.g., $\theta_k = (-\infty, \infty)$, one may believe that a $2n$-dimensional Euclidean space, whose mutually orthogonal axes conform to different values of the index k, is adequate to the whole set of the corresponding concrete objects. In this case dv in (36) is of the form

$$dv = dx_1 \ldots dx_n \, dx_{n+1} \ldots dx_{2n}. \tag{6.38}$$

The validity of such a supposition, however, may be verified only by statistical experiments, as always. But if it is true, Eqn. (36) yields

$$\rho(x_1, \ldots, x_n, x_{n+1}, \ldots, x_{2n}) = c\, \delta(H'). \tag{6.36$'$}$$

It follows from (36′) that, if H' does not depend explicitly on a variable x_l, probability distributions for any functions of the Hamiltonian random variables cannot depend on it either. In particular, the probability distribution for the random variable t_l itself is independent of it and, hence, the probability density of t_l is constant in every state of the corresponding abstract object. In this case, the value of the random variable t_{n+l} canonically conjugate to t_l turns out to be one of the constants determining a state of the abstract object. All of this leads to the possibility of omitting integration over the variables x_l and x_{n+l} when utilizing Eqn. (36′).

Hence, it follows further that if Eqns. (1) or (20) do not depend explicitly on x_n, the probability density of t_n is constant in any state of the corresponding abstract object, the value x_{2n} of the random variable t_{2n} being one of the constants determining a state. Now, let the set of Hamiltonian random variables be separated into subsets in accordance with the separation of Eqns. (1) and (20). Then the joint probability density of the random variables belonging to each such subset is determined by an expression of the form (36′), and the joint probability density of all these random variables takes the form

$$\rho(x_1, \ldots, x_n, x_{n+1}, \ldots, x_{2n}) = c \prod_k \delta(x_{2n}^{(k)} + H_k). \tag{6.36$''$}$$

In this case the joint probability densities of the t_j and of the t_{n+j} are

$$\rho(x_1, \ldots, x_{n-1}) = \prod_k \rho(x_j^{(k)}); \qquad \rho(x_{n+1}, \ldots, x_{2n-1}) = \prod_k \rho(x_{n+j}^{(k)}), \tag{6.39}$$

where $\rho(x_j^{(k)})$ and $\rho(x_{n+j}^{(k)})$ are the joint probability densities of those t_j and t_{n+j} ($j \leqslant n-1$), that belong to the kth subset of the Hamiltonian random variables.

Suppose the solution of Eqn. (20) is represented explicitly, that is, as

$$x_l = f_l(x_n, C_1, \ldots, C_{2n}), \quad l = 1, \ldots, n-1, n+1, \ldots, 2n-1 \tag{6.40}$$

and that $\rho(x_n)$ is known. Then one can, making use of (3.32), immediately obtain the expression for $\rho(x_l)$, namely,

$$\rho(x_l) = \sum_k |dx_n^{(k)}/dx_l| \rho[\varphi_k(x_l)], \tag{6.41}$$

where $x_n^{(k)} = \varphi_k(x_l)$ are the roots of the equation $x_l - f_l(x_n, C_1, \ldots, C_{2n}) = 0$. If in Eqn. (20) R does not depend on x_n explicitly, $\rho(x_n) = \text{const}$ and, therefore,

$$\rho(x_l) = a_l \sum_k |\mathrm{d}x_n^{(k)}/\mathrm{d}x_l|, \tag{6.42}$$

the normalization constant a_l being determined from the condition

$$\int \rho(x_l)\,\mathrm{d}x_l = 1. \tag{6.43}$$

As follows from the preceding section, the random variables t_{n+j} are functions of some random variables t_j and some of the constfants C_k determining a state. Further, one may assert that a joint probability distribution exists for any number of functions of one multidimensional random variable, because the corresponding probability density can be formed by means of Eqn. (3.34), being determined obviously by the probability density of the argument of those functions. Hence, joint probability distributions for functions of Hamiltonian random variables exist as well and are fully determined by the probability distribution for the first n of these, that is, for the n-dimensional random variable $t_{1\ldots n} = (t_1, \ldots, t_n)$. This fact is highly significant; it shows, in particular, that often there is actually no necessity at all either to know or to use such joint probability distribution for the whole set of Hamiltonian random variables.

Let us illustrate this by an example: Eqn. (32). The joint probability density of t_3 and t_1 is given by

$$\rho(x_3, x_1) = c\,\delta\,[x_3^2 + 2\varphi(x_1) - E^0], \tag{6.44}$$

where c is the normalization constant, and $E^0 = -2x_4^0$ is the constant determining a state. Hence, the probability density of t_1 is

$$\rho(x_1) = c\,[E^0 - 2\varphi(x_1)]^{-1/2}. \tag{6.45}$$

To write the probability density of t_3 explicitly it is necessary to know the function $\varphi(x_1)$ definitely. In particular, if $q(x_1) = x_1$, then $\varphi(x_1) = 2^{-1}x_1^2$, and, by means of (44), one gets

$$\rho(x_3) = \pi^{-1}(E^0 - x_3^2)^{-1/2}. \tag{6.46}$$

In this case, the normalization constant in both (44) and (45) proves to be $c = \pi^{-1}$ (cf. Eqn. (3.45)). Further, the solution of Eqn. (32) for this case gives

$$x_1 = C_2 \sin(x_2 + C_1), \qquad x_3 = C_2 \cos(x_2 + C_1), \tag{6.47}$$

whence, first, (35) follows again with $C_2^2 = E^0$ and, second, by means of Eqn. (42) we immediately obtain the same expression (45) and (46) for the probability densities $\rho(x_1)$ and $\rho(x_3)$.

To calculate the joint mathematical expectation $Mt_1^m t_3^k$, that is, the mathematical expectation of the random variable $t_1^m t_3^k$, one can use Eqn. (44). However, it follows from the form of the right-hand side of (44) that, if k is odd $Mt_1^m t_3^k = 0$ for any m. Hence, it is sufficient to consider the case of even k. Introducing the notation $t_3^2 = g$ and $E^0 - 2\varphi(t_1) = r$, we obtain from (34)

$$g - r = 0. \tag{6.48}$$

Now, we utilize the random variable

$$P = F(t_1) \sum_{l=0}^{k-1} g^l r^{k-l-1}(g - r). \tag{6.49}$$

Since

$$\sum_{l=0}^{k-1} g^l r^{k-l-1}(g - r) = 0, \tag{6.50}$$

the mathematical expectation of P is

$$MP = MFg^k - MFr^k = 0, \tag{6.51}$$

whence

$$MFg^k = MFr^k = MF(t_1)[E^0 - 2\varphi(t_1)]^k \tag{6.52}$$

and, in particular,

$$Mt_1^m t_3^{2l} = Mt_1^m [E^0 - 2\varphi(t_1)]^l = \int x_1^m [E^0 - 2\varphi(x_1)]^l \rho(x_1)\, dx_1. \tag{6.53}$$

Thus, calculation of the mathematical expectation of the random variable $t_1^m t_3^{2l}$ reduces, as it should, to that of the random variable $t_1^m [E^0 - 2\varphi(t_1)]^l$, the probability density of the random variable t_1 alone being used. This confirms that the joint probability distribution for t_3 and t_1 is not needed when calculating joint mathematical expectations of the form $Mt_1^m t_3^k$. But it is needed, of course, for calculation of conditional probability distributions and conditional mathematical expectations (cf. Sect. 3.4).

6.3. Canonically Conjugate Operators

Later some results concerning canonically conjugate operators will prove to be helpful. That term is used for any operators $\hat{a}$ and $\hat{b}$ (the operator $[\hat{a}\hat{b}]$ is called the commutator of the operators $\hat{a}$ and $\hat{b}$) obeying

$$\hat{a}\hat{b} - \hat{b}\hat{a} \equiv [\hat{a}\hat{b}] = c\hat{I}, \tag{6.54}$$

where c is a constant and $\hat{I}$ is the unit operator (in what follows we will designate the unit operator by $\hat{I}$, not by $\hat{E}$ as in Chap. 5). If $\hat{a}$ and $\hat{b}$ are Hermitian operators, c must obviously be pure imaginary: $c = -i\alpha$, where α is a real constant. It follows from (54) that

$$[\hat{b}\hat{a}] = (-c)\hat{I}. \tag{6.54$'$}$$

If the operator $\hat{a}$ is given, all operators $\hat{b}'$ canonically conjugate to it are of the form $\hat{b}' = \gamma_1 \hat{b} + f_1(\hat{a})$, where $\hat{b}$ is one of them, f_1 is an arbitrary function, and γ_1 is an arbitrary constant. On the other hand, if the operator $\hat{b}$ is given, all operators $\hat{a}'$ canonically conjugate to it are of the form $\hat{a}' = \gamma_2 \hat{a} + f_2(\hat{b})$, where $\hat{a}$ is one of them, while γ_2 and f_2 are a constant and a function, both also arbitrary.

Let us assume that in linear space of functions $\varphi(x_1, \ldots, x_n)$ a certain linear operator $\hat{b}$ is defined which likewise generally depends explicitly on the variables x_j. In particular, this may be the operator of multiplication by a certain function of these variables. Supposing that in the equality

$$F\varphi = \psi, \tag{6.55}$$

where F, φ, and ψ are functions of the x_j, φ is transformed into ψ by multiplication by F, we can rewrite (55) as

$$\hat{F}\varphi = \psi. \tag{6.55$'$}$$

Here

$$\hat{F} = F(x_1, \ldots, x_n)\hat{I} \tag{6.56}$$

is the operator: multiplication by F.

We also retain the form similar to (55′) in order to write that a certain operator $\hat{b}$ (not necessarily the operator of multiplication by a function of the x_j) transforms the function φ into the function ψ:

$$\hat{b}\varphi = \psi. \tag{6.57}$$

Since φ and ψ are vectors of a linear space, Eqns. (55′) and (57) should be considered the particular representation of certain more general vector equations. This representation, which corresponds to a choice of a basis in a vector space such that vectors are represented by functions of the x_j, will be called the x-representation (cf. Sect. 5.3).

In the x-representation the partial derivative of the operator $\hat{b}$ with respect to a variable x_m is defined by

$$\frac{\partial \hat{b}}{\partial x_m}\varphi = \lim_{\Delta x_m \to 0} \frac{\hat{b}(x_m + \Delta x_m)\varphi - \hat{b}(x_m)\varphi}{\Delta x_m}. \tag{6.58}$$

If the partial derivatives $\partial\varphi/\partial x_m$, $\partial\psi/\partial x_m$, and $\partial\hat{b}/\partial x_m$ exist, then, in the x-representation,

$$\frac{\partial(\hat{b}\varphi)}{\partial x_m} = \frac{\partial\hat{b}}{\partial x_m}\varphi + \hat{b}\,\frac{\partial\varphi}{\partial x_m} \tag{6.59}$$

whence

$$\frac{\partial\hat{b}}{\partial x_m}\varphi = \left(\frac{\partial}{\partial x_m}\,\hat{b} - \hat{b}\,\frac{\partial}{\partial x_m}\right)\varphi. \tag{6.60}$$

Introducing in the x-representation the operator of differentiation with respect to x_m,

$$\hat{D}_m = \partial/\partial x_m, \tag{6.61}$$

we get from (60)

$$\partial\hat{b}/\partial x_m = [\hat{D}_m\hat{b}]. \tag{6.62}$$

In particular, if $\hat{b} = \hat{x}_m$, Eqns. (62), (58), and (56) yield

$$[\hat{D}_m\hat{x}_m] = \hat{I}. \tag{6.63}$$

Thus, the operators $\hat{D}_m$ and $\hat{x}_m$ turn out to be canonically conjugate.

Now, let us state some results immediately following from the definition (54) of canonically conjugate operators (their deduction is given in detail, among other related topics, in Appendix 2). The formal derivative of $\hat{F} = F(\hat{A})$ with respect to A is defined as

$$\frac{\mathrm{d}\hat{F}}{\mathrm{d}\hat{A}} \equiv \hat{F}^{(1)} = \lim_{\epsilon\to 0}\frac{F(\hat{A}+\epsilon\hat{I}) - F(\hat{A})}{\epsilon}. \tag{6.64}$$

This definition of $\hat{F}^{(1)}$ fully conforms to the definition of the operator $\hat{F} = F(\hat{x}_1, \ldots, \hat{x}_n)$ as the operator that, in the x-representation, is of the form given in (56).

If

$$\hat{\varphi} = \varphi(\hat{a}) = \sum_{n=0}^{\infty} \alpha_n\hat{a}^n, \qquad \hat{f} = f(\hat{b}) = \sum_{n=0}^{\infty} \beta_n\hat{b}^n, \tag{6.65}$$

it follows from Eqns. (54) and (54′), respectively, that

$$[\hat{a}\hat{f}] = c\hat{f}^{(1)}, \qquad [\hat{b}\hat{\varphi}] = (-c)\hat{\varphi}^{(1)}. \tag{6.66}$$

Comparing (66) with (62), we note that, irrespective of what the canonically conjugate operators $\hat{a}$ and $\hat{b}$ are, in the representation where one of them is the operator of multiplication by the corresponding variable, the other is the operator of differentiation with respect to the same variable (accurate to an arbitrary constant factor and a summand which is an arbitrary function of this variable).

Further, for the operators $\hat{\varphi} = \varphi(\hat{a})$ and $\hat{f} = f(\hat{b})$ having the necessary derivatives, we obtain

$$[\hat{\varphi}\hat{f}] = \sum_{l=1} (l!)^{-1} c^l \hat{f}^{(l)} \hat{\varphi}^{(l)}, \tag{6.67}$$

$$[\hat{\varphi}\hat{f}] = \sum_{l=1} (-1)^{l+1} (l!)^{-1} c^l \hat{\varphi}^{(l)} \hat{f}^{(l)}, \tag{6.67'}$$

where $\hat{\varphi}^{(l)}$ and $\hat{f}^{(l)}$ are the lth derivatives of the corresponding operators with respect to their arguments. (Absence of an upper summation limit shows that the number of the summands may be finite or infinite.)

For the operators $\hat{\varphi} = \varphi(\hat{a}, \hat{b})$ and $\hat{f} = f(\hat{a}, \hat{b})$ one gets

$$[\hat{\varphi}\hat{f}] = \sum_{l=1} (l!)^{-1} c^l \left(\frac{\partial^l \hat{f}}{\partial \hat{b}^l} \frac{\partial^l \hat{\varphi}}{\partial \hat{a}^l} - \frac{\partial^l \hat{\varphi}}{\partial \hat{b}^l} \frac{\partial^l \hat{f}}{\partial \hat{a}^l} \right), \tag{6.68}$$

$$[\hat{\varphi}\hat{f}] = \sum_{l=1} (-1)^{l+1} (l!)^{-1} c^l \left(\frac{\partial^l \hat{\varphi}}{\partial \hat{a}^l} \frac{\partial^l \hat{f}}{\partial \hat{b}^l} - \frac{\partial^l \hat{f}}{\partial \hat{a}^l} \frac{\partial^l \hat{\varphi}}{\partial \hat{b}^l} \right). \tag{6.68'}$$

If there are n mutually independent pairs of canonically conjugate operators $\hat{a}_j$ and $\hat{b}_j$ ($j = 1, \ldots, n$) for which the relations

$$[\hat{a}_j \hat{b}_k] = c\, \delta_{jk} \hat{I} \quad \text{and} \quad [\hat{a}_j \hat{a}_k] = [\hat{b}_j \hat{b}_k] = 0 \tag{6.69}$$

are valid, then for the operators $\hat{\varphi} = \varphi(\hat{a}_1, \ldots, \hat{a}_n, \hat{b}_1, \ldots, \hat{b}_n)$ and $\hat{f} = f(\hat{a}_1, \ldots, \hat{a}_n, \hat{b}_1, \ldots, \hat{b}_n)$ we obtain

$$[\hat{\varphi}\hat{f}] = \sum_{l=1} (l!)^{-1} c^l \sum_{k=1}^{n} \left(\frac{\partial^l \hat{f}}{\partial \hat{b}_k^l} \frac{\partial^l \hat{\varphi}}{\partial \hat{a}_k^l} - \frac{\partial^l \hat{\varphi}}{\partial \hat{b}_k^l} \frac{\partial^l \hat{f}}{\partial \hat{a}_k^l} \right) + \ldots \tag{6.70}$$

$$[\hat{\varphi}\hat{f}] = \sum_{l=1} (-1)^{l+1} (l!)^{-1} c^l \sum_{k=1}^{n} \left(\frac{\partial^l \hat{\varphi}}{\partial \hat{a}_k^l} \frac{\partial^l \hat{f}}{\partial \hat{b}_k^l} - \frac{\partial^l \hat{f}}{\partial \hat{a}_k^l} \frac{\partial^l \hat{\varphi}}{\partial \hat{b}_k^l} \right) + \ldots \tag{6.70'}$$

In Eqns. (70) and (70′) the carets denote terms including cross-derivatives which are of no interest here. Taking $\hat{a}_j$ or $\hat{b}_j$ instead of $\hat{\varphi}$, one obtains from (70) and (70′) the same results

$$[\hat{a}_k \hat{f}] = c\, \partial \hat{f}/\partial \hat{b}_k; \qquad [\hat{b}_k \hat{f}] = -c\, \partial \hat{f}/\partial \hat{a}_k \tag{6.71}$$

as should be the case.

Pairs of canonically conjugate operators $\hat{a}_j$ and $\hat{b}_j$ ($j = 1, \ldots, n$) are mutually independent, provided there are no conditions imposed on the set of vectors on which operators are defined. Suppose, however, that every vector ψ of this set must obey

$$-\hat{a}_k \psi = \hat{G}_{n+k} \psi, \tag{6.72}$$

where $\hat{G}_{n+k} = G_{n+k}(\hat{a}_j, \hat{b}_j)$ is a function of all $\hat{a}_j$ and $\hat{b}_j$, except for $\hat{a}_k$. Then all the operators $\hat{a}_j$ and $\hat{b}_j$ defined on this subspace (of the whole space on which they are originally defined), which satisfies (72), prove to be interrelated; there arises generally an implicit dependence of $\hat{a}_j$ ($j = 1, \ldots, n$) and $\hat{b}_j$ ($j \neq k$) on $\hat{b}_k$.

To demonstrate this we make use of the b-representation. In this representation a vector ψ is represented by a function of the b_j, while $\hat{a}_k$ may be represented (accurate to a factor) as the operator of differentiation with respect to b_k:

$$\hat{a}_k = c\, \partial/\partial b_k. \tag{6.73}$$

Considering all operators $\hat{a}_j$ ($j = 1, \ldots, n$) to be Hermitian, we can put in (73) $c = -i\alpha$, where α is a real number. Now, let us calculate in the b-representation the derivative of an *incomplete scalar product* $(\hat{F}\psi, \psi)_{b_k}$, where $\hat{F}$ is a certain operator (generally a function of all the $\hat{a}_j$ and $\hat{b}_j$), with respect to b_k. The index b_k (or any other index x) attached to a scalar product will always denote an incomplete scalar product at the value b_k given (respectively, at the value x given). This is the result of integration (or summation) over all variables except b_k (except x) having a certain definite value. Thus, the incomplete scalar product $(\varphi, \psi)_x$ is a function of x, whereas the scalar product (φ, ψ) is a number. The incomplete scalar product obviously has all the properties of the scalar product (cf. Sect. 5.3).

Calculating the derivative mentioned above, we get

$$\frac{\mathrm{d}}{\mathrm{d}b_k}(\hat{F}\psi, \psi)_{b_k} = \left(\frac{\partial \hat{F}}{\partial \hat{b}_k}\psi, \psi\right)_{b_k} + \left(\hat{F}\frac{\partial \psi}{\partial b_k}, \psi\right)_{b_k} + \left(\hat{F}\psi, \frac{\partial \psi}{\partial b_k}\right)_{b_k}. \tag{6.74}$$

Replacing in (74) $\partial/\partial b_k$ by $(i/\alpha)\hat{a}_k$ and utilizing (72), we find

$$\frac{\mathrm{d}}{\mathrm{d}b_k}(\hat{F}\psi, \psi)_{b_k} = \left(\frac{\partial \hat{F}}{\partial \hat{b}_k}\psi, \psi\right)_{b_k} + \left(-\frac{i}{\alpha}\hat{F}\hat{G}_{n+k}\psi, \psi\right)_{b_k} + \left(\hat{F}\psi, -\frac{i}{\alpha}\hat{G}_{n+k}\psi\right)_{b_k}. \tag{6.75}$$

Considering $\hat{G}_{n+k}$ the Hermitian operator on the set of vectors involved, we can rewrite (75) as follows

$$\frac{\mathrm{d}}{\mathrm{d}b_k}(\hat{F}\psi, \psi)_{b_k} = \left(\left\{\frac{\partial \hat{F}}{\partial \hat{b}_k} + \frac{i}{\alpha}[\hat{G}_{n+k}\hat{F}]\right\}\psi, \psi\right)_{b_k}. \tag{6.76}$$

Equation (76) shows that in the subspace satisfying (72) the incomplete scalar product $(\hat{F}\psi, \psi)_{b_k}$ can depend on b_k even if the operator $\hat{F}$ does not depend on $\hat{b}_k$ explicitly. This is possible when $\hat{F}$ does not commute with $\hat{G}_{n+k}$, and may be ascribed to the implicit dependence of $\hat{F}$ on $\hat{b}_k$. Having defined the operator $\mathrm{d}\hat{F}/\mathrm{d}\hat{b}_k$ as obeying in the b-representation

$$\left(\frac{\mathrm{d}\hat{F}}{\mathrm{d}\hat{b}_k}\psi, \psi\right)_{b_k} = \frac{\mathrm{d}}{\mathrm{d}b_k}(\hat{F}\psi, \psi)_{b_k}, \tag{6.77}$$

where ψ is a vector of the subspace satisfying (72), one gets from (76),

$$\frac{\mathrm{d}\hat{F}}{\mathrm{d}\hat{b}_k} = \frac{\partial\hat{F}}{\partial\hat{b}_k} + \frac{i}{\alpha}[\hat{G}_{n+k}\hat{F}]. \tag{6.78}$$

Replacing in (71) c by $-i\alpha$ and $\hat{f}$ by $\hat{F}$, we obtain

$$\frac{\partial\hat{F}}{\partial\hat{b}_k} = \frac{i}{\alpha}[\hat{a}_k\hat{F}] \tag{6.79}$$

and finally,

$$\frac{\mathrm{d}\hat{F}}{\mathrm{d}\hat{b}_k} = \frac{i}{\alpha}[\hat{a}_k\hat{F}] + \frac{i}{\alpha}[\hat{G}_{n+k}\hat{F}] = \frac{i}{\alpha}[\hat{G}'_{n+k}\hat{F}], \tag{6.80}$$

where

$$G'_{n+k} = \hat{a}_k + \hat{G}_{n+k}. \tag{6.81}$$

In Eqn. (80), $[\hat{a}_k\hat{F}]$ conforms to the explicit dependence of $\hat{F}$ on $\hat{b}_k$, whereas $[\hat{G}_{n+k}\hat{F}]$ conforms to the implicit one.

Since there are cases in which Eqn. (75) does not lead to Eqn. (76) (cf. Sect. 5.5), it is precisely Eqn. (80) which should be considered the definition of $\mathrm{d}\hat{F}/\mathrm{d}\hat{b}_k$ in the general case.

In the cases $\hat{F} = \hat{a}_j$ or $\hat{F} = \hat{b}_j$ ($j = 1, \ldots, n$), Eqn. (80), in view of (70) and (70′), yields

$$\mathrm{d}\hat{b}_j/\mathrm{d}\hat{b}_k = \partial\hat{G}'_{n+k}/\partial\hat{a}_j; \qquad \mathrm{d}\hat{a}_j/\mathrm{d}\hat{b}_k = -\partial\hat{G}'_{n+k}/\partial\hat{b}_j. \tag{6.82}$$

Similar formulas hold if the vector ψ satisfy the equation

$$\hat{b}_k\psi = \hat{G}_k\psi, \tag{6.72′}$$

where $\hat{G}_k$ is the Hermitian operator – a function of all $\hat{a}_j$ and $\hat{b}_j$ except for $\hat{b}_k$. All of this confirms indeed the above statement that the operators $\hat{a}_j$ and $\hat{b}_j$ ($j = 1, \ldots, n$) are interrelated, provided the set of vectors on which they are defined obeys equations of the form given in (72) or (72′).

If $\hat{F} = \hat{I}$, it follows from (80) that

$$d\hat{I}/d\hat{b}_k = 0, \tag{6.83}$$

which is immediately evident as well. But then in the b-representation it follows from (77) that

$$\frac{d}{db_k}(\psi, \psi)_{b_k} = 0 \tag{6.84}$$

if ψ satisfies (72). So, fulfillment of (84) is the necessary condition for a vector ψ to belong to the subspace defined by (72), $\hat{G}_{n+k}$ being a Hermitian operator on that.

6.4. Quantum Approach

This is the common name for the approach in physics which, from the standpoint of probabilistics, is a specific mode of probabilistic treatment utilizing the methods of linear algebra (cf. Sect. 5.4). When employing it, a state of an abstract object is described with respect to some random variables by a state vector or a density operator defined in the corresponding space, that is, the probability distributions for those random variables are uniquely determined in the state under consideration by those mathematical quantities. This requires forming an appropriate Hermitian operator for every random variable. In probabilistics those operators are auxiliary mathematical quantities and appear originally in connection with the calculation of mathematical expectations of random variables. Therefore, one may generally take, as an operator for a random variable t having a set θ_t of values, any Hermitian operator $\hat{T}$ such that the set θ_T of its eigenvalues satisfies the condition $\theta_T \supseteq \theta_t$. But then any state of the corresponding abstract object is described by a linear combination of only those eigenvectors of $\hat{T}$ which refer to its eigenvalues $\lambda \in \theta_t$. However, it is most expedient, of course, to choose a Hermitian operator for a random variable such that its spectrum coincides with the set of values of the latter.

Therefore, in the space of functions of all x_j ($j = 1, \ldots, n$), that is, in the x-representation, the operator of multiplication by x_l,

$$\hat{t}_l = \hat{x}_l = x_l\hat{I} \tag{6.85}$$

should be taken as the Hermitian operator for every Hamiltonian random variable t_l having values $x_l \in \theta_l$ ($l = 1, \ldots, n$). But when choosing Hermitian operators for t_{n+l} ($l = 1, \ldots, n$) one has to reckon with the fact noted in Section 2 that the joint probability distribution for any number of functions of all Hamiltonian random variables is fully determined by the probability distribution for the n-dimensional random variable $t_{1\ldots n}$ in any state of the corresponding abstract object. This means that the description of a state with respect to the first n Hamiltonian random variables $t_1, \ldots, t_n$ completely determines probability distributions for any random variables that are functions of all the Hamiltonian random variables $t_1, \ldots, t_n, t_{n+1}, \ldots, t_{2n}$; that is, any state of the corresponding abstract object is entirely described by a state vector which in the x_j-representation is of the form $\psi(x_1, \ldots, x_n)$. Because of this, in any state, the probability distributions for the random variables t_{n+l} $(l = 1, \ldots, n)$ are fully determined by the corresponding state function $\psi(x_1, \ldots, x_n)$. Hence, the operators for t_{n+l} cannot be formed irrespective of the choice of the operators for the t_l, e.g., as the operators of multiplication by the x_{n+l}. They should be chosen in the space of functions of only the x_l if the operators for the t_l are taken of the form given in (85). For a number of reasons the operator for t_{n+l} in this case has to be chosen in the form

$$\hat{t}_{n+l} = -i\alpha\left[\frac{\partial}{\partial x_l} + \varphi_l(x_l)\hat{I}\right]. \tag{6.86}$$

The direct reason for such a choice is as follows. An eigenfunction of the operator $\hat{t}_{n+l}$ determines a state where t_{n+l} has a definite value. However, if H does not depend explicitly on values of t_l, then in any state, t_{n+l} has a quite definite value, the probability density of t_l being constant in the state. Therefore, if the scalar product (φ, ψ) is defined by

$$\int \varphi(x_l)\bar{\psi}(x_l)f(x_l)\,\mathrm{d}x_l, \tag{6.87}$$

where $f(x_l)$ is a weight function, then a ψ-function describing a state with a definite value of t_{n+l} has to be of the form

$$\psi(x_l) = \text{const}\,\exp\,[if_1(x_l)]\,f(x_l)^{-1/2}, \tag{6.88}$$

where f_1 is a real function, since only in this case $\rho(x_l) = |\psi(x_l)|^2 f(x_l)$ does not depend on x_l. But such a ψ-function may be the necessary eigenfunction of a Hermitian operator $\hat{t}_{n+l}$ if the operator is of the form given in (86), while

$$\varphi_l(x_l) = \frac{1}{2}\,\frac{\mathrm{d}\ln f(x_l)}{\mathrm{d}x_l}, \qquad f_1(x_l) = \alpha^{-1}\gamma x_l \tag{6.89}$$

and γ is the value of t_{n+l} in the state under consideration. The constant α can be determined from statistical experiments only, but we are not interested in its value here.

The correctness of such a choice of operators for the t_{n+l} is confirmed, in particular, by the correspondence between Eqns. (6′) and (82), since in this case the operators $\hat{t}_{n+l}$ and $\hat{t}_l$ prove to be canonically conjugate; the former being similar to $\hat{a}_l$, and the latter to $\hat{b}_l$. Now, it follows from the preceding section that in the x_{n+j}-representation the operator of multiplication by x_{n+l} corresponds to the random variable t_{n+l}, and the operator of differentiation with respect to x_{n+l} corresponds to the random variable t_l ($l = 1, \ldots, n$).

The noncommutativity of the operators $\hat{t}_{n+l}$ and $\hat{t}_l$ leads to the well-known difficulties in forming operators for functions of t_l and t_{n+l}. In particular, an operator $\hat{H}'$ for the parametric Hamiltonian function H' can be formed easily and uniquely only if $H'(t)$ is a sum of terms including functions of the t_l and functions of the t_{n+l} separately: $H'(t) = f_1(t_1, \ldots, t_n) + f_2(t_{n+1}, \ldots, t_{2n})$. In this case, in order to form $\hat{H}'$ one has only to replace in $H'(t)$ the t_j and t_{n+j} by the operators for them, that is, to write $\hat{H}' = f_1(\hat{t}_1, \ldots, \hat{t}_n) + f_2(\hat{t}_{n+1}, \ldots, \hat{t}_{2n})$. This way of forming $\hat{H}'$ follows directly from the approach leading to the introduction of Hermitian operators for random variables.

If H' does not depend on x_n explicitly, in a state where the random variable t_{2n} has a definite value x^0_{2n}, the random variable $H(t)$ has the negative of the same value. Hence, such a state is described by a common eigenvector of the operators $\hat{H}$ and $\hat{t}_{2n}$, satisfying simultaneously the equations

$$\hat{t}_{2n}\psi = x^0_{2n}\psi; \qquad \hat{H}\psi = -x^0_{2n}\psi \tag{6.90}$$

and, therefore, also the equation

$$-\hat{t}_{2n}\psi = \hat{H}\psi \tag{6.91}$$

which in the x-representation takes the form

$$i\alpha\, \partial\psi/\partial x_n = \hat{H}\psi. \tag{6.92}$$

Equation (91) keeps the same form if H depends on x_n explicitly. This follows from the fact that the mathematical expectation of the random variable $H'(t) = t_{2n} + H(t)$ must be equal to zero, whence

$$-(\hat{t}_{2n}\psi, \psi) = (\hat{H}\psi, \psi). \tag{6.93}$$

But Eqn. (93) is valid if ψ obeys (91).

Since Eqn. (91) is of the same form as Eqn. (72), the formulas

$$\hat{\dot{t}}_j = \frac{d\hat{t}_j}{d\hat{t}_n} = \frac{\partial \hat{H}'}{\partial \hat{t}_{n+j}}, \quad \text{and} \quad \hat{\dot{t}}_{n+j} = \frac{d\hat{t}_{n+j}}{d\hat{t}_n} = -\frac{\partial \hat{H}'}{\partial \hat{t}_j} \tag{6.94}$$

are valid; they define on the set of vectors satisfying (91) the operators for the random variables $\dot{t}_j$ and $\dot{t}_{n+j}$. More generally, if a random variable F is a function of t_k ($k = 1, \ldots, n, n+1, \ldots, 2n$), the operator for the random variable $\dot{F}$ is defined by

$$\hat{\dot{F}} = \frac{d\hat{F}}{d\hat{t}_n} = \frac{i}{\alpha}[\hat{H}'\hat{F}] = \frac{i}{\alpha}[\hat{t}_{2n}\hat{F}] + \frac{i}{\alpha}[\hat{H}\hat{F}], \tag{6.95}$$

where $\hat{F}$ is the operator for F. Equations defining operators for the random variables $\mathrm{d}F/\mathrm{d}t_k$ and $\mathrm{d}F/\mathrm{d}t_{n+k}$ are, in view of (80) and (72′),

$$\frac{\mathrm{d}\hat{F}}{\mathrm{d}\hat{t}_k} = \frac{i}{\alpha}\,[\hat{G}'_{n+k}\hat{F}] = \frac{i}{\alpha}\,[\hat{t}_{n+k}\hat{F}] + \frac{i}{\alpha}\,[\hat{G}_{n+k}\hat{F}] \tag{6.96}$$

and

$$\frac{\mathrm{d}\hat{F}}{\mathrm{d}\hat{t}_{n+k}} = \frac{i}{\alpha}\,[\hat{G}'_k\hat{F}] = -\frac{i}{\alpha}\,[\hat{t}_k\hat{F}] + \frac{i}{\alpha}\,[\hat{G}_k\hat{F}], \tag{6.97}$$

where $\hat{G}_k$ and $\hat{G}_{n+k}$ are the same as in Section 6.3, the t_{n+k} replacing $\hat{a}_k$, and the $\hat{t}_k - \hat{b}_k$.

Let us now discuss some peculiarities which reveal themselves when one applies the quantum approach to objects possessing Hamiltonian random variables. One is *quantization* of certain random variables, that is, appearance of a denumerable set θ_d of discrete values for a random variable instead of the continuum θ of its original values, the discreteness of θ_d being due to that of the spectrum of the Hermitian operator corresponding to the random variable involved. This means that the set of eigenvectors of this operator is denumerable as well. The only reason why such a situation arises in a functional space is the requirement that the operator for a random variable should be Hermitian. This requirement is, as has already been mentioned in Section 5.4, very strong – it singles out in the space, where an operator of the formally necessary appearance is defined, the subspace in which this operator is Hermitian. The totality of the eigenvectors of this Hermitian operator, which forms the basis of this subspace, can be denumerable.

Let us illustrate this by the example of a random variable t_l possessing values $x_l \in (a, b)$. In the x-representation t_{n+l} is of the form given in (86), while $\varphi_l(x_l)$ is given by (89). But among all the eigenfunctions

$$\psi_\beta = (b - a)^{-1/2} \exp(i\beta x_l) f(x_l)^{-1/2} \tag{6.98}$$

of this operator, the only ones that belong to the domain of definition of the Hermitian operator t_{n+l} are those that can be paired such that

$$\psi_{\beta_1}(b)\bar{\psi}_{\beta_2}(b)f(b) = \psi_{\beta_1}(a)\bar{\psi}_{\beta_2}(a)f(a). \tag{6.99}$$

This statement results from the following considerations. Let us calculate $(\hat{t}_{n+l}\psi_1, \psi_2)$, making use of (87). In view of

$$\hat{t}_{n+l} = -i\alpha\left[\frac{\partial}{\partial x_l} + \frac{1}{2}\,\frac{\mathrm{d}\ln f(x_l)}{\mathrm{d}x_l}\,\hat{I}\right] \tag{6.100}$$

we get

$$(\hat{t}_{n+l}\psi_1, \psi_2) = -i\alpha\int_a^b \frac{\mathrm{d}\psi_1}{\mathrm{d}x_l}\,\bar{\psi}_2 f\,\mathrm{d}x_l - \frac{i\alpha}{2}\int_a^b \psi_1\bar{\psi}_2\,\frac{\mathrm{d}f}{\mathrm{d}x_l}\,\mathrm{d}x_l. \tag{6.101}$$

Integration by parts yields, after simple rearrangements,

$$(\hat{t}_{n+l}\psi_1, \psi_2) = (\psi_1, \hat{t}_{n+l}\psi_2) - i\alpha[\psi_1(b)\overline{\psi}_2(b)f(b) - \psi_1(a)\overline{\psi}_2(a)f(a)], \qquad (6.102)$$

which proves the statement made.

As is easily seen, the eigenfunctions needed are

$$\psi_m = (b-a)^{-1/2} \exp\,[im(b-a)^{-1}2\pi x_l]\, f(x_l)^{-1/2}, \qquad (6.103)$$

the respective eigenvalues being $x^m_{n+l} = m2\pi(b-a)^{-1}\alpha$ ($m = 0, \pm1, \pm2, \ldots$). Thus, it is the requirement that the operator t_{n+l} of Eqn. (100) be Hermitian that leads, in the last analysis, to quantization of the values of t_{n+l} in case the values of t_l fully occupy a certain finite interval.

The same also holds in other cases when a set of values of a certain random variable which is a function of Hamiltonian random variables becomes denumerable when using the quantum approach. In every such case the reason for this is that the set of eigenvectors of the corresponding Hermitian operator, which must belong to its domain of definition, proves to be countable. For instance, for Eqn. (32) with $q(x_1) = x_1$,

$$\hat{H} = 2^{-1}\hat{t}_3^2 + 2^{-1}\hat{t}_1^2; \qquad \hat{H}' = \hat{t}_4 + \hat{H}. \qquad (6.104)$$

In the x_1-representation, provided $\alpha = 1$,

$$\hat{H} = -2^{-1}\,\mathrm{d}^2/\mathrm{d}x_1^2 + 2^{-1}x_1^2\hat{I}. \qquad (6.105)$$

A state with a definite value of $x_4^0 = -2^{-1}E^0$ of the random variable t_4 is described by a ψ-function obeying

$$-\mathrm{d}^2\psi(x)/\mathrm{d}x^2 + x^2\psi(x) = E^0\psi(x) \qquad (6.106)$$

(for the sake of simplicity the index '1' is omitted). The requirement that $\hat{H}$ should be a Hermitian operator, provided in (87) $f(x_1) = 1$, leads to the condition

$$\left(\frac{\mathrm{d}\psi_1}{\mathrm{d}x}\psi_2 - \frac{\mathrm{d}\psi_2}{\mathrm{d}x}\psi_1\right)\Bigg|_{-\infty}^{\infty} = 0 \qquad (6.107)$$

for any pair of functions ψ_1 and ψ_2 satisfying Eqn. (106). This condition is fulfilled only for a denumerable set of functions

$$\psi_n(x) = \pi^{-1/4}(2^n n!)^{-1/2} \exp\,(-x^2/2)H_n(x), \qquad (6.108)$$

where $H_n(x)$ is the nth Hermite polynomial, the respective eigenvalues of H being

$$2^{-1}E_n^0 = n + \tfrac{1}{2}, \qquad n = 0, 1, 2, \ldots \qquad (6.109)$$

Let us elucidate a misunderstanding which may arise due to the apparent quantization of t_{2n}, whereas the set of values of t_n is either an infinite or a half-infinite interval. As follows from (103), in this case an operator $\hat{t}_{2n}$ of the form (100) is Hermitian for any values of β in (98) and, hence, there is no reason for direct quantization of the values of t_{2n}. Meanwhile, in many cases, in particular in the instance just considered, these values prove to be quantized. The contradiction is eliminated as follows. The requirement of hermiticity of the operator $\hat{H}$ actually leads to direct quantization of the random variable $H(t)$, whereas the values of t_{2n} are compelled to be quantized by virtue of the necessity that Eqn. (29) be satisfied.

The above example also reveals another peculiarity related to the quantum approach. In a state with a definite value of $H(t)$, say, $H^0 = 2^{-1}$, values of t_1 can exceed in magnitude $x_1^0 = 1$ as much as one wants. Since the values $2^{-1}x_3^2$ of the random variable $2^{-1}t_3^2$ are nonnegative, they cannot compensate for the excess of the summand $2^{-1}x_1^2$ over 2^{-1} [cf. Eqn. (34) with $q(x_1) = x_1$]. But the probability density of a value of t_1 such that the summand $2^{-1}x_1^2$ alone is greater than 2^{-1} is nonzero, as follows from Eqn. (108). So, a curious situation arises: For concrete objects corresponding to a state with a definite value $2^{-1}E^0$ of the property $-t_4$, the property $H(t)$ has the same value, whereas for $q(x_1) = x_1$, each one of the two nonnegative summands in the first of Eqns. (34) can exceed $2^{-1}E^0$.

To explain this peculiarity it is necessary to take into account the following circumstances. The ranges of probabilistic problems solvable in principle do not coincide entirely for cases in which one makes use of the classical approach (that is, of the approach utilizing Eqns. (1) and (20)) and of the quantum approach. The range of problems solvable on the basis of the classical approach is wider, and not all these problems can be solved with the quantum approach. For instance, the problem of finding the joint probability density of canonically conjugate random variables, which is easily solvable when applying the classical approach, cannot be generally solved in principle when one utilizes the quantum approach. Therefore, the quantum approach does not allow one to find for these random variables any conditional probabilities and, hence, conditional mathematical expectations. But this approach reveals peculiarities of the probability distributions for Hamiltonian random variables that cannot be found when applying the classical approach. At the same time, a number of probabilistic problems can be solved on the basis of both approaches. For such problems the quantum approach is more general, and the probability distributions obtained on applying it either coincide with or go over with sufficient accuracy into those obtained by use of the classical approach when the quantum numbers become large enough.

The necessity of coincidence of the probability densities of the t_l, for both approaches, in states with definite values of the corresponding t_{n+l} has even been utilized as a direct reason for the choice of the Hermitian operators for the t_{n+l} in the form given in (100). And the practical coincidence of the probability density obtained when using the quantum approach with the classical one for large enough quantum numbers can be demonstrated on the interesting example of Hamiltonian random variables related by Eqn. (32) with $q(x_1) = x_1$. In this case, the quantum approach gives for the probability density of t_1 in a state with a definite value $x_4 = -(n + \frac{1}{2})$ of t_4, in view of (108), the following expression

$$\rho(x) = |\psi_n(x)|^2 = \pi^{-1/2}(2^n n!)^{-1} \exp(-x^2) H_n^2(x). \tag{6.110}$$

When using asymptotic formulas for $H_n(x)$ at large n (see Ref. [5]) one obtains from (110), in view of (109),

$$\rho(x) = [\pi(E^0 - x^2)^{1/2}]^{-1} [1 + \sin E^0(\varphi_1 - \tfrac{1}{2} \sin 2\varphi_1) + O(E^{0-1})] \tag{6.111}$$

if $1 > xE^{0-1/2} > -1$, and

$$\rho(x) = [2\pi(x^2 - E^0)^{1/2}]^{-1} \exp[-E^0(\tfrac{1}{2} \sinh 2\varphi_2 - \varphi_2)] [1 + O(E^{0-1})] \tag{6.112}$$

if $|x|E^{0-1/2} > 1$, where φ_1 is positive and is given by $x = E^{0\,(1/2)} \cos \varphi_1$, and φ_2 is positive and is given by $|x| = E^{0\,(1/2)} \cosh \varphi_2$. If n is large enough, (111) does not differ practically from the classical expressions (45) and (46), while the right-hand side of (112) vanishes.

One may conclude from the above that in the general case, that is, for any quantized values of Hamiltonian random variables, the connection between these random variables is not determined by Eqn. (20), nor therefore, is it expressed by its solution explicitly. The question, what actually is this connection, is apparently unanswerable at present. But since it is determined by Eqn. (20) (with sufficient accuracy) for high quantum numbers only, the relation (34) as well as the more general one expressing $H'(t)$ as a sum of terms depending on only the t_l and only the t_{n+l} separately will not necessarily be fulfilled in the general case. In this way one may explain generally the noncompliance with equations of the form (34) shown by concrete objects corresponding to states with definite values of t_{2n}. In connection with this it is necessary to note one more circumstance, we called only the random variables obeying (20′) the Hamiltonian ones, but it actually turned out that Eqn. (20′) fits such random variables in the transition to the classical limit only. Despite this, however, the term 'Hamiltonian random variables' keeps its sense in the general case as well, because the operators for these random variables are connected by Eqn. (94), which is quite similar to Eqn. (6′).

Since the random variables t_{n+l} must be functions of the random variables t_l ($l = 1, \ldots, n$) in the general case as well, any functions of the random variables t_l and t_{n+l} always are, in the last analysis, functions of the t_l only. In any state, therefore, both probability distributions and mathematical expectations for functions of the t_l and t_{n+l} have to be uniquely determined by the probability distributions for t_l ($l = 1, \ldots, n$). And, although, as noted above, it is impossible to form the joint probability distribution for t_l and t_{n+l} when using the quantum approach, one can obtain in principle the probability distributions for any functions of them as well as the mathematical expectations of these functions, provided the respective Hermitian operators are chosen correctly. Thus, the problem reduces to a correct choice of a Hermitian operator for the random variable under consideration. As soon as the necessary operator is found, all the values we are interested in can be obtained in principle by means of the standard mathematical techniques. As an illustrative example, let us consider the joint mathematical expectation $Mt_1^m t_3^k$ for the random variables t_1 and t_3 corresponding to Eqn. (32) in a state with a definite discrete value $H^0 = 2^{-1}E^0$ of the random variable $H(t)$, the state function ψ^0 describing this state being considered real.

The operator $\hat{H}$ in this case is obtained from Eqn. (34). As an operator for the random variable $t_1^m t_3^k$ we take the Hermitian operator

$$\hat{t}_{(1,3)}^{(m,k)} = 2^{-1}(\hat{t}_1^m \hat{t}_3^k + \hat{t}_3^k \hat{t}_1^m). \tag{6.113}$$

This choice seems to be the simplest, while its correctness is confirmed as follows. If k is an odd integer, then $Mt_1^m t_3^k = 0$ for any m, as it should be, because this result holds in the classical limit. Further, it follows from (48) that

$$(\hat{g} - \hat{r})\psi^0 = 0, \tag{6.114}$$

where $\hat{g}$ and $\hat{r}$ are the Hermitian operators corresponding to g and r in (48). The Hermitian operator

$$\hat{P} = 2^{-1} \sum_{j=0}^{k-1} [\hat{t}_1^m \hat{r}^j \hat{g}^{k-j-1}(\hat{g} - \hat{r}) + (\hat{g} - \hat{r})\hat{g}^{k-j-1}\hat{r}^j \hat{t}_1^m] \tag{6.115}$$

corresponds to the random variable P given by Eqn. (49) for $F = t_1^m$. Since in the state involved, $MP = 0$, we get for the mathematical expectation of the random variable $t_1^m t_3^{2l}$, after a simple calculation

$$Mt_1^m t_3^{2l} = \int x_1^m [E^0 - 2\varphi(x_1)]^l \rho(x_1)\,\mathrm{d}x_1 + \sum_{j=1}^{l-1} ([[\hat{g}^j \hat{r}]\hat{r}^{l-j-1}\hat{t}_1^m]\psi^0, \psi^0), \tag{6.116}$$

where $\rho(x_1) = |\psi^0(x_1)|^2$. Hence it follows for $l = 1$

$$Mt_1^m t_3^2 = \int x_1^m [E^0 - 2\varphi(x_1)]\rho(x_1)\,\mathrm{d}x_1, \tag{6.117}$$

and for $l = 2$

$$Mt_3^4 = \int [E^0 - 2\varphi(x_1)]^2 \rho(x_1)\,\mathrm{d}x_1. \tag{6.118}$$

These equations coincide in form with the classical ones given by (53), the coincidence becoming numerical when the classical limit is approached. Finally, if $q(x_1) = x_1$, then also for m odd, $Mt_1^m t_3^k = 0$, and in the classical limit (53) follows from (116) for any m and k.

Let us note once more, that one also may try some other choice of Hermitian operator for $t_1^m t_3^k$. If it turns out that in the classical limit $Mt_1^m t_3^k$ does not satisfy the necessary conditions, the choice is incorrect. But if all these conditions are satisfied, one has to consider as correct also such a choice of the Hermitian operator for $t_1^m t_3^k$. The possible absence of uniqueness in the choice of an operator for $t_1^m t_3^k$ conforms to the fact that the operators for random variables are auxiliary mathematical quantities only and have no independent meaning in probabilistics.

6.5. Standard Deviations of Canonically Conjugate Random Variables

The standard deviation of a random variable t, which will be denoted by Δt, is defined as the positive square root of the variance $\mathrm{D}t$ of this random variable in the corresponding state of an abstract object (see Sect. 5.4):

$$\Delta t = (\mathrm{D}t)^{1/2} \tag{6.119}$$

If a state is described with the aid of the state vector ψ normalized to unity, then in view of Eqns. (3.59) and (5.109),

$$\mathrm{D}t = ([\hat{t} - (\hat{t}\psi, \psi)\hat{I}]^2 \psi, \psi), \tag{6.120}$$

where $\hat{t}$ is the Hermitian operator corresponding to t, the vector ψ belonging to the domain of definition of this Hermitian operator (briefly, to the hermiticity domain of this operator). Since $\hat{t} - (\hat{t}\psi, \psi)\hat{I}$ is also a Hermitian operator on the same set of vectors as $\hat{t}$, we can rewrite (120) in the form

$$\mathrm{D}t = ([\hat{t} - (\hat{t}\psi, \psi)\hat{I}]\, \psi, [\hat{t} - (\hat{t}\psi, \psi)\hat{I}]\, \psi) \tag{6.121}$$

whence, in view of (119),

$$\Delta t = |[\hat{t} - (\hat{t}\psi, \psi)\hat{I}]\, \psi|. \tag{6.122}$$

The standard deviation of a random variable t is a quite definite nonnegative number for each state of the corresponding abstract object, which can be calculated by means of (122) if the operator $\hat{t}$ and the ψ-vector are known. For instance, the standard deviations of t_1 and t_3 (see Eqn. (104)) in the nth state of the corresponding abstract object determined by the definite value $x_4^0 = -2^{-1}E_n^0$ of the random variable t_4, calculated with the aid of (108), are

$$\Delta t_1 = \Delta t_3 = (2^{-1}E_n^0)^{1/2} = (n + 2^{-1})^{1/2}, \tag{6.123}$$

whereas in any state described by (103), which corresponds to a definite value of t_{n+l},

$$\Delta t_l = \frac{|b - a|}{2\sqrt{3}}, \quad \Delta t_{n+l} = 0. \tag{6.124}$$

The product of the standard deviations of any two random variables $t_{(1)}$ and $t_{(2)}$ is also a quite definite nonnegative number for each state of the corresponding abstract object. The lower limit of this product can be estimated by making use of Schwarz's inequality (5.77). By replacing the x and y by $[\hat{t}_{(1)} - (\hat{t}_{(1)}\psi, \psi)\hat{I}]\,\psi$ and $[\hat{t}_{(2)} - (\hat{t}_{(2)}\psi, \psi)\hat{I}]\,\psi$, respectively, we get, in view of (122) and $\operatorname{Im} z \leqslant |z|$ for any z,

$$\Delta t_{(1)}\,\Delta t_{(2)} \geqslant |\operatorname{Im}\,([\hat{t}_{(1)} - (\hat{t}_{(1)}\psi, \psi)\hat{I}]\,\psi, [\hat{t}_2 - (\hat{t}_{(2)}\psi, \psi)\hat{I}]\,\psi)|. \tag{6.125}$$

Since $(\hat{t}_{(1)}\psi, \psi)$ and $(\hat{t}_{(2)}\psi, \psi)$ are real numbers, this gives

$$\begin{aligned}\Delta t_{(1)}\,\Delta t_{(2)} &\geqslant \operatorname{Im}\,(\hat{t}_{(1)}\psi, \hat{t}_{(2)}\psi)\\ &= -\frac{i}{2}\,[(\hat{t}_{(1)}\psi, \hat{t}_{(2)}\psi) - (\hat{t}_{(2)}\psi, \hat{t}_{(1)}\psi)].\end{aligned} \tag{6.126}$$

Let us calculate the right-hand side of (126) for the case when $t_{(1)}$ and $t_{(2)}$ are canonically conjugate random variables: $t_{(1)} = t_l$ and $t_{(2)} = t_{n+l}$. Making use of (87) and (100), we get, after simple calculations,

$$\operatorname{Im}\,(\hat{t}_{n+l}\psi, \hat{t}_l\psi) = \frac{\alpha}{2}\left[1 - (|\psi|^2 x_l f)\Big|_a^b\right]. \tag{6.127}$$

We thus finally obtain from (126)

$$\Delta t_{n+l}\,\Delta t_l \geqslant \left|\frac{\alpha}{2}\left[1 - (|\psi|^2 x_l f)\Big|_a^b\right]\right|. \tag{6.128}$$

Since $x_l\psi$ was not required to belong to the hermiticity domain of the operator $\hat{t}_{n+l}$ when deriving Eqn. (128), this is a universal inequality valid for any state of the corresponding abstract object. But if $x_l\psi$ does belong to that domain, then the second term in the parentheses on the right-hand side of (128) should vanish, in view of (99). Therefore, in this particular case, Eqn. (128) goes over into

$$\Delta t_{n+l}\,\Delta t_l \geqslant \left|\frac{\alpha}{2}\right|. \tag{6.129}$$

For this case, however,

$$-\frac{i}{2}\,[(\hat{t}_{n+l}\psi, \hat{t}_l\psi) - (\hat{t}_l\psi, \hat{t}_{n+l}\psi)] = \frac{i}{2}\,([\hat{t}_{n+l}\hat{t}_l]\,\psi, \psi) \tag{6.130}$$

and (129) follows immediately from (126), because of

$$[\hat{t}_{n+l}\hat{t}_l] = -i\alpha\hat{I}. \tag{6.131}$$

Since ψ in (123) belongs to the hermiticity domain of the operator t_{n+l}, it obeys (99). Hence, in order for $x_l\psi$ to belong to that domain, it should satisfy the condition

$$|\psi(b)|^2 f(b)(b-a) = |\psi(a)|^2 f(a)(b-a) = 0. \tag{6.132}$$

But, in view of $b - a \neq 0$, this can be valid only when

$$|\psi(b)|^2 f(b) = |\psi(a)|^2 f(a) = 0. \tag{6.133}$$

In Section 5.5 it has been proven that if for two Hermitian operators $\hat{A}$ and $\hat{B}$ Eqn. (5.126) is fulfilled, that is, if the operators are canonically conjugate, and if x_α is an eigenvector of $\hat{A}$, then the vector $\hat{B}x_\alpha$ does not belong to the hermiticity domain of $\hat{A}$. But as stated in Section 3, in the representation in which $\hat{B}$ is the operator of multiplication by its eigenvalues, $\hat{A}$ is the operator of the form given in (100). Therefore, not only $\hat{B}x_\alpha$, but also any $\hat{B}\psi$ is not found in the hermiticity domain of $\hat{A}$ if ψ is a linear combination of eigenvectors of $\hat{A}$, which does not satisfy Eqn. (133).

It is necessary, finally, to stress some results following from the above. First, in contrast to Eqn. (128), Eqn. (129) is not universal. Second, neither from (128) nor from (129) does it follow that the larger Δt_{n+l}, the smaller Δt_l and conversely. The incorrectness of such a conclusion is illustrated by the examples considered above [see Eqns. (123) and (124)]. Third, it is quite possible that Eqn. (129) is really valid, although ψ in it does not satisfy (133), that is, although Eqn. (129) cannot be obtained from Eqn. (126) because $x\psi$ does not belong to the hermiticity domain of the operator $\hat{t}_{n+l}$. As an example, let us consider the case: $a = 0$, $b = 2\pi$, $f(x_l) = 1$, and $\psi(x_l) = \beta\psi_m(x_l) + (1-\beta)\psi_n(x_l)$, where β is a positive number, and $\psi_m(x)$ and $\psi_n(x)$ are of the form as given in (103). It is obvious that $\psi(x_l)$ does not satisfy (133) for any value of $\beta \neq 0$ or 1, and hence, Eqn. (129) cannot be obtained for this function from Eqn. (126). At the same time, simple calculations show that

$$\mathrm{D}t_{n+l} = \alpha^2\gamma^2(m-n)^2; \qquad \mathrm{D}t_l = \frac{\pi^2}{3} + \frac{4\gamma}{(n-m)^2}, \tag{6.134}$$

where $\gamma = \beta(1-\beta)/[\beta^2 + (1-\beta)^2]$, and therefore

$$\Delta t_{n+l}\,\Delta t_l = |\alpha|\gamma\left[\frac{\pi^2(m-n)^2}{3} + 4\gamma\right]^{1/2}. \tag{6.135}$$

It is now easily seen that in two definite intervals of β depending on $|n-m|$, namely $0 \leqslant \beta < \beta_1$ and $\beta_2 < \beta \leqslant 1$, $\Delta t_{n+l}\,\Delta t_l < |\alpha|/2$, while in the interval $\beta_1 \leqslant \beta \leqslant \beta_2$, $\Delta t_{n+l}\,\Delta t_l \geqslant |\alpha|/2$ is really valid.

Problems

6.1. Verify that Eqn. (1) with A_0, A_1, and Q determined by Eqn. (27) can be represented in the form of Eqn. (2) with L determined by Eqn. (28).

6.2. Can a state of an abstract object be described by a ψ-function of the form $\psi = f(x_n)\psi_0(x_1, x_2, \ldots, x_{n-1})$, where f is a function such that $f(x_n)\bar{f}(x_n) = \varphi(x_n) \neq \text{const}$?

6.3. Find the intervals of β, in which the right-hand side of Eqn. (135) does not satisfy Eqn. (129) for $|m - n| = 1$.

Suggested References

1. Mayants, L. S.: 1967, *Quantum Physics and Transformations of Matter* (in Russian), VINITI, Moscow. This monograph contains detailed considerations of generalized Hamiltonian equations, canonically conjugate operators, and the transition from quantum to classical probability distributions.
2. Mayants, L. S.: *Found. Phys.* 3, 413. This paper contains a brief consideration of the transition from quantum to classical probability distributions for the particular case discussed in this chapter.
3. Mayants, L. S.: *Found. Phys.* 4, 335. In this paper the inequalities for the product of the standard deviations of canonically conjugate random variables and related problems are discussed for the first time.
4. Mayants, L. S.: *Found. Phys.* 7, 3. This paper introduces for the first time the Hamiltonian random variables and discusses their peculiarities.
5. Szego, G.: 1975, *Orthogonal Polynomials*, 4th edn., Amer. Math. Soc., New York. This book contains the asymptotic formulas for the Hermite polynomials.

Chapter 7

Random Fields

7.1. Definition

Let us remember that the starting point of probabilistics is the notion of a set A of concrete objects which is capable of being subjected to random tests. The concrete objects $a \in A$ have a finite set T of properties t_j ($j = 0, 1, \ldots, n$), every concrete object has one value $f_j \in \theta_j$ of the property t_j, and for all these concrete objects one and the same value f_0 of the property t_0 defines that they belong to set A (see Chap. 1). The structure of the concrete objects was of no interest to us. We now proceed to a consideration of continuous concrete objects, a case which must be treated specially.

Some properties t_α ($\alpha = 1', 2', \ldots, m'$) of a continuous concrete object can have values that are generally different for its distinct points and are, hence, definite functions $f_\alpha(r)$ of the totality r of the space coordinates determining a point in this object. The function $f_\alpha(r)$ is usually called the field of the property t_α. Since values of t_α may depend on the time instant τ, too, in the general case fields have the form of definite functions $f_\alpha(r, \tau)$.

As an example of a continuous object, let us consider a string. If the X-axis of a Cartesian coordinate system is directed along the equilibrium position of the string vibrating in the XY-plane, then for a concrete string the value y of the property 'string deformation' is a definite function of the x-coordinate of a string point and the time instant τ: $y = f(x, \tau)$.

For every continuous concrete object the field $f_\alpha(r, \tau)$ may be regarded as one value of the property t_α for this concrete object. The set $\theta^{(\alpha)}(r, \tau) \ni f_\alpha(r, \tau)$ of the values of t_α is determined, first of all, by the specificity of the corresponding continuous objects $a \in A$. For instance, if the free vibrations of a string in the XY-plane are small, then the equation

$$\frac{\partial^2 y}{\partial \tau^2} - \gamma \frac{\partial^2 y}{\partial x^2} = 0 \tag{7.1}$$

is valid, where $\gamma = T_0 \eta^{-1}$, while T_0 and η are the tension and linear mass density of the string, respectively. Therefore, any function $y = f(x, \tau) \in \theta(x, \tau)$, which describes such vibrations, is a time-dependent field – a value of the property 'string deformation' – that must obey Eqn. (1). If, in addition, a string is fastened at the points, say, $x_1 = 0$ and $x_2 = 1$, then for any τ the conditions

$$f(0, \tau) = f(1, \tau) = 0 \tag{7.2}$$

also have to be fulfilled.

For a set A of continuous concrete objects the abstract object $f_{0\alpha} \equiv f_\alpha$ corresponds to the class $A(f_\alpha) \subset A$ of elements characterized by one and the same field $f_\alpha(r, \tau) \in \theta^{(\alpha)}$, that is, by one and the same value of the property t_α. Let us remember that f_α is the image of the class $A(f_\alpha)$ on the set $\theta^{(\alpha)}$ of the values of the property $t_{0\alpha} = (t_0, t_\alpha)$ inherent in concrete objects $a \in A$; since the only value f_0 of the property t_0 defines that continuous objects under consideration belong to set A, the index '0' in what follows will be omitted.

In accordance with the definition of random variables in probabilistics (see Sect. 3.1) it seems natural to define a one-dimensional random field as a property t_α of continuous objects, whose values $f_\alpha(r, \tau) \in \theta^{(\alpha)}$ are real functions. A set of k one-dimensional random fields will be called a k-dimensional random field.

A definition of a random field may also be approached from a more general point of view. Suppose that in a set A of concrete objects each element $a \equiv B_a \in A$ is, in turn, a set of concrete objects $b_a \in B_a$, the b_a being distinguished by different values $f_{lm\ldots p} \in \theta_{lm\ldots p}$ of an s-dimensional random variable $t_{lm\ldots p}$, one and the same for all the B_a. Let a random variable $f_a = f_a(t_{lm\ldots p})$, which is a definite real function of the random variable $t_{lm\ldots p}$, correspond to every $B_a \in A$, and let $f_a \in \theta$, where θ is the totality of these functions that corresponds to set A. Then the property t_α of the concrete objects $B_a \in A$, whose values are random variables f_a, will be called one-dimensional random function (of the random variable $t_{lm\ldots p}$). A set of k one-dimensional random functions will be called a k-dimensional random function.

A concrete mechanical system of N particles bound harmonically may serve as an example of a concrete object $B_a \in A$ possessing a random function. Each concrete particle of this system is an element $b_{aj} \in B_a$ ($j = 1, \ldots, N$). Let us choose $t_{14} = (t_1, t_4)$ as a random variable for b_{aj} at any a. Here t_4 is time, while t_1 is any random variable, whose values uniquely determine what a particle is meant; it may be, for instance, the particle number (the numeration is supposed, of course, to be the same for each concrete system B_a) or any other quantity determined by this number, say, the radius vector of the particle's equilibrium position. The set $\theta_1 = \{1, \ldots, N\}$ is discrete, and set θ_4 is an interval. The deformation of each concrete system B_a, that is, its deviation from the initial equilibrium position, at the time zero, is a definite three-dimensional random variable – a totality of three definite functions of the independent random variables t_1 and t_4. Values of these functions, which may be written as: $x_k^{(a1)}, x_k^{(a2)}, x_k^{(a3)}$, are the components of the displacement of the kth particle in B_a along the respective axes X_1, X_2, and X_3 of a Cartesian system common for all the B_a. So, in the example given the property 'system deformation' for an element $B_a \in A$ is a three-dimensional random function of the two-dimensional independent random variable t_{14}, that is, of particle number and time, which may also be regarded as a $3N$-dimensional random function of time. The set θ of this random function's values are determined, first of all, by the equations of motion for a system B_a and corresponding initial conditions.

For a set A of continuous objects every concrete object $B_a \in A$ is a set of concrete points $b_a \in B_a$ which are distinguished by different values $f_{1234} = (x_1, x_2, x_3, x_4)$ of one and the same for all the B_a independent random variable $t_{1234} = (t_1, t_2, t_3, t_4)$. Values $r = (x_1, x_2, x_3)$ of the random variable $t_{123} = (t_1, t_2, t_3)$ determine the spatial positions of continuous object points, while values $\tau = x_4$ of the random variable t_4 give the corresponding time instants. Hence, a random field is a random function of the independent random variable t_{1234}, whose values $f_{1234} = (r, \tau)$ determine the time–space positions of continuous object points. The value of a random field on every continuous concrete object is, thus, a definite random variable, which is a definite function of the independent random variable t_{1234}. This definition of a random field obviously is equivalent to that given above.

7.2. Two Types of Related Finite-Dimensional Random Variables

'Randomness' of a random field (more general, of a random function) reveals itself, as for any random variables, only when random tests are carried out. But random tests can give in principle either a value of the property t_α on a concrete object B_a, that is, the corresponding function of the independent random variable $t_{lm\ldots p}$ on this concrete object, or values of this function only for a finite number of concrete objects $b_a \in B_a$. For continuous objects, however, only the second possibility can be realized, as a rule. This circumstance conforms to the specific difficulties of classical probabilistic treatment of fields, which results from the impossibility of generally forming a set $A^{(a)}$ adequate (see Chap. 1) to a set A of continuous concrete objects. This becomes obvius from the following reasoning.

In accordance with the definition, a set $A^{(a)}$ should be measurable and has to satisfy, therefore, the condition: $A^{(a)} \subseteq R_n$, where R_n is some n-dimensional Euclidean space. But in the case under consideration this condition cannot generally be satisfied for any fixed number n. Indeed, a random field t_α gives rise to an infinite set of k-dimensional random variables of at least two types. The first-type k-dimensional random variables $t_{(\alpha)l_1l_2\ldots l_k}$ $(k = 1, 2, \ldots)$ have values $f_{(\alpha)l_1l_2\ldots l_k}$, which are the sets of values of the field $f_\alpha(r, \tau)$ for k different values of the independent random variable t_{123}, that is, at k distinct space points: $r_{l_1}, r_{l_2}, \ldots, r_{l_k}$.

Another type of random variable is obtained as follows: If the set $\theta^{(\alpha)} \ni f_\alpha(r, \tau)$ is a function space, a complete set of real orthonormal functions $f_j^{(\alpha)}(r, \tau)$ forming its basis, then any field $f_\alpha \in \theta^{(\alpha)}$ can be represented as

$$f_\alpha = \sum_j \beta_j^{(\alpha)} f_j^{(\alpha)}(r, \tau), \tag{7.3}$$

where $\beta_j^{(\alpha)} \in \theta_j^{(\alpha)}$ are real numbers. One may regard sets $\{\beta_{l_1}^{(\alpha)}, \beta_{l_2}^{(\alpha)}, \ldots, \beta_{lk}^{(\alpha)}\} \equiv f_{l_1 l_2 \ldots lk}^{(\alpha)}$ as the values of the corresponding second-type k-dimensional random variable $t_{l_1 l_2 \ldots lk}^{(\alpha)}$ $(k = 1, 2, \ldots)$ which are created by the same random field t_α.

Further, the definition of a set $A^{(a)}$ requires the existence of classes $A^{(a)}(f_{lm \ldots p}^{(a)}) \subseteq A^{(a)}$ for any values of the indices $l, m, \ldots, p$. And the dimension of a space R_n including the set $A^{(a)}$ cannot be less than the dimension of any $\theta_{(\alpha)l_1 l_2 \ldots lk} \ni f_{(\alpha)l_1 l_2 \ldots lk}$ or $\theta_{l_1 l_2 \ldots lk}^{(\alpha)} \ni f_{l_1 l_2 \ldots lk}^{(\alpha)}$, since at least one continuous concrete object, for which the element of $A^{(a)}$ conforming to it is in a class $A^{(a)}(f_{(\alpha)l_1 l_2 \ldots lk}^{(a)})$ or in a class $A^{(a)}(f_{l_1 l_2 \ldots lk}^{(\alpha, a)})$, may correspond to each value $f_{(\alpha)l_1 l_2 \ldots lk}$ or $f_{l_1 l_2 \ldots lk}^{(\alpha)}$, respectively. Meanwhile, for k-dimensional random variables of both types the number k and, therefore, the dimensions of $\theta_{(\alpha)l_1 l_2 \ldots lk}$ and $\theta_{l_1 l_2 \ldots lk}^{(\alpha)}$ can exceed any given number n.

Hence, in the general case there are no sets $A^{(a)}$ adequate to a set A of continuous concrete objects possessing a random field t_α. Let us note, by the way, that this conclusion does not contradict Axiom 2 of probabilistics asserting the existence of at least one set $A^{(a)}$ adequate to any set A of concrete objects which is capable of being subjected to random tests. Indeed, in the general case, the results of random tests carried out on continuous concrete objects would have to be concrete values of the random field t_α, that is, concrete functions $f_\alpha(r, \tau)$ wholly, which obviusly is unachievable.

Thus, the usual classical way of the probabilistic treatment – the utilization of a set $A^{(a)}$ adequate to set A of concrete objects – is inapplicable to continuous objects possessing random fields. Therefore, one generally has to be satisfied with the consideration of the finite-dimensional random variables of both types mentioned above. In a number of cases, this allows one to obtain approximate solutions of certain probabilistic problems. The choice of a type of random variable, which is suitable for solving various probabilistic problems, is determined by both the problems themselves and the essential differences between the random variables of these two types. By virtue of the continuity of the objects under consideration, the first-type one-dimensional random variable, $t_{(\alpha)l_j} (j = 1, 2, \ldots)$, can generally have neither mutually uncorrelated nor, all the more, independent probability distributions. This becomes especially clear for the case when the corresponding space points, r_{l_j} $(j = 1, 2, \ldots)$, lie close enough to each other. But the continuity of objects imposes no restrictions as such on the second-type one-dimensional random variable, $t_{l_j}^{(\alpha)}$ $(j = 1, 2, \ldots)$, since the functions $f_j^{(\alpha)}$ themselves take account of it. Therefore, the random variables of this type can have even mutually independent probability distributions.

7.3. Lagrangian and Hamiltonian Partial Equations

Lagrangian and Hamiltonian equations for continuous objects can be obtained from the corresponding sets of ordinary differential equations for discrete objects when proceeding to the limit. Since this procedure, being independent of the specificity of objects, is purely mathematical, one may ignore it and get the results wanted in another formally mathematical way.

Let $Q = Q(u, \ldots, w)$ be a functional, and $u, \ldots, w$ be functions of independent variables x_1, x_2, x_3, x_4. The formal functional (variational) derivative, $\delta Q/\delta u(r')$, of Q with respect to u at the point $r' = (x_1', x_2', x_3')$ can be defined as follows:

$$\frac{\delta Q}{\delta u(r')} = \lim_{\epsilon \to 0} \frac{\delta_{u(r')}^{(\epsilon)} Q}{\epsilon}, \tag{7.4}$$

where

$$\delta_{u(r')}^{(\epsilon)} Q = Q(u + \delta_{r'}^{(\epsilon)}, \ldots, w) - Q(u, \ldots, w) \tag{7.5}$$

and

$$\delta_{r'}^{(\epsilon)} = \epsilon\, \delta(x_1 - x_1')\, \delta(x_2 - x_2')\, \delta(x_3 - x_3') \equiv \epsilon\, \delta(r - r'). \tag{7.6}$$

If, for instance, Q is of the form

$$Q = \int D^{(Q)}\, \mathrm{d}v, \tag{7.7}$$

where

$$D^{(Q)} = D^{(Q)}(u, P(D_j)u, \ldots, w, \ldots, x_4) \quad (j = 1, 2, 3), \tag{7.8}$$

$P(D_j)$ is a polynomial differential operator with constant coefficients, D_j is the operator $\partial/\partial x_j$, and $\mathrm{d}v = \mathrm{d}x_1\, \mathrm{d}x_2\, \mathrm{d}x_3$, then

$$\begin{aligned} \delta_{u(r')}^{(\epsilon)} Q = \int [D^{(Q)}(u + \delta_{r'}^{(\epsilon)}, P(D_j)(u + \delta_{r'}^{(\epsilon)}), \ldots, w, \ldots, x_4) - \\ - D^{(Q)}(u, P(D_j)u, \ldots, w, \ldots, x_4)]\, \mathrm{d}v \equiv \\ \equiv \left(\frac{\partial D^{(Q)}}{\partial u}, \delta_{r'}^{(\epsilon)} \right) + \sum_{j=1}^{3} \left(\frac{\partial D^{(Q)}}{\partial P(D_j)u}, P(D_j)\delta_{r'}^{(\epsilon)} \right). \end{aligned} \tag{7.9}$$

In this section we make use of the scalar product

$$(u, w) = \int u w\, \mathrm{d}v \tag{7.10}$$

of the functions u and w, considering these real, which replaces the matrix product of row and column matrices in Section 6.1.

It follows from (6) that

$$P(D_j)\, \delta_{r'}^{(\epsilon)} = \epsilon P(D_j)\, \delta(r - r'). \tag{7.11}$$

In view of (9) and (11), Eqn. (4) yields

$$\frac{\delta Q}{\delta u(r')} = \left[\frac{\partial D^{(Q)}}{\partial u} + \sum_{j=1}^{3} {}^{t}P(D_j)\frac{\partial D^{(Q)}}{\partial P(D_j)u}\right]_{r'}, \tag{7.12}$$

where ${}^{t}P(D_j)$ is the operator adjoint to $P(D_j)$ and differing from the latter only by the signs at odd degrees of D (see, for instance, Ref. [4]). The expression obtained shows that $\delta Q/\delta u(r')$ is a function of r' and x_4.

The definition of a functional derivative given here is a natural expansion of the definition of a partial derivative of a function with respect to its argument, the δ-function playing the role of a unit. This definition allows one to calculate functional derivatives for functionals of more general form and various type functional derivatives of higher order. If a certain variable x_l takes on only discrete values, then the corresponding Eqn. (7) includes no partial derivatives with respect to it and integration is replaced by summation over the values of x_l, the respective terms vanishing in Eqn. (12).

Let a functional $S = S(u^{(1)})$ be of the form

$$S = \int_{\tau'}^{\tau''} L \, d\tau, \tag{7.13}$$

where

$$L = L(u^{(1)}, \dot{u}^{(1)}, \tau) = \int D^{(L)}(u^{(1)}, P(D_j)u^{(1)}, \dot{u}^{(1)}, \tau) \, dv \quad (j = 1, 2, 3) \tag{7.14}$$

is a functional of $u^{(1)}$ and $\dot{u}^{(1)}$, and generally also an explicit function of τ, called in what follows Lagrangian function; $u^{(1)} = u^{(1)}(x_1, x_2, x_3, \tau)$; $dv = dx_1 \, dx_2 \, dx_3$; τ is independent of the x_j variable singled out, a total τ-derivative being denoted from now on by a dot above the corresponding letter. Then it follows from (12) that

$$\frac{\delta L}{\delta u^{(1)}} = \frac{\partial D^{(L)}}{\partial u^{(1)}} + \sum_{j=1}^{3} {}^{t}P(D_j) \frac{\partial D^{(L)}}{\partial P(D_j)u^{(1)}} \tag{7.15}$$

and

$$\frac{\delta L}{\partial \dot{u}^{(1)}} = \frac{\partial D^{(L)}}{\partial \dot{u}^{(1)}} \tag{7.16}$$

(for simplicity the denotation of the variables, whose functions are these functional derivatives, is omitted).

The condition for an extremum of S has the familiar form:

$$\delta S = 0 \tag{7.17}$$

for any variation of $u^{(1)}$. This is fulfilled when the equation

$$\frac{\delta S}{\delta u^{(1)}} = 0 \tag{7.18}$$

is satisfied. But in view of (14), S takes on the form

$$S = \int_{\tau'}^{\tau''} \int D^{(L)}(u^{(1)}, P(D_j)u^{(1)}, \dot{u}^{(1)}, \tau)\, \mathrm{d}v\, \mathrm{d}\tau \quad (j = 1, 2, 3). \tag{7.19}$$

Therefore,

$$\frac{\delta S}{\delta u^{(1)}} = \frac{\partial D^{(L)}}{\partial u^{(1)}} + \sum_{j=1}^{3} {}^{t}P(D_j) \frac{\partial D^{(L)}}{\partial P(D_j)u^{(1)}} - \frac{\mathrm{d}}{\mathrm{d}\tau} \frac{\partial D^{(L)}}{\partial \dot{u}^{(1)}}. \tag{7.20}$$

In view of (15) and (16), Eqn. (18) yields

$$\frac{\delta L}{\delta u^{(1)}} - \frac{\mathrm{d}}{\mathrm{d}\tau} \frac{\delta L}{\delta \dot{u}^{(1)}} = 0. \tag{7.21}$$

Equation (21) seemingly differs from the respective Eqn. (6.2) only by the substitution of the function $u^{(1)}$ for the column matrix $x^{(1)}$ and of the functional derivative $\delta/\delta u^{(1)}$ for the derivative $\partial/\partial\tilde{x}^{(1)}$. But essentially these two equations do not differ at all. Therefore, if a partial equation can be written in the form given in (21), it seems natural to call it a Lagrangian partial equation.

By analogy with the reasoning stated in Section 6.1, one can introduce the auxiliary function

$$u^{(2)} = \frac{\delta L}{\delta \dot{u}^{(1)}} \tag{7.22}$$

and call the pair of functions $u^{(1)}$ and $u^{(2)}$ canonically conjugate functions. It is possible now to form the Hamiltonian function

$$H = H(u^{(1)}, u^{(2)}, \tau) = (u^{(2)}, \dot{u}^{(1)}) - L, \tag{7.23}$$

where $\dot{u}^{(1)}$ is supposed to be expressed in terms of $u^{(2)}$ and $u^{(1)}$. The Hamiltonian function H obviously is a functional of $u^{(1)}$ and $u^{(2)}$, generally depending explicitly on τ also.

Further, it follows from (23) that

$$\delta H = (\delta u^{(2)}, \dot{u}^{(1)}) + (u^{(2)}, \delta\dot{u}^{(1)}) - \left(\frac{\delta L}{\delta u^{(1)}}, \delta u^{(1)} \right). \tag{7.24}$$

Since on the other hand

$$\delta H = \left(\frac{\delta H}{\delta u^{(1)}}, \delta u^{(1)} \right) + \left(\frac{\delta H}{\delta u^{(2)}}, \delta u^{(2)} \right), \tag{7.25}$$

we can, by making use of Eqn. (21) rewritten as

$$\frac{\delta L}{\delta u^{(1)}} = \dot{u}^{(2)} \tag{7.21a}$$

and utilizing Eqn. (22), get the corresponding Hamiltonian equations:

$$\dot{u}^{(1)} = \frac{\delta H}{\delta u^{(2)}}; \qquad \dot{u}^{(2)} = -\frac{\delta H}{\delta u^{(1)}}. \tag{7.26}$$

Besides, it follows from (23) that

$$\frac{\partial H}{\partial \tau} = -\frac{\partial L}{\partial \tau}. \tag{7.27}$$

Equations (26) and (27) are analogous to Eqns. (6.6) and (6.7), respectively. By making use of the reasoning similar to that in Section 6.1 for the case of ordinary differential equations, we find that the auxiliary variable P canonically conjugate to τ is connected with H by the relation

$$H' = P + H = 0, \tag{7.28}$$

where H' is the parametric Hamiltonian function – a functional of $u^{(1)}$ and $u^{(2)}$, and generally an explicit function of τ and P. When H' is used instead of H, Eqns. (26) retain their form and, besides, by virtue of (27) one more equation

$$\dot{P} = -\frac{\partial H}{\partial \tau} \tag{7.29}$$

appears.

If some partial equations can be written in the form given in (26), it also seems natural to call them Hamiltonian partial equations.

Now let $F = F(u^{(1)}, u^{(2)}, \tau, P)$ be a functional of $u^{(1)}$ and $u^{(2)}$, and generally an explicit function of τ and P, the functions $u^{(1)}$ and $u^{(2)}$ obeying (26). Then, also like the case of ordinary differential equations, one gets

$$\dot{F} = \frac{\partial F}{\partial \tau} + \left(\frac{\delta F}{\delta u^{(1)}}, \dot{u}^{(1)} \right) + \left(\frac{\delta F}{\delta u^{(2)}}, \dot{u}^{(2)} \right) + \frac{\partial F}{\partial P} \dot{P}, \tag{7.30}$$

which in view of (26), (28), and (29), can be rewritten as

$$\dot{F} = \left(\frac{\delta H'}{\delta u^{(2)}}, \frac{\delta F}{\delta u^{(1)}} \right) - \left(\frac{\delta F}{\delta u^{(2)}}, \frac{\delta H'}{\delta u^{(1)}} \right) + \frac{\partial H'}{\partial P} \frac{\partial F}{\partial \tau} - \frac{\partial F}{\partial P} \frac{\partial H'}{\partial \tau}. \tag{7.31}$$

By means of the expanded Poisson bracket, $[Q, R]'$, defined for functionals $Q = Q(u^{(1)}, u^{(2)}, \tau, P)$ and $R = R(u^{(1)}, u^{(2)}, \tau, P)$ by

$$[Q, R]' = \left(\frac{\delta Q}{\delta u^{(2)}}, \frac{\delta R}{\delta u^{(1)}}\right) - \left(\frac{\delta R}{\delta u^{(2)}}, \frac{\delta Q}{\delta u^{(1)}}\right) + \frac{\partial Q}{\partial P}\frac{\partial R}{\partial \tau} - \frac{\partial R}{\partial P}\frac{\partial Q}{\partial \tau}$$

$$= \frac{\partial Q}{\partial P}\frac{\partial R}{\partial \tau} - \frac{\partial R}{\partial P}\frac{\partial Q}{\partial \tau} + [Q, R], \tag{7.32}$$

where

$$[Q, R] = \left(\frac{\delta Q}{\delta u^{(2)}}, \frac{\delta R}{\delta u^{(1)}}\right) - \left(\frac{\delta R}{\delta u^{(2)}}, \frac{\delta Q}{\delta u^{(1)}}\right) \tag{7.33}$$

is the conventional Poisson bracket, Eqn. (31) can be rewritten as

$$\dot{F} = [H', F]' = [H, F]' + [P, F]'. \tag{7.34}$$

On the right of Eqn. (34) the first term is nonzero when F depends on τ implicitly, the second one, when the dependence is explicit.

Let us apply Eqn. (34) to the calculation of the total derivative with respect to τ for the value $u^{(1)}(r')$ of the function $u^{(1)}$. Since $u^{(1)}(r')$, which is the value of $u^{(1)}$ at the point $r = r'$, is a functional of $u^{(1)}$, we can write

$$u^{(1)}(r') = \int u^{(1)}(r)\, \delta(r - r')\, dv. \tag{7.35}$$

Then (12) yields the functional derivative

$$\frac{\delta u^{(1)}(r')}{\delta u^{(1)}(r)} = \delta(r - r') \tag{7.36}$$

as a function of r. Substituting it in (34), we find

$$\dot{u}^{(1)}(r') = \left(\frac{\delta H}{\delta u^{(2)}}, \delta(r - r')\right) = \frac{\delta H}{\delta u^{(2)}(r')},$$

that is, the first equation in (26). The functional derivative

$$\frac{\delta u^{(2)}(r')}{\delta u^{(2)}(r)} = \delta(r - r') \tag{7.37}$$

as a function of r and the second equation in (26) is obtained analogously.

Substituting P in (34) instead of F, we get Eqn. (29). If H is taken as F, Eqn. (34) yields

$$\dot{H} = \frac{\partial H}{\partial \tau} \tag{7.38}$$

because of which $\dot{P} = -\dot{H}$. However, it should be emphasized that P and $-H$ are different quantities (cf. Sect. 6.1). This is confirmed, in particular, by the fact that (38) results from the second term on the right of Eqn. (34), whereas (29) comes from the first term. If H does not depend on τ explicitly, then P is a constant.

By making use of (32), (36), and (37), we easily find

$$[u^{(1)}(r'), u^{(1)}(r'')]' = [u^{(2)}(r'), u^{(2)}(r'')]' = 0;$$
$$[u^{(2)}(r'), u^{(1)}(r'')]' = \delta(r' - r''). \tag{7.39}$$

The relations

$$[u^{(1)}, \tau]' = [u^{(2)}, \tau]' = [u^{(1)}, P]' = [u^{(2)}, P]' = 0; \qquad [P, \tau]' = 1 \tag{7.40}$$

obviously are also valid.

So, canonically conjugate quantities $u^{(1)}$ and $u^{(2)}$ as well as τ and P satisfy relations similar to those in the case of ordinary differential equations.

Let the Fourier method (the method of separation of variables) be applicable to a certain Lagrangian partial equation, that is, let a solution to the equation be representable as

$$u^{(1)} = \sum_j q_j(\tau)\varphi_j(x_1, x_2, x_3), \tag{7.41}$$

then sets of ordinary differential equations for q_j and φ_j separately result from this partial equation, while the corresponding Lagrangian function (14), provided the φ_j satisfy the equations obtained, takes on the form

$$L = L(u^{(1)}, \dot{u}^{(1)}, \tau) = L'(q_j, \dot{q}_j, \tau) \equiv L'. \tag{7.42}$$

Thereby the functional (13) goes over into

$$S' = \int_{\tau'}^{\tau''} L' \,\mathrm{d}\tau \tag{7.43}$$

and the ordinary differential equations for the q_j should result from the variational condition

$$\delta S' = 0, \tag{7.44}$$

that is, they can be written as

$$\frac{\partial L'}{\partial q_j} - \frac{\mathrm{d}}{\mathrm{d}\tau}\frac{\partial L'}{\partial \dot{q}_j} = 0. \tag{7.45}$$

Thus, if the Fourier method can be used for solving a Lagrangian partial equation, then the ordinary differential equations for the q_j in (41) also prove to be Lagrangian.

To illustrate the above, let us consider some examples. We start with an equation of the form

$$\frac{\partial^2 u^{(1)}}{\partial \tau^2} + Gu^{(1)} = 0, \tag{7.46}$$

where G is an independent of τ Hermitian operator. Equation (46) obviously is Lagrangian and, since $G^{1/2}$ also is a Hermitian operator, both

$$L = 2^{-1} [(\dot{u}^{(1)}, \dot{u}^{(1)}) - (G^{1/2} u^{(1)}, G^{1/2} u^{(1)})] \tag{7.47}$$

and

$$L = 2^{-1} [(\dot{u}^{(1)}, \dot{u}^{(1)}) - (u^{(1)}, Gu^{(1)})] \tag{7.48}$$

can be taken, for instance, as the corresponding Lagrangian function. On the other hand, substituting (41) into (46), we get for any j

$$\ddot{q}_j \varphi_j + q_j G \varphi_j = 0 \tag{7.49}$$

whence it follows that, as is generally known,

$$\ddot{q}_j + \lambda_j q_j = 0 \tag{7.50}$$

and

$$G\varphi_j = \lambda_j \varphi_j, \tag{7.51}$$

where λ_j is a constant – an eigenvalue of the operator G, as shown in Eqn. (51). And since Eqn. (51) is solvable, the function $u^{(1)}$ can be represented in the form given in (47), and Eqns. (50) are Lagrangian.

Let us note that the example considered also includes the case when $u^{(1)}$ is a function of variables which take on only denumerable sets of values. In this case $u^{(1)}$ can be written, for instance, in the form of either the column matrix $u^{(1)}$ or row matrix $\tilde{u}^{(1)}$ with elements $u_j^{(1)} = u^{(1)}(j)$, or in the form of the diagonal matrix $\|u_j^{(1)} \delta_{ij}\|$ with the same elements. For this case it is convenient to write Eqn. (46) in the form

$$\ddot{u}^{(1)} + Gu^{(1)} = 0, \tag{7.52}$$

where G is a symmetrical matrix representing the corresponding Hermitian operator on the basis chosen. As a Lagrangian function resulting in (52), one can take, for instance,

$$L = 2^{-1} (\tilde{\dot{u}}^{(1)} \dot{u} - \tilde{u}^{(1)} G u^{(1)}) \tag{7.53}$$

(cf. also Sect. 6.1).

Let us scrutinize how the 'Fourier method' can be applied to the case under consideration. To that end we write

$$u^{(1)} = \sum_l q_l \varphi_l, \tag{7.54}$$

where φ_l is a column matrix independent of τ, and $q_l = q_l(\tau)$ is a scalar function of τ. Substituting (54) into (52), one gets for every summand separately

$$\ddot{q}_l \varphi_l + q_l G \varphi_l = 0. \tag{7.55}$$

Introducing the diagonal matrix $D_{(l)}$, whose elements are the corresponding inverse elements of φ_l, one finds

$$D_{(l)} \varphi_l = e, \tag{7.56}$$

where e is the column matrix, all of whose elements are units. In view of (56), and multiplying Eqn. (55) from the left by $q_l^{-1} D_{(l)}$, one obtains

$$q_l^{-1} \ddot{q}_l e + D_{(l)} G \varphi_l = 0. \tag{7.57}$$

Equation (57) can obviously be satisfied only in case the equalities

$$q_l^{-1} \ddot{q}_l = -\lambda_l \tag{7.58}$$

and

$$D_{(l)} G \varphi_l = \lambda_l e \tag{7.59}$$

are fulfilled simultaneously, where λ_l is a constant. From (58) it follows that

$$\ddot{q}_l + \lambda_l q_l = 0. \tag{7.58a}$$

Multiplying (59) from the left by $D_{(l)}^{-1}$ and using (56), one gets

$$G \varphi_l = \lambda_l \varphi_l. \tag{7.59a}$$

Equations (58a) are obviously Lagrangian, as indeed should be the case, while the constants λ_l are eigenvalues of the matrix G.

Let us now consider one more partial equation

$$i \frac{\partial u^{(1)}}{\partial \tau} = G u^{(1)}, \tag{7.60}$$

where $u^{(1)} = u^{(1)}(x_1, x_2, x_3, \tau)$, and G is a real Hermitian operator independent of τ. Having written $u^{(1)} = u_1^{(1)} + i u_2^{(1)}$, where $u_1^{(1)}$ and $u_2^{(1)}$ are real functions, we find, using (60), the equations

$$\frac{\partial u_1^{(1)}}{\partial \tau} = G u_2^{(1)}; \qquad \frac{\partial u_2^{(1)}}{\partial \tau} = -G u_1^{(1)}. \tag{7.61}$$

These equations are clearly Hamiltonian ones, and one can choose $u_2^{(1)}$ as the auxiliary function $u_1^{(2)}$ canonically conjugate to $u_1^{(1)}$. Equations (61) are of the form as given in (26) if, for instance,

$$H = 2^{-1} [(u_1^{(1)}, Gu_1^{(1)}) - (u_1^{(2)}, Gu_1^{(2)})]. \tag{7.62}$$

On the other hand, representing in (61) $u_1^{(1)}$ and $u_2^{(1)}$ as

$$u_1^{(1)} = \sum_j q_j'(\tau)\varphi_j; \qquad u_2^{(1)} = \sum_j q_j''(\tau)\varphi_j, \tag{7.63}$$

where $\varphi_j = \varphi_j(x_1, x_2, x_3)$ are eigenfunctions of G, we get

$$\dot{q}_j' = \lambda_j q_j'', \qquad \dot{q}_j'' = -\lambda_j q_j', \tag{7.64}$$

the λ_j being the eigenvalues of G conforming to the respective φ_j. Equations (64) are also Hamiltonian, q'' turning out to be the auxiliary variable canonically conjugate to q', and the corresponding Hamiltonian function H having the form

$$H_j = 2^{-1}\lambda_j(q_j'^2 + q_j''^2). \tag{7.65}$$

In the examples considered H does not depend on τ explicitly, because of which the auxiliary variable P canonically conjugate to τ is constant. When using the Fourier method in these examples, the Lagrangian and Hamiltonian ordinary differential equations for the functions $q_j(\tau)$ prove to be completely separated into single equations for each j, the corresponding variables P_j canonically conjugate to τ being constants satisfying the relation

$$P = \sum_j P_j. \tag{7.66}$$

Let us emphasize once again that all the results obtained are purely mathematical and concern differential equations of the forms involved, irrespective of the origin of those equations.

7.4. Hamiltonian Random Fields

For the same reasons as those stated in Section 6.2, random fields (more generally, random functions) will be called Hamiltonian random fields (or Hamiltonian random functions, respectively), if they (that is, their values) obey Lagrangian or Hamiltonian equations.

It is easy to make sure that to random variables of the first and second type created by a Hamiltonian random field (or by a Hamiltonian random function) there correspond analogous Hamiltonian random variables (see Sect. 6.2). For the first-type one-dimensional random variables, $t_{(\alpha)l}$, resulting from the random field t_α at the points r_l this becomes especially clear if one introduces random variables of the same type, ${}^c t_{(\alpha)l}$, created at those points by the auxiliary random field ${}^c t_\alpha$ defined by the relation

$$ {}^c t_\alpha = \frac{\delta L(t_\alpha, \dot{t}_\alpha, t_4)}{\delta t_\alpha} \tag{7.67}$$

resulting from (22). On the basis of Eqn. (26), $t_{(\alpha)l}$ and ${}^c t_{(\alpha)l}$ obey

$$ \dot{t}_{(\alpha)l} = \left[\frac{\delta H(t_\alpha, {}^c t_\alpha, t_4)}{\delta^c t_\alpha}\right]_l, \qquad {}^c\dot{t}_{(\alpha)l} = -\left[\frac{\delta H(t_\alpha, {}^c t_\alpha, t_4)}{\delta t_\alpha}\right]_l \tag{7.68}$$

and, hence, are canonically conjugate Hamiltonian random variables (of the first type).

The connection between the second-type one-dimensional random variables, $t_l^{(\alpha)}$, and the corresponding Hamiltonian random variables is revealed by the following reasoning. Suppose that the Hamiltonian random field can be represented as

$$ t_\alpha = \sum_j t_j^{(4)}(t_4)\varphi_j^{(\alpha)}, \tag{7.69}$$

where $\varphi_j^{(\alpha)} = \varphi_j^{(\alpha)}(t_{123})$ are functions of the independent random variable t_{123}. Then, as follows from the preceding section, $t_j^{(4)}(t_4)$ are Hamiltonian random variables (of the second type). If the functions $\varphi_j^{(\alpha)}$ are normalized, then in view of (3), one can write

$$ t_j^{(4)}(t_4) = t_j^{(\alpha)} q_j^{(\alpha)}, \tag{7.70}$$

where $q_j^{(\alpha)} = q_j^{(\alpha)}(t_4)$ are normalized functions of t_4, and $t_j^{(\alpha)}$ are the second-type one-dimensional random variables, whose values $\beta_j^{(\alpha)} \in \theta_j^{(\alpha)}$ are those of Eqn. (3). Different sets of $\beta_j^{(\alpha)}$ generally refer to distinct continuous concrete objects $B_a \in A$.

In the example of a mechanical system consisting of N particles bound harmonically, 'system deformation' t is a Hamiltonian random function, since it obeys an equation of the form given in (52), namely

$$ \ddot{t} + Wt = 0, \tag{7.71}$$

where a certain $3N$-by-$3N$ symmetrical matrix, whose elements are uniquely determined by the particles' masses, force coefficients, and the equilibrium configuration of the system, can be chosen as W. From (54), (58a), and (59a) it follows for this case that

$$ t = \sum_j t_j^{(4)}\varphi_j, \tag{7.72}$$

where φ_j is one-by-3N column matrix, whose elements are independent of t_4. Then

$$\ddot{t}_j^{(4)} + \lambda_j t_j^{(4)} = 0 \tag{7.73}$$

and

$$W\varphi_j = \lambda_j \varphi_j. \tag{7.74}$$

The $t_j^{(4)}$ obviously are second-type Hamiltonian random variables, whereas the elements of the matrix t themselves may be regarded as 'first-type' Hamiltonian random variables.

As is generally known, six zero-eigenvalues, λ_j, of the matrix W (five in case the equilibrium configuration is linear) refer to the translations and rotations of the whole system. The rest of its eigenvalues corresponds to the so-called normal vibrations of the system; the mode of the jth normal vibration, that is, the relation between the elements in the matrix φ_j, being determined by Eqn. (74). In particular, if all the particles are of the same mass and form a crystal grating, then the dependence of the magnitudes of elements in each column matrix φ_j on the numbers of the elements proves to be harmonic. This conforms to the fact that combined Eqns. (74) for a crystal grating can be written as a set of second-order difference equations similar to the corresponding set of three ordinary second-order differential equations.

In the example of a string, the random field 'string deformation' t obeys the equation

$$\frac{\partial^2 t}{\partial t_4^2} - \gamma \frac{\partial^2 t}{\partial t_1^2} = 0 \tag{7.75}$$

following from Eqn. (1), as well as the boundary conditions

$$t(0, t_4) = t(1, t_4) = 0 \tag{7.76}$$

resulting from Eqn. (2). One can easily make sure that t is a Hamiltonian random field creating Hamiltonian random variables of both types, since the Fourier method is applicable to Eqn. (1).

All the results obtained in Chapter 6 for the Hamiltonian random variables are directly applicable to the second-type Hamiltonian random variables created by Hamiltonian random functions and, in particular, by Hamiltonian random fields. Hence, the quantum approach is also applicable to consideration of such random variables. When considering Hamiltonian random functions obeying equations of the form given in (46) or (60), one can immediately utilize some general results contained in that chapter. For instance, in the respective equations of the form given in (50) and (64) there are two initial variables for each j, that is, $n = 2$. Since the equations do not explicitly contain t_4, the auxiliary random variables $t_{(4)+2}^{(j)}$ canonically conjugate to t_4 are constants. The totality of these constants for all the j determines a state of the corresponding abstract object, while probability distributions for any functions of $t_j^{(4)}$ and t_4 in the state are independent of t_4. So, the values $-x_{(4)+2}^{(j)} = E^{(j)}$ of the random variables $-t_{(4)+2}^{(j)}$ are constant in this state, and the value

$$E = \sum_j E^{(j)} \tag{7.77}$$

of the corresponding summary random variable is one and the same for each concrete object $B_a \in A$ conforming to the abstract object occupying the state with a definite set of $E^{(j)}$-values obeying (77).

When one makes use of the quantum approach, the α-constant in the expressions for the operators corresponding to the auxiliary random variables canonically conjugate to $t_j^{(4)}$ and t_4 must be determined from the appropriate statistical experiments (cf. Sect. 6.4). Whenever, in equations of the form given in (50), the λ_j are positive, the corresponding quantum equations has the form of Eqn. (6.106), and for each j the set of $E^{(j)}$-values is discrete, these values being given by

$$E_n^{(j)} = (n + \tfrac{1}{2})\alpha\lambda_j^{1/2}. \tag{7.78}$$

In case of Eqn. (60) it follows from (64) that for each j the Hamiltonian function can be taken in the form

$$H_j(t) = 2^{-1}\lambda_j(t_j'^{(4)2} + t_j''^{(4)2}). \tag{7.79}$$

Therefore, for Eqn. (60)

$$E_n^{(j)} = (n + \tfrac{1}{2})\alpha\lambda_j \tag{7.78a}$$

is valid, regardless of whether the λ_j are positive or negative.

In conclusion it is necessary to note the following important facts: The Lagrangian and Hamiltonian partial equations may be regarded as obtained from the corresponding sets of ordinary differential equations by proceeding to the limit. This stipulates the different mathematical meanings of the independent variables x_1, x_2, and x_3, on the one hand, and the variable τ, on the other hand. The totality $r = (x_1, x_2, x_3)$ conforms to the order-number of a variable in a set of ordinary differential equations. But the variable τ singled out plays one and the same role both in a set of ordinary differential equations and in the corresponding partial equation, and mathematically it is similar to a value at any point r of the function $u^{(1)}$, whose derivatives go into the partial equation. The difference between r and τ becomes apparent in the fact, too, that for τ, as also for $u^{(1)}(r)$, there exist an auxiliary variable P canonically conjugate to it, whereas for x_1, x_2, and x_3 such variables do not exist at all.

For Hamiltonian random fields an analogous difference exists also between the independent random variable $t_{123} = (t_1, t_2, t_3)$, on the one hand, and the random variable t_4, on the one hand. Besides, each r-value of the random variable t_{123} determines spatial coordinates of a certain point of a continuous concrete object $B_a \in A$, which are independent of τ in a proper coordinate system. These circumstances should be taken into consideration when discussing physical random fields.

Suggested References

1. Goldstein, H.: 1980, *Classical Mechanics*, 2nd edn., Addison-Wesley, Reading. This excellent book contains the consideration of the transition from sets of ordinary differential equations for discrete objects to the corresponding partial differential equations for continous ones.
2. Mayants, L. S.: 1981, *Found. Phys.* **11**, 57. In this paper the probabilistic treatment of fields in probabilistics was given for the first time.
3. Schiff, L. I.: 1968, *Quantum Mechanics*, 3rd edn., McGraw-Hill, New York. This comprehensive book also considers the transition from ordinary differential equation sets for discrete objects to the partial differential equations for continuous objects.
4. Schwartz, L.: 1966, *Mathematics for the Physical Sciences*, Addison-Wesley, Reading. This book contains some mathematics utilized in obtaining Eqn. (12).

Part II

Fundamentals of Probabilistic Physics

Chapter 8

General Considerations

8.1. Preliminaries

Probabilistic physics is the name of a unified and vast science which deals with probabilistic treatment of physical systems. In other words, probabilistic physics is the science that emerges as a result of application of probabilistics to consideration of any kind of physical systems. Various separate disciplines such as, say, classical statistical physics, quantum physics, statistical hydrodynamics, kinetics of physical and chemical processes, etc., are thus particular domains of probabilistic physics.

In probabilistic physics concrete objects and abstract objects are concrete physical systems and abstract physical systems, respectively. Properties of objects are, in particular, various physical quantities including coordinates and time. Since values of these properties are real numbers, physical quantities prove to be random variables by definition (see Sect. 3.1).

A state of an abstract physical system is given by definite values of certain physical quantities, and it can be described, in particular, by a distribution function, probability density, a characteristic function, a state vector, or a density operator, etc.

One of the main tasks of probabilistic physics is to find probability distributions of certain physical quantities and to calculate mathematical expectations for some of their functions in states of an abstract physical system. The problem can be solved in principle by at least two different ways. The first that has been called classical approach is a trivial way of solving common probabilistic problems, the starting point of which is, as a matter of fact, the construction of a set $A^{(a)}$, adequate to the corresponding set A of concrete physical systems. This method was actually used by Gibbs in constructing classical statistical physics (ensemble method).

The starting point of the second method, named the quantum approach (see Sect. 6.4), is finding Hermitian operators for various physical quantities and calculating their eigenvectors and eigenvalues.

When it is possible to use the classical method, description of a state by either a state vector or a density operator is unnecessary and seems therefore quite artificial. Thus, it is natural to use the quantum approach only when it is impossible to make use of the classical method. However, when both are suitable, they should lead to the same results.

One can verify the correctness of both the construction of a set $A^{(a)}$ adequate to the corresponding set of concrete physical systems and the choice of operators for various physical quantities by comparing theoretically obtained results with those of statistical experiments.

It follows from all the statistical experiments we know that the classical method is suitable for mechanical systems studied in classical statistical physics, that is, for 'clsssical' systems consisting of particles of large enough mass, whereas it is unsuitable for mechanical systems studied in quantum mechanics, that is, for 'microsystems'. We find that the quantum approach alone works for microsystems, but it should also be applicable, of course, to classical systems. Since operators for random variables are the same for both microsystems and classical systems, quantum mechanical probability distributions must go over into the corresponding classical statistical probability distributions when passing from the former to the latter. Whenever this assertion is capable of being checked it proves to be true.

8.2. Classification of Physical Systems

The conventional classification of physical systems relative to their interreaction with other physical systems is as follows.

A concrete physical system is called closed if it does not interact with any other one. If the forces acting on a concrete physical system do not depend on time, the system is called a conservative system; a closed system obviously is a conservative system too. All other concrete physical systems, on which time-dependent forces do act, may hence be termed non-conservative systems.

Later on we will need some refinement of this classification, but in order to get to it we must consider, first, the methods of the description of the motion of mechanical systems by means of certain mathematical spaces. There are different kinds of mathematical spaces allowing one to describe the motion of a concrete mechanical system consisting of particles. If the number of the spatial coordinates characterizing the system is n, an n-dimensional space can be introduced to that end, each coordinate x_j of which being uniquely assigned to each coordinate q_j of the system. Then any configuration $q = (q_1, q_2, \ldots, q_n)$ of the system in the real three-dimensional physical space is represented by a point $x^{(1)}(x_1, x_2, \ldots, x_n)$ in that n-dimensional mathematical space called configuration space. The real motion of a concrete mechanical system, that is, the change in time in the positions of the particles it consists of in real physical space, is represented by the motion of the cooresponding point in the configuration space, which depicts a certain concrete curve there.

To avoid any terminological muddle, we will call curves along which particles move in real physical space particle paths. A curve on which a point moves in the configuration space will be termed a C-trajectory. Thus, to every C-trajectory for a mechanical system there uniquely corresponds the totality of the paths of the particles it consists of. The path of every particle can be described parametrically, for instance, by indicating the dependence of the coordinates of the particle on time. The C-trajectories may also be described in a similar way, but they are highly tangled even for the simplest cases and are therefore very hard to deal with.

As is well known, the generalized coordinates q_j of a mechanical system consisting of particles obey the corresponding Lagrangian equations of the form given in (6.2), where the x_j are replaced by the q_j, and x_n by t (time). In this case the Lagrangian function has the form

$$L(q_j, \dot{q}_j, t) = T - V, \tag{8.1}$$

where T and V are the kinetic and the potential energy of the system, respectively. The generalized momenta p_j, being defined by the corresponding equations similar to (6.4), are canonically conjugate to the respective coordinates. The Hamiltonian function H is obtained from Eqn. (6.5) by replacing $x^{(1)}$ by $q = (q_1, q_2, \ldots, q_n)$, $x^{(2)}$ by $p = (p_1, p_2, \ldots, p_n)$, and x_n by t, and has consequently the form

$$H = \sum_{j=1}^{n} p_j \dot{q}_j - L. \tag{8.2}$$

The Hamiltonian equations for a mechanical system are of the form as given in (6.6), after the necessary changes in designation of the variables have been made. They can be written, hence, as

$$\dot{q}_j = \frac{\partial H}{\partial p_j}; \qquad \dot{p}_j = -\frac{\partial H}{\partial q_j}. \tag{8.3}$$

The so-called phase space can now be introduced for the description of the motion of a mechanical system, which proves to be much more convenient than the configuration space. In that $2n$-dimensional space, n coordinates (say, first) are assigned to the generalized coordinates q, and n more coordinates to the corresponding generalized momenta p of the system. Thus, a point in the phase space determines simultaneously both the totality of coordinates characterizing the configuration of a system in real physical space and the totality of momenta, which determine the changes in this configuration at the next close enough instant of time. The real motion of a concrete system is represented by the motion of the corresponding point in the phase space along a curve which will be called P-trajectory of the system. If the P-trajectory of a system is known, one can obviously find both its C-trajectory (and therefore the paths of all its particles) and changes in the momenta of the system in time. Also a P-trajectory can be described parametrically by indicating the dependence of all the coordinates and momenta of a system on a certain parameter, in particular, time.

For a conservative system, only one P-trajectory can pass through each point of the phase space. This follows from the fact that for a conservative system H does not depend on time and, hence, Eqns. (3) uniquely determine the inclination of the tangent to the P-trajectory at each point, which does not depend on time either. Therefore, if at some time instant a conservative system is characterized by a definite point in the phase space, all its consequent motion, as well as the previous one, is uniquely determined by the only P-trajectory passing through this point.

Supplementing the phase space with two more coordinate axes corresponding to time, t, and canonically conjugate to it (with the sign reversed) energy, E, we obtain the $2n+2$-dimensional complete phase space (CPS). In this space suitable for the consideration of any mechanical systems (both conservative and non-conservative) the real motion of a concrete system is also represented by a curve called CP-trajectory of the system. Since $-E$ replaces x_{2n} in Eqn. (6.9), we have

$$E = H. \tag{8.4}$$

Every CP-trajectory obviously must satisfy this equation, that is, it should have been disposed upon the $2n+1$-dimensional hypersurface in CPS determined by Eqn. (4), which will be denoted by HCPS. It follows from the parametric Hamiltonian equations (6.6′), which in the case under consideration add to Eqns. (3) one more equation

$$E = \partial H/\partial t \tag{8.5}$$

and the identity $\dot{t} = \dot{t}$, that one and only one CP-trajectory passes through each point of HCPS. The motion of any concrete mechanical system is thus uniquely determined by any point of the corresponding CP-trajectory. Therefore the question of what is the concrete behavior of a concrete mechanical system, that is, what are the values of all the physical properties it has as functions of time, is quite sensible.

The solution of the problem is equivalent to finding the concrete CP-trajectory for this system in the corresponding CPS, and this obviously is possible in principle when one point on the CP-trajectory is given. In other words, the concrete behavior of any mechanical system is completely determined, both in the future and the past, by the so-called initial conditions, which are values of coordinates and momenta at a certain time instant; the initial value of energy is given by Eqn. (4).

We should now, however, refine the conventional classification of physical systems given above, since no reservations have been made there, whether 'closed' and 'conservative' systems, respectively, are believed to be of that kind in the entire infinite interval of time $(-\infty, \infty)$ or in a certain time interval (t_1, t_2), of which at least one end is at a finite distance from the zero. Meanwhile, the behavior of a concrete system depends essentially on whether it is closed or conservative, respectively, at any time instant of the interval $(-\infty, \infty)$ or only in a certain more restricted time interval.

For instance, if a particle is not subjected to any action of forces in the time interval $(-\infty, \infty)$, it moves freely in this interval, and its path is fully determined by the corresponding initial conditions not only in the future, but also in the past.

However, if forces do not act on a particle only in the interval (t_1, t_2) with either $t_1 \neq -\infty$ or $t_2 \neq \infty$, then it moves freely in this time interval only, and its path is fully determined only within this interval and by the initial conditions at some time instant of the same interval. Those initial conditions cannot certainly determine the motion of the particle outside the indicated interval.

If a system is closed or conservative, respectively, in a certain interval (t_1, t_2) which does not coincide with $(-\infty, \infty)$, then in any time intervals containing the former it is no longer closed or conservative. We will therefore call such a system pseudoclosed and pseudoconservative, respectively, keeping the terms 'closed' and 'conservative' for those systems only, which obey the necessary conditions in the time interval $(-\infty, \infty)$.

Let us note, by the way, that here time does not play any particular role differing from that of the spatial coordinates. If a system is closed, for instance, in some time interval, it also is closed on the corresponding segment of the path along which it, as a single whole, moves freely. To the entire infinite time interval of closedness there corresponds the whole infinite spatial closedness interval, to a part of the former there corresponds a part of the latter, and conversely. But for simplicity, we will often speak of time intervals only.

Let us now consider a system I which is a part (subsystem) of a certain closed system. We will denote by II the rest of this closed system. Some forces act, generally speaking, on system I from the side of system II and therefore generally it is not closed. If the forces do not explicitly depend on time, it is conservative. If they do, then system I is nonconservative. The same is true for system II. It can happen that the interaction of systems I and II ends off from a certain time instant t_1. Since in this case the forces acting upon the systems depend on time explicitly, both systems are nonconservative. But from the time instant t_1 they become pseudoclosed, while the entire system consisting of systems I and II remains closed.

Under some conditions, a nonclosed system can be considered a part of the corresponding closed system. The latter is obtained by addition to the former all the physical objects it interacts with. Let us note, by the way, that a subsystem of a closed system can be considered some autonomous conservative system in the limiting case only, when its mass is infinitesimally small comperatively to that of the rest of the system. Indeed, only in such a case is it possible to describe the motion of this subsystem relatively an inertial frame of reference connected with the rest of the system. The motion of a conservative system coincides thus with the motion of the corresponding closed system of the indicated type considered in the reference frame connected with the center of mass of the latter. In all the other cases, if one wants to exclude the rest of the system from consideration, it is necessary to introduce forces explicitly depending on time.

So, strictly speaking, a subsystem of a closed system can, as a rule, be only nonconservative (in particular, pseudoclosed), the dependence of the forces acting upon it on spatial coordinates and time being determined by the entire closed system on the whole. In other words, a subsystem of a closed system is not, as a rule, autonomous, and its motion is determined by the motion of the entire closed system. Exceptions are, to say nothing about the indicated limiting case (conservative systems), closed systems consisting of closed subsystems. They are, however, of no special interest.

If a system becomes pseudoclosed from a certain time instant t_1 and remains such to $t = \infty$, then its CP-trajectory (on HCPS for the system) is determined by any of its points, but only 'ahead' – in the positive direction of time axis. 'Back' – in the negative direction of time axis, this CP-trajectory is determined by any of its points to time instant t_1 only. It could seem that a pseudoclosed system is fully autonomous within the interval of its closedness, because at any time instant from this interval, its motion is determined by seemingly arbitrary initial conditions for some time instant from the same interval. But this is not the case, since these initial conditions are not, as a matter of fact, arbitrary and are determined by the initial conditions for the entire closed system on the whole.

For a pseudoclosed system, there exist its own complete phase space (CPS) and HCPS in it, but both the latter and the concrete CP-trajectory on it are determined by the motion of the entire closed system containing this pseudoclosed system. The concrete motion of a pseudoclosed system can be considered both proceeding from the entire closed system, whose subsystem it is, and proceeding from itself. In the second case one might regard the pseudoclosed system as fully autonomous, if not the fact that the initial conditions for it are determined by the whole closed system which contains it. If those initial conditions become anyway known, the pseudoclosed system can be treated as a closed system, and it is possible (at least in principle) to find the concrete CP-trajectories in its CPS.

Let us now consider the question whether there generally exist closed systems in Nature, and whether it is possible to speak of such systems, at least approximately, in some cases. The first part of the question should be answered in the negative. Indeed, a closed system must not, by definition, undergo any action of external forces within the entire infinite time interval $(-\infty, \infty)$. But this is, strictly speaking, impossible for any finite system (that is, one consisting of a finite number) of concrete physical objects.

The second part of the question can be answered in the affirmative. It is possible to say beforehand that the less the external forces acting on some system in comparison with those acting inside it, the more accurately the system can be regarded as closed. For instance it is, to a very great accuracy, possible to consider as closed our solar system, or rather the collection of bodies of large enough mass it contains, since the forces exerted on these bodies by the heavenly bodies being outside the solar system are very small in comparison with those acting inside this system.

Matters stand differently for particles of small mass. For such particles, even the solar system can no longer be considered closed, since their motion is determined not so much by small gravitational forces acting on them, as by forces arising, for instance, when colliding with other particles. The motion of such particles is affected by both direct collisions with particles coming from any parts of the Universe and collisions with other particles which have changed their own motion due to the collisions with these cosmic 'strangers'. It is impossible to isolate in principle such particles from the influence of the Universe, since the walls even of any hermetic vessels, in which the particles can be put, consist of the same kind of particles themselves. Therefore even if these walls do not immediately let 'strange' particles pass into the vessel, they nevertheless transfer their influence; the motion of particles inside a hermetic vessel changes when colliding with particles its walls consist of, the changes being dependent on the motion of these particles, which in turn changes as a result of collisions with the strangers from the Universe.

Thus, no finite closed system exists, such that particles of small enough mass could be its subsystems even approximately. This fact makes the motion of such particles in definite paths for a long enough time impossible in principle. Nevertheless, they can be considered pseudoclosed systems within the interims between collisions with other particles.

8.3. Two Possible Types of Problems

Till now we dealt with concrete physical systems, in particular, concrete mechanical systems. Any questions concerning concrete behavior of concrete physical systems are relevant for them in principle. It is quite sensible, therefore, to try to answer questions of the type: What are the values of the physical quantities the concrete system is possessed of? For a concrete closed mechanical system, in order to answer such a question, it is necessary and sufficient to know one point on the CP-trajectory in CPS for the corresponding concrete system, as has been noted in the preceding section. For a concrete pseudoclosed mechanical system, it is possible to solve the problem of finding its CP-trajectory in the case only, when the initial conditions are given within the interval of closedness, since this trajectory is fully determined by these conditions in that interval only, in other words, the motion of a concrete pseudoclosed system is uniquely determined only in its interval of closedness and by any point on its CP-trajectory within this interval.

If the initial conditions are not given, we cannot solve the problem of finding the values of various physical quantities for any mechanical system in principle. In the case of a free nonrelativistic motion of a particle of mass m along x axis of a certain inertial coordinate system, for instance, it is impossible to answer the question of what is, say, the coordinate x of the particle at time instant t, unless the initial conditions are given. It is also impossible to answer the question of what are the values of its energy and momentum. The only thing one can assert in this case is that its energy, momentum, and coordinate can have any values. However, if it is known that energy of such a particle has the value E, then the magnitude of its momentum can have only the value $|p| = (2mE)^{1/2}$. The direction of its motion remains, however, undetermined thereby (it can move either in the positive or the negative direction of the x axis), and the value of its coordinate x at the time instant t is completely uncertain too [x can have any value within the interval $(-\infty, \infty)$].

So, if the initial conditions are not given, it is impossible to get a definite answer to the question of what are values of physical quantities in a mechanical system. At best, one may make only, in this case, probabilistic statements about the values of various physical quantities. But such kind of statements can refer to abstract physical system only, which have no definite values of properties they may possess (cf. Chap. 1). Hence, the possibility arises to raise probabilistic problems concerning the values of physical quantities for physical systems, which appear as random variables (see Sect. 3.1). In order for such a probabilistic problem to be sensible, the corresponding set of concrete physical systems should be defined distinctly, and the check-up of the calculated probabilities by a statistical method has to be feasible.

In the case of closed mechanical systems, to every concrete system there corresponds a point in the respective complete phase space (CPS). As we know, different concrete systems differ in at least one value of their properties (see Sect. 1.1) and, therefore, the points corresponding to them in CPS are different as well. This fact makes it possible to use one common CPS for all the concrete mechanical systems belonging to one set. The CP-trajectories for different concrete systems can never intersect, because otherwise one point in CPS would correspond to different concrete mechanical systems, which contradicts the very notion of concrete objects. It is obvious that any set of concrete mechanical systems is in such a one-to-one correspondence with the respective set of concrete points in their common CPS that both sets are mutually adequate (see Sect. 1.8). Therefore, probabilistic problems concerning mechanical systems can be solved in principle by making use of the respective sets of points in CPS conforming to these systems. The well-known Gibbs' ensemble method is actually based on the similar considerations with utilization of the phase space of mechanical systems for that purpose.

The particular choice of the set of concrete systems and, hence, that of the set of points in the corresponding CPS depends on what probabilistic problem is to be solved. In the above example of a nonrelativistic free particle of mass m moving along the x-axis of a certain inertial coordinate system, the following problem may be posed: What is the probability that its Cartesian coordinates x, y, and z take values within the intervals $(x_0 - \frac{1}{2}\Delta x, x_0 + \frac{1}{2}\Delta x)$, $(y_0 - \frac{1}{2}\Delta y, y_0 + \frac{1}{2}\Delta y)$, and $(z_0 - \frac{1}{2}\Delta z, z_0 + \frac{1}{2}\Delta z)$, respectively? Since no reservations have been made for other peculiarities of the motion, it should be meant that a particle can have, in this case, any values of coordinates and momenta at any time instant. But such a particle is thus an element of an infinite set of concrete particles of the indicated type which have any values of coordinates and momenta at any time instant. And the problem posed actually concerns the corresponding abstract particle. The set of points in the properly chosen respective eight-dimensional CPS, which is adequate to the set of concrete particles under consideration, is, hence, this entire infinite space itself. If the space is Euclidean, then the sought-for probability, being proportional to the measure of the corresponding subset of points, is proportional to $dv = \Delta x\, \Delta y\, \Delta z$ and does not depend on x_0, y_0, and z_0.

One more example. For a one-dimensional classical harmonic oscillator of mass

m vibrating with angular frequency ω and amplitude A along the x-axis, a problem can be posed: What is the probability distribution of its coordinate?

In accordance with this statement of the problem, the set of concrete oscillators corresponding to the abstract oscillator under consideration consists of all the oscillators of the indicated type vibrating with amplitude A and angular frequency ω at any initial conditions. The original CPS in this case is also the corresponding infinite eight-dimensional Euclidean space. But, since the coordinates y and z are not indicated in the statement of the problem, neither they nor the corresponding momenta are of importance, and we may confine ourselves with the consideration of its four-dimensional subspace corresponding to x, $p_x = p$, E, and t, singled out by the conditions: y = const, z = const, p_y = const, p_z = const.

The Hamiltonian function for a concrete oscillator now is

$$H = P^2/2m + m\omega^2 x^2/2. \tag{8.6}$$

We introduce for simplicity dimensionless coordinate, η, and momentum, ϵ, defined by the relations

$$\eta = x\sqrt{(m\omega^2/2E_0)} \tag{8.7}$$

and

$$\epsilon = P/\sqrt{(2mE_0)} \tag{8.8}$$

respectively, where E_0 denotes a certain energy unity. Then we obtain for the dimensionless Hamiltonian, $H_d = H/E_0$,

$$H_d = \epsilon^2 + \eta^2. \tag{8.9}$$

Now the condition of the reality of motion (4) leads to

$$\epsilon^2 + \eta^2 = E_d \tag{8.10}$$

where $E_d = E/E_0$ is the dimensionless energy. This equation picks out the three-dimensional hypersurface in that four-dimensional space.

The value

$$E_d = A^2 m\omega^2/2E_0 \tag{8.11}$$

of the dimensionless energy corresponds to the value A of the amplitude of the oscillators we should consider. Now the intersection of that hypersurface with the planes $E_d = E_{d_1}$ = const and t = const gives (at any t) the circumference

$$\epsilon^2 + \eta^2 = E_{d_1} \tag{8.12}$$

a set of points which corresponds to the set of concrete oscillators we should take into consideration.

In order to find the sought-for probability distribution it is now sufficient to calculate the relative measure of the set of points on the circle (12) whose coordinate η has values within an interval $(\eta_1, \eta_1 + \Delta\eta)$. This measure is proportional to the length 2 ds of the two segments of the circle contained between the straight lines $\eta = \eta_1$ and $\eta = \eta_1 + \Delta\eta$, and we can write

$$\rho(\eta)\,\mathrm{d}\eta = \text{const } 2\,\mathrm{d}s = \text{const}\left[1 + \left(\frac{\mathrm{d}\epsilon}{\mathrm{d}\eta}\right)^2\right]^{1/2}_{\eta=\eta_1} \mathrm{d}\eta. \tag{8.13}$$

In view of (12) and the fact that η takes values on the interval $(-\sqrt{E_{d_1}}, \sqrt{E_{d_1}})$, we obtain from (13)

$$\rho(\eta) = \pi^{-1}(E_{d_1} - \eta^2)^{-1/2}. \tag{8.14}$$

This result coincides, as it should, with Eqn. (3.45).

When a probabilistic problem is posed for a pseudoclosed system, one can also proceed in principle both from the entire closed system, whose part this system is, and from the pseudoclosed system itself. The nonautonomousness of the pseudoclosed system now is displayed particularly distinctly in the first case. This can be seen, in particular, from the fact that if one proceeds from the consideration of the entire closed system, then a state of any pseudoclosed system it contains cannot be characterized by a definite value of energy, irrespective of the energy value of the rest of the closed system (which is a pseudoclosed system too). Indeed, if a state of the closed system on the whole is characterized by a certain definite energy value E, then energy value E_1 of its pseudoclosed subsystem I is no longer definite and is connected with the energy value E_2 of the subsystem II of this closed system by the condition $E_1 + E_2 = E$.

When a probabilistic problem is posed for a pseudoclosed system proceeding from the system itself, its nonautonomousness within its interval of closedness is not revealed at all, since no initial conditions exist (and are required) in such kind of problems. Therefore, for probabilistic problems, a pseudoclosed system, when considered proceeding from itself, does not differ in its closedness interval from a closed system. Any pseudoclosed system can thus be regarded as a closed one in its closedness interval (when considering it proceeding from itself) if probabilistic problems are posed for it. The nonautonomousness of a pseudoclosed system considered proceeding from itself, for probabilistic problems, is displayed only in the distinction of its interval of closedness from the entire infinite interval.

It is necessary to note that, in the case when pseudoclosed systems are considered proceeding from themselves, it is impossible to extrapolate both trajectories of concrete systems and the probability distributions for the physical quantities of abstract systems outside the closedness interval. This follows, in particular, from the fact that a pseudoclosed system cannot by itself determine the entire closed system whose part it had been (or will have been outside the interval of its closedness.

In the preceding section it has been shown that, for particles of small enough mass, there does not exist at all such a closed system whose part they could be. Nevertheless, setting up of probabilistic problems is quite possible also for them, since in the interims between collisions these particles are pseudoclosed systems and, as follows from the above, can be treated as closed systems.

8.4. Conservation Laws

There exist a number of conservation laws in physics, and the role they play can hardly be exaggerated. The universal validity of the laws has never been and even cannot be proven theoretically. But the large number of experimental data now accumulated in the real sciences do confirm their correctness. So, we can only postulate the conservation laws of physics to be universal, proceeding from all the experiments we know. One may, however, believe that these laws reflect a universal principle of Nature – the principle of unannihilatability of matter (that is, of the impossibility for matter to be annihilated; here matter is understood in the most abstract sense as corresponding to any physical objects).

It is usually believed, however, that for mechanical systems some conservation laws can be obtained theoretically, and they are due to the particular properties of physical space and time. The reasoning is as follows: If a coordinate is not contained in the Lagrange function (cyclic coordinate), neither is it contained in the Hamiltonian function. Then, in view of Eqn. (3), the respective momentum remains unchanged (conserved) during the process of motion. For a closed system, physical space where it moves as a single whole is homogeneous, because no external forces act on it, and its motion does not depend, hence, on the choice of the origin of the inertial frame of reference which the motion is treated in. Therefore the coordinates of the center of mass of a closed system are not contained in its Lagrange function, and the corresponding momenta – the components of the momentum of the system along the corresponding coordinate axes – are conserved. In connection with this fact, the momentum of a system can be defined as a physical quantity (vector) which is conserved for a closed system in virtue of the homogeneity of physical space.

By a similar reasoning, time is not contained in the Lagrange function of a closed system, and therefore the corresponding momentum also is a conservable physical quantity. This physical quantity is the energy of the system with the sign reversed. Hence, energy can be defined as a physical quantity which is conserved for a closed system in virtue of the homogeneity of time.

Thus, the conservation laws for the momentum and energy of a closed system can be thought to be connected with the homogeneity of space and time, respectively. But when there are no external forces acting on a system (closed system) the physical space it moves in also is isotropic, that is, has equal properties in all directions. Therefore, the angle of rotation of a closed system, as a single whole, about any axis is not contained in Lagrange function, and the generalized momentum corresponding to this angle, which is known to be the component of the angular momentum of the system along that axis, is conserved as well. Thus, with the isotropy of physical space there is connected the law of conservation of the component of the angular momentum of a closed system along an arbitrary axis. And the angular momentum component itself can be defined as a physical quantity which is conserved for a closed system in connection with the isotropy of physical space.

Under certain conditions, conservation laws can hold (maybe, partly) for non-closed systems too. If the potential energy of a system does not depend on time explicitly, that is, the system is conservative, the homogeneity of time is not disturbed and the energy of the system is conserved. If the external forces are central, the isotropy of physical space relative to the center of forces holds, and the component of the angular momentum of the system along any axis is conserved. If the external forces are such that in physical space there exists one singled out axis, then this space is isotropic in the directions making equal angles with the axis, and the component of the angular momentum of the system along this axis is conserved. If the external forces are such that physical space is homogeneous in one or two mutually perpendicular directions, then the components of the momentum of the system along these directions are conserved.

Let us dwell now on the question whether different conservation laws can be fulfilled in common. The energy conservation law is connected with the homogeneity of time only and is fulfilled for conservative systems irrespective of the possibility of the fulfillment of other conservation laws. Therefore energy conservation law is compatible with all the other conservation laws. Since the homogeneity of physical space in a certain direction is compatible with that in any other one, the components of the momentum of a closed system along any directions are conserved in common. Hence, the homogeneity of physical space results in the law of conservation of the momentum as a vector quantity.

Matters stand differently for the angular momentum of a system. For a concrete mechanical system, this physical quantity is originally defined as a vector referring to some spatial point chosen, in particular, center of mass of the system. But when deducing the corresponding conservation law proceeding from the isotropy of physical space, one should choose some definite axis and make use of the isotropy of space in the directions perpendicular to it. Thereby, merely the conservation law for the angular momentum referring to the axis chosen, not to the original point chosen, actually is proven. Although the axis can be arbitrary due to the isotropy of space, it must be considered quite definite whenever the deduction is performed. It is impossible to perform this deduction for two different axes simultaneously, since introduction of the second axis disturbs the isotropy of space relative to the first one. Therefore, the deduction based on the consideration of the isotropy of physical space only yields the law of conservation of the component of the angular momentum along some one arbitrary axis. This deduction does not give the law of conservation of components of the angular momentum along different axes in common, that is, the law of conservation of the angular momentum as a vector.

And what is more, the choice of a definite axis in the deduction of the conservation law for the component of the angular momentum along the axis disturbs the homogeneity of the space in the directions perpendicular to this axis, and only retains it in the direction of the axis itself. Therefore, it is possible to deduce in common merely laws of conservation for the component of the angular momentum along an arbitrary axis and for the component of the momentum along the same axis.

Thus, the deduction of conservations laws proceeding from the general properties of physical space and time leads to the compatibility of the energy conservation law with either the conservation law for the momentum or the conservations laws for the components of the angular momentum and the momentum along one and the same axis. To prove in classical mechanics the validity of the conservation laws for the energy, momentum, and angular momentum jointly, one has to make use of some additional reasoning proceeding from the consideration of the forces acting on a mechanical system, which is relevant for concrete systems only. At the same time, the law of conservation of the angular momentum magnitude can be proven to follow immediately from the isotropy of the physical space, since this physical quantity equals to the maximum component of the angular momentum along a certain definite axis chosen.

Consider now concrete physical systems having a certain totality of physical quantities conserved in common. Each concrete system has a definite value of each physical quantity conserved. To a set of concrete systems having one and the same definite value of a physical quantity conserved, there corresponds an abstract system with the same value of the quantity (cf. Sect. 1.1). In accordance with the reasoning of Section 5.1, we may say that this abstract system is in the state determined by this value of the quantity conserved. To a set of concrete systems having a totality of definite values of all the quantities conserved in common, there corresponds an abstract system with the same totality of values of these quantities. And we may say that this abstract system is in the state determined by the totality of the definite values of all the quantities conserved in common.

The above statements formulate in probabilistic physics the conservation laws for both classical abstract systems and abstract microsystems and refer only to the physical quantities which are conserved due to the general properties of space and time, as follows from the foregoing reasoning. We will still revert to this matter, whenever it is necessary.

Suggested References

1. Mayants, L. S.: 1967, *Quantum Physics and Transformations of Matter* (in Russian), VINITI, Moscow. This monograph contains the first version of the statements of Sections 8.2–8.4.
2. Mayants, L. S.: 1973, *Found. Phys.* 3, 413. This paper introduces for the first time the main principles of probabilistic physics.

Chapter 9

Equilibrium Classical Statistical Mechanics

9.1. Microcanonical Distribution

Statistical mechanics deals, as a matter of fact, with probabilistic treatment of physical systems considered from the point of view of their mechanical properties. The denomination 'Probabilistic Mechanics' would, of course, be more pertinent to this discipline. But its usual name is so firmly entrenched in the literature that we will also use it, always keeping in mind, however, its real meaning.

A probabilistic problem can be posed for any abstract mechanical system in principle, irrespective of what are the masses of the bodies it consists of, even though statistical mechanics is usually thought to deal merely with physical systems consisting of microparticles, such as molecules, atoms, etc.

The properties of mechanical systems considered in statistical mechanics obey Lagrangian and Hamiltonian equations. Therefore, from the standpoint of probabilistics they are Hamiltonian random variables (cf. Sect. 6.2), and Eqn. (6.36) is immediately applicable to solving any probabilistic problems posed for such systems in principle. The infinitesimal volume, dv, in the complete phase space (CPS) for a mechanical system having n degrees of freedom may be chosen of the form given in (6.38), and this choice is confirmed by experiment. Therefore, from Eqn. (6.36′) we get for the joint probability density of all the generalized coordinates q and momenta p (including time t and, respectively, energy E)

$$\rho(q_1, \ldots, q_n, t, p_1 \ldots, p_n, E) \equiv \rho = c\,\delta(H') \tag{9.1}$$

where H' is the parametric Hamiltonian function of the form given in (6.11), namely,

$$H' = H - E. \tag{9.2}$$

Equation (1) is valid for any mechanical system, irrespective of whether its Hamiltonian function H depends on t explicitly. In case it does not, neither does the probability density ρ, as is seen from (1), and all the probability distributions for any functions of generalized coordinates and momenta are independent of time, too, since they are all obtained from ρ merely by integration over some of the variables involved (cf. Sects. 6.2 and 3.3). In this case E is a certain constant value of energy determining a state of the corresponding abstract mechanical system.

Let us consider a separate system, say, a molecule in a closed vessel. For the reasons indicated in Section 8.2, it is not a subsystem of any finite closed system. But in the intervals between collisions, it is a pseudoclosed system itself and can be treated as a closed system having constant values of energy E in those intervals.

To a set A_E of concrete molecules of one species, which have one and the same value E of energy, there corresponds an abstract molecule with the same value E of energy determining its state. Equation (1) describing the state takes, in view of (2), the form

$$\rho = c\,\delta(H - E) \tag{9.3}$$

that is known to be a Gibbs' or microcanonical distribution. Since the distribution does not depend on time, it may be regarded as an equilibrium one.

The set A_E consists of all the molecules having the value E of energy and any values of coordinates and momenta compatible with this energy value and the shape of the vessel, while different concrete molecules of the set should differ in the values of at least one coordinate or one momentum. Of course, the set is never in the hands of an experimenter as a whole. However, this fact does not prevent the experimenter from carrying out the statistical experiment on it, even a mental one, since every random test can reveal only one concrete object of a set under examination, and there is no need at all in dealing with the entire set A_E at once. This set is a subset of the whole set A of concrete molecules having values of energy, which can occur under the conditions of the experiment.

Consider now again a single molecule being in a closed vessel. If it has the energy value E at a certain time instant, then it forms consecutively, due to its development in time stipulated by the laws of mechanics for a closed system, elements of a subset of the set A_E (from the point of view of an experimental statistician), before it collides at random with some other particle either being also inside the vessel or in its walls. Thus, between every two random collisions, either elastic or inelastic, the molecule forms some subset of the entire set A mentioned above. And the totality of all these subsets obtained as a result of a very large number of random collisions coincides approximately with the entire set A, the set A_E being contained in it. Hence, carrying out the (mental) experiment on a single molecule at a large number of different time instants, the experimenter can get the necessary statistical data. Indeed, the experimental statistician would record as results of each test only the values of the coordinates and momenta of the molecule under examination, regardless of the time instant the obtained data are related to. Therefore, the data obtained in every test can be regarded as referring to different concrete molecules – elements of set A.

As follows from the Note at the end of Section 1.6, there is one more way of carrying out the statistical experiment in the case under consideration. One may take a very large number of identical closed vessels, each containing only one molecule of the same species, and carry out one test on each of them. The results obtained will give the same statistical distribution, that is, the same approximate probability distribution for coordinates and momenta of the corresponding abstract molecule as in the preceding case, provided each molecule has undergone at least one collision with the walls of the vessel it is contained in.

It will be recalled that the randomness of the tests and, hence, the statistical character of the experiment in both cases are actually ensured by Nature itself (cf. Sects. 1.6 and 4.2).

9.2. Canonical Distribution

The microcanonical distribution given by Eqn. (3) should be valid for any abstract mechanical system, irrespective of how large it is, provided the corresponding separate systems can be regarded as pseudoclosed. Suppose that a large separate pseudoclosed system with energy value E contains a relatively small pseudoclosed subsystem I whose energy value ϵ is much less than E. In different intervals of its closedness (cf. Sect. 8.2) the subsystem I has different energy values ϵ, while the entire system containing it is supposed to have one and the same constant energy value E in the course of testing.

For the reasons explained in the preceding section, by testing the subsystem I a large number of times, one obtains statistical results which can be referred to different concrete subsystems of type I that are in equilibrium with the rest of the large system. The same statistical results may be obtained by testing once a large number of concrete subsystems I, each of which is contained in its own concrete large system of one and the same type, which has the same energy value E.

A probabilistic problem may be posed now: What are the equilibrium probability distributions for various physical quantities of the corresponding abstract subsystem I under the circumstances indicated, that is, when the energy value E of the large system containing it is constant?

To solve this problem theoretically, it is convenient to proceed from Eqn. (3) for the entire pseudoclosed system, in order to find, first, the probability density $p_1(q, p)$ of the coordinates and momenta of subsystem I. Then it is possible to calculate the probability densities of any functions of these quantities, and in particular, the probability density $\rho_1(\epsilon)$ of energy.

Denoting by $H_2(Q, P)$ the Hamiltonian function of the rest of the large pseudoclosed system, which may be called a thermostat, we get

$$H = H_1(q, p) + H_2(Q, P), \tag{9.4}$$

where Q and P are the sets of the coordinates and momenta of the thermostat, respectively, while $H_1(q, p)$ obviously is the Hamiltonian function of subsystem I. Equation (3) which holds for the entire large pseudoclosed system can be rewritten in the form

$$\rho(q, p, Q, P) = c\,\delta\,[H_2(Q, P) - (E - H_1(q, p))] \tag{9.5}$$

and to obtain the probability density $\rho_1(q, p)$ one should integrate Eqn. (5) over all Q and P. Hence we have, in view of $\epsilon = H_1(q, p)$,

$$\rho_1(q, p) = C \int_{\Gamma_2} \delta\,[H_2(Q, P) - (E - H_1(q, p))]\ \mathrm{d}Q\,\mathrm{d}P$$

$$= C \int_{\Gamma_2} \delta\,[H_2(Q, P) - (E - \epsilon)]\ \mathrm{d}Q\,\mathrm{d}P, \tag{9.6}$$

where the integration is over all the values of the variables involved.

Equation (6) shows that $\rho_1(q, p)$ should be a function of only the energy ϵ of subsystem I, which depends on E, as on a parameter, and on the form of the function $H(Q, P)$. It is impossible to calculate the right-hand side of (6) in the general case, and we shall consider, first, the following rather simple example.

Let the thermostat consist of a large number N of noninteracting atoms of any kind, that is, be an ideal monatomic gas. Then

$$H_2(Q, P) = \sum_{i=1}^{N} \frac{p_{ix}^2 + p_{iy}^2 + p_{iz}^2}{2m_i} \tag{9.7}$$

where m_i is the mass of ith atom, and $p_{i\alpha}$ is its momentum along the α-axis of the inertial coordinate system chosen ($\alpha = x, y, z$). The integration over the Cartesian coordinates of the atoms in (6) can be performed immediately, since $H_2(Q, P)$ does not depend on them, and the result obtained, namely V^N, where V is the volume of the thermostat, may be incorporated in the constant C before the integral. Thus the integral on the right-hand side of (6) reduces to

$$\int \delta \left(\sum_{i=1}^{N} \frac{P_{ix}^2 + P_{iy}^2 + P_{iz}^2}{2m_i} - E_2 \right) \prod_{i=1}^{N} \mathrm{d}p_{ix}\,\mathrm{d}p_{iy}\,\mathrm{d}p_{iz}, \tag{9.8}$$

where

$$E_2 = E - H_1(p, q) = E - \epsilon. \tag{9.9}$$

Introducing dimensionless momenta

$$\pi_{i\alpha} = \frac{p_{i\alpha}}{(2m_i E_0)^{1/2}} \qquad (i = 1, \ldots, N; \alpha = x, y, z) \tag{9.10}$$

and dimensionless energy

$$\epsilon_2 = \frac{E_2}{E_0}, \tag{9.11}$$

where E_0 is some unit of energy, we get, up to a constant factor incorporated into the constant C in (6),

$$\int \delta\left(\sum_{j=1}^{3N} \pi_j^2 - \epsilon_2\right) \prod_{j=1}^{3N} \mathrm{d}\pi_j \tag{9.12}$$

instead of (8), the π_j ($j = 1, 2, \ldots, 3N$) replacing the $\pi_{i\alpha}$ ($i = 1, 2, \ldots, N$; $\alpha = x, y, z$).

To calculate (12), let us introduce $3N$-dimensional polar coordinates r and φ_j ($j = 1, 2, \ldots, 3N-1$), whose ranges are:

$$0 \leqslant r < \infty; \qquad 0 \leqslant \varphi_j \leqslant \pi \qquad (j = 1, 2, \ldots, 3N-2); \qquad 0 \leqslant \varphi_{3N-1} < 2\pi,$$

related to π_j by

$$\pi_1 = r\cos\varphi_1; \qquad \pi_j = r\cos\varphi_j \prod_{i=1}^{j-1} \sin\varphi_i \quad (j = 2, \ldots, 3N-1);$$
$$\pi_{3N} = r \prod_{i=1}^{3N-1} \sin\varphi_i. \tag{9.13}$$

Thereby (12) is replaced by

$$\int \delta(r^2 - \epsilon_2) r^{3N-1}\, \mathrm{d}r \prod_{j=1}^{3N-1} (\sin\varphi_j)^{3N-1-j}\, \mathrm{d}\varphi_j. \tag{9.14}$$

Integration over the φ_j yields a constant factor which can also be incorporated in the constant C in (6), and it remains only to calculate

$$\int_0^\infty \delta(r^2 - \epsilon_2) r^{3N-1}\, \mathrm{d}r. \tag{9.15}$$

This, in view of the positivity of $\epsilon_2^{1/2}$, gives

$$\frac{1}{2\epsilon_2^{1/2}} \int_0^\infty \delta(r - \epsilon_2^{1/2}) r^{3N-1}\, \mathrm{d}r = \tfrac{1}{2}\, \epsilon_2^{3N/2-1}. \tag{9.16}$$

We now obtain from (6), in view of (9) and (11),

$$\rho_1(q, p) = C[E - H_1(q, p)]^\alpha = C(E - \epsilon)^\alpha, \tag{9.17}$$

where $\alpha = 3N/2 - 1$. The right-hand side of (17) can be represented in the form

$$\text{const}\left(1 - \frac{\epsilon}{E}\right)^\alpha. \tag{9.18}$$

But ϵ is supposed to be small relative to E. Therefore

$$\left(1-\frac{\epsilon}{E}\right) \cong e^{-\epsilon/E} + O\left(\frac{\epsilon^2}{E^2}\right). \tag{9.19}$$

Now, (18) can be rewritten approximately as

$$\text{const } e^{-\alpha\epsilon/E} = \text{const } e^{-\beta\epsilon}, \tag{9.20}$$

where

$$\beta = \left(\frac{E}{\alpha}\right)^{-1} \tag{9.21}$$

is a parameter having the dimension of inverse energy. So we finally obtain from (17)

$$\rho(q, p) = \text{const } e^{-\beta H_1(q,\, p)} = \text{const } e^{-\beta\epsilon}. \tag{9.22}$$

Since N is a large number, α in (21) can be set equal to $3N/2$, instead of $3N/2 - 1$, and because of $\epsilon \ll E$, we may assign E to the thermostat itself. And from the deduction of Eqn. (22) it is seen that to a very great accuracy α is one-half of the number of the quadratic terms in the expression of the form (7) for the Hamiltonian function $H_2(Q, P)$ of the thermostat. Indeed, any term of this kind can be represented as one of the π_j^2 by introducing appropriate dimensionless coordinates and momenta, irrespective of whether it belongs to the kinetic or potential part of the Hamiltonian function, while the result of the calculation of expressions of form (12) obviously depends on the number of these terms only. This fact allows us to calculate easily the right-hand side of (6) in any case where $H_2(Q, P)$ consists merely of quadratic terms. If $H_2(Q, P)$ consists of M quadratic terms, then $\rho_1(q, p)$ has the form (22) with $\alpha = M/2$ in (21). We shall apply these considerations to one more example.

Let the thermostat be a classical giant N-atomic molecule oscillating harmonically. In this case the Hamiltonian function $H_2(Q, P)$ of the thermostat also consists of only quadratic terms. And to a very great accuracy the number of these can be set equal to $6N$ because to each one of the $3N - 6$ degrees of freedom possessed by the molecule there correspond two quadratic terms in $H_2(Q, P)$ – one in its kinetic and another in its potential parts, where N is a large number. Thus, in the case under consideration, we come again to Eqn. (22) for $\rho_1(q, p)$, but with $\alpha = 3N$ in (21), instead of $\alpha = 3N/2$ for the preceding case of monoatomic ideal gas.

But what about the general case when it is impossible to calculate the right-hand side of Eqn. (6)? The examples considered suggest that $\rho_1(q, p)$ will always have the form given in Eqn. (22) with the parameter β of the form given in (21), but with the value of α depending on what is the fucntion $H_2(Q, P)$. Hence, one may expect Eqn. (22) to be valid for any thermostat, with β having the form as given in (21), the α being a functional of $H_2(Q, P)$. One may consider 2α to have the meaning of an 'effective number' of quadratic terms in the corresponding Hamiltonian function $H_2(Q, P)$.

We can now try to elucidate the meaning of the parameter β in Eqn. (22). Suppose there are two thermostats, whose values of E, α, and β are: E_1, α_1, β_1, and E_2, α_2, β_2, respectively. Let us combine both into one thermostat. The total energy E of that will be, in view of (21)

$$E = E_1 + E_2 = \alpha_1 \beta_1^{-1} + \alpha_2 \beta_2^{-1}. \tag{9.23}$$

If the thermostats do not interact, the Hamiltonian function of the united one will be equal to the sum of their Hamiltonian functions, and therefore, in view of the meaning of the α, the value of α for it is

$$\alpha = \alpha_1 + \alpha_2. \tag{9.24}$$

For the value of β^{-1} of the united thermostat we have

$$\beta^{-1} = \frac{E}{\alpha} = \frac{\alpha_1 \beta_1^{-1} + \alpha_2 \beta_2^{-1}}{\alpha_1 + \alpha_2}. \tag{9.25}$$

If $\beta_1^{-1} = \beta_2^{-1}$, then (25) gives

$$\beta^{-1} = \beta_1^{-1} = \beta_2^{-1}. \tag{9.26}$$

When $\beta_1^{-1} > \beta_2^{-1}$, then from (25) it follows that

$$\beta^{-1} = \beta_1^{-1} - \frac{\alpha_2(\beta_1^{-1} - \beta_2^{-1})}{\alpha_1 + \alpha_2} = \beta_2^{-1} + \frac{\alpha_1(\beta_1^{-1} - \beta_2^{-1})}{\alpha_1 + \alpha_2}, \tag{9.27}$$

that is,

$$\beta_2^{-1} < \beta^{-1} < \beta_1^{-1}. \tag{9.28}$$

As a result of combining the thermostats, a part of the energy of the thermostat with higher β^{-1}-value is transferred to that with the lower one. Since the α values for the thermostats do not change after uniting them, the amount of the energy transferred is

$$\begin{aligned} \Delta E &= E_1 - E_1' = E_2' - E_2 = \alpha_1(\beta_1^{-1} - \beta^{-1}) \\ &= \alpha_2(\beta^{-1} - \beta_2^{-1}) = \frac{\alpha_1 \alpha_2(\beta_1^{-1} - \beta_2^{-1})}{\alpha_1 + \alpha_2}, \end{aligned} \tag{9.29}$$

where E_1' and E_2' are the new energy values for the corresponding thermostats.

We thus see that the peculiarity of β^{-1} does not differ from that of temperature and, therefore, β^{-1} should have the form

$$\beta^{-1} = kT, \tag{9.30}$$

where T is absolute temperature and k is a constant factor to be determined. This can be done, in particular, by considering the ideal monatomic gas thermostat. In this case, because $\alpha = 3N/2$, Eqn. (21) gives

$$E = \frac{3N}{2} kT. \tag{9.31}$$

But, as is well-known from thermodynamics, E is the internal energy of the ideal monatomic gas thermostat, and the factor before T is equal to $\frac{3}{2}R$ per mole, where R is the gas constant per mole. Since the number of molecules in a mole is N_a (Avogadro constant), we finally obtain

$$k = \frac{R}{N_a} \tag{9.32}$$

which is known as Boltzmann's constant.

Equation (22) can now be written in the form

$$\rho(q, p) = Z^{-1}\, \mathrm{e}^{-H(q,p)/kT} = Z^{-1}\, \mathrm{e}^{-\epsilon/kT}, \tag{9.33}$$

where, in view of

$$\int_\Gamma \rho(q, p)\, \mathrm{d}q\, \mathrm{d}p = 1, \tag{9.34}$$

$$Z = \int_\Gamma \mathrm{e}^{-H(q,p)/kT}\, \mathrm{d}q\, \mathrm{d}p = \int_\Gamma \mathrm{e}^{-\epsilon/kT}\, \mathrm{d}\Gamma \tag{9.35}$$

(all marks denoting subsystem I are henceforth omitted).

Equation (33) is known as the canonical distribution, and Z is called the partition function. The canonical distribution describes an equilibrium state of any physical system, irrespective of its complexity, which is determined by a definite value T of the temperature of a thermostat being in contact with the system. This distribution is related, of course, to the corresponding abstract physical system, whereas all the concrete physical systems of the respective set are formed, as was shown before, due to the contact of any of them with the thermostat in particular.

Let us immediately make two practically important remarks. (1) At the beginning of this section it was supposed that a thermostat is a pseudoclosed system. But for the derivation of Eqn. (22) this requirement actually is unnecessary. Indeed, we not only made some approximations when passing from (17) to (22), but also tacitly assumed that changes in the energy value of the subsystem do not change the energy value of the thermostat. Therefore, if the interaction of the thermostat with the surroundings causes only relatively small changes in its energy E, then this cannot practically change its temperature T and, hence, the distribution given in Eqn. (33). In real experiments, thermostats are never pseudoclosed systems. On the contrary, they are supplied with special equipment – heating or cooling devices – which ensure the approximate stability of their temperatures.

(2) The integration over the momenta in Eqn. (35) and others usually is over the infinite intervals $(-\infty, \infty)$, which obviously does not meet the real conditions. But this can be done, since the relative difference in the results of integration within the real and infinite limits is negligible.

Equation (33) allows one to obtain the probability distributions and, hence, mathematical expectations of any functions of coordinates and momenta of the system under consideration, and thereby to build a bridge between statistical mechanics and thermodynamics. In particular, the probability density $\rho(\epsilon)$ of energy ϵ has, in view of (33) and $\epsilon = H(q, p)$, the form

$$\rho(\epsilon) = \int_\Gamma \delta\,[\epsilon - H(q, p)]\,\rho(q, p)\,\mathrm{d}q\,\mathrm{d}p$$

$$= Z^{-1}\,\mathrm{e}^{-\epsilon/kT}\,\omega(\epsilon), \tag{9.36}$$

where $\omega(\epsilon)$ is proportional to the volume of that part of the phase space, where $H(q, p)$ has the value ϵ.

Making use of Eqn. (36), we can also rewrite the partition function Z as

$$Z = \int_\epsilon \mathrm{e}^{-\epsilon/kT}\,\omega(\epsilon)\,\mathrm{d}\epsilon. \tag{9.37}$$

9.3. A Separate Atom in a Thermostat

By way of example we will consider a separate free atom being in a container of volume V, which contacts a thermostat of temperature T. The probability distributions for any functions of its coordinates and momenta, and hence, the mathematical expectations of all those quantities can be obtained from Eqn. (33). In this particular case, we have

$$H(x, y, z, p_x, p_y, p_z) = \frac{p_x^2 + p_y^2 + p_z^2}{2m}, \tag{9.38}$$

where x, y, and z are the Cartesian coordinates of the atom; p_x, p_y, and p_z its corresponding momenta, and m its mass. Therefore, Eqn. (33) takes the form

$$\rho(x, y, z, p_x, p_y, p_z) = Z^{-1}\exp\left(-\frac{p_x^2 + p_y^2 + p_z^2}{2mkT}\right). \tag{9.39}$$

As seen from Eqn. (39), all the Cartesian coordinates and momenta are mutually independent random variables, and, in accordance with the results of Section 6.2, the probability densities of coordinates are constants. From Eqn. (35) it follows that

$$Z = \int_{-\infty}^{\infty} \exp\left(-\frac{p_x^2 + p_y^2 + p_z^2}{2mkT}\right)\mathrm{d}x\,\mathrm{d}y\,\mathrm{d}z\,\mathrm{d}p_x\,\mathrm{d}p_y\,\mathrm{d}p_z = (2\pi mkT)^{3/2}\,V, \tag{9.40}$$

whence

$$\rho(p_x, p_y, p_z) = \int \rho(x, y, z, p_x, p_y, p_z)\, \mathrm{d}x\, \mathrm{d}y\, \mathrm{d}z$$

$$= (2\pi mkT)^{-3/2} \exp\left(-\frac{p_x^2 + p_y^2 + p_z^2}{2mkT}\right). \tag{9.41}$$

Introducing

$$p_\alpha = mv_\alpha \quad (\alpha = x, y, z) \tag{9.42}$$

we get from (41)

$$\rho(v_x, v_y, v_z) = \left(\frac{m}{2\pi kT}\right)^{3/2} \exp\left[-\frac{m(v_x^2 + v_y^2 + v_z^2)}{2kT}\right], \tag{9.43}$$

that is, the joint probability density of the velocity components of an atom in a thermostat. It is easily seen that the probability density of a separate velocity component v_α $(\alpha = x, y, z)$ is given by

$$\rho(v_\alpha) = \left(\frac{m}{2\pi kT}\right)^{1/2} \mathrm{e}^{-mv_\alpha^2/2kT}. \tag{9.44}$$

The probability density $\rho(\epsilon)$ of energy is, in view of (36), (41), and

$$\epsilon = \frac{p_x^2 + p_y^2 + p_z^2}{2m} = \frac{p^2}{2m} \tag{9.45}$$

of the form

$$\rho(\epsilon) = (2\pi mkT)^{-3/2} \int_0^\infty \mathrm{e}^{-\epsilon/kT}\, \delta\left(\epsilon - \frac{p^2}{2m}\right) 4\pi p\, \mathrm{d}p$$

$$= 2\pi^{-1/2} (kT)^{-3/2}\, \mathrm{e}^{-\epsilon/kT} \epsilon^{1/2}. \tag{9.46}$$

In view of

$$v = (v_x^2 + v_y^2 + v_z^2)^{1/2} = (2\epsilon/m)^{1/2}, \tag{9.47}$$

Eqn. (46) leads to

$$\rho(v) = \left(\frac{2}{\pi}\right)^{1/2} \left(\frac{m}{kT}\right)^{3/2} \mathrm{e}^{-mv^2/2kT} v^2, \tag{9.48}$$

which is the well-known Maxwell velocity distribution.

For the mathematical expectation of energy, $\overline{\epsilon}$, we obtain from (46)

$$\overline{\epsilon} = \int_0^\infty \rho(\epsilon)\epsilon \, d\epsilon = 2\pi^{-1/2}(kT)^{-3/2} \int_0^\infty e^{-\epsilon/kT}\epsilon^{3/2} \, d\epsilon = \tfrac{3}{2}kT. \tag{9.49}$$

(In statistical physics, the mathematical expectation is indicated by a bar over the symbol denoting the quantity.) From

$$\overline{\epsilon} = \frac{m}{2}\overline{v^2} \tag{9.50}$$

we now get

$$\overline{v^2} = \frac{3kT}{m}. \tag{9.51}$$

When a concrete atom moves in a certain direction, which can be chosen as the positive direction of the z-axis of a Cartesian coordinate system, with velocity v_{z+}, it carries the momentum mv_{z+}. The mathematical expectation of the momentum transported by an atom in that direction through unit area of the xy-plane per unit time is

$$\rho_n \overline{|v_{z+}|mv_{z+}} = V^{-1}m\overline{v_{z+}^2} = \frac{m}{V}\int_0^\infty \rho(v_z)v_z^2 \, dv_z, \tag{9.52}$$

where ρ_n is the number of particles in unit volume (in the case under consideration $\rho_n = V^{-1}$). It is obvious that $\overline{v_{z+}^2} = \frac{1}{2}\overline{v_\alpha^2}$ $(\alpha = x, y, z)$. Therefore, in view of (50) and $\overline{v_\alpha^2} = \frac{1}{3}\overline{v^2}$,

$$m\overline{v_{z+}^2} = \tfrac{1}{3}\overline{\epsilon}. \tag{9.53}$$

The mathematical expectation of the force exerted by an atom on unit area of the plane perpendicular to the direction of its motion, that is, the 'pressure', P, exerted by an atom, is, in view of $v_{z-} = -v_{z+}$ and (49),

$$P = V^{-1}\left[|v_{z+}|mv_{z+} - |v_{z-}|mv_{z-}\right] = 2m\overline{v_{z+}^2}V^{-1} = kTV^{-1}. \tag{9.54}$$

Equation (54), rewritten in the form

$$PV = kT \tag{9.55}$$

is the ideal gas equation of state for one atom. It confirms once again that the parameter k in (30) is the Boltzmann constant.

As is generally known, the notions of temperature and pressure are pertinent to thermodynamics, which deals with macroscopic physical systems. Therefore, a question may arise whether it is relevant to use these notions when considering a particle, say, an atom, which obviously is a microscopic physical system. The answer is as follows.

Temperature and pressure are, indeed, properties of concrete macroscopic physical systems. But, at the same time, these variables have definite probabilistic meanings related to the distributions for the corresponding abstract systems. Therefore, it is possible to use the notions when dealing even with the canonical distribution for an abstract particle, bearing in mind their probabilistic sense only. For the same reason, and with the same reserve, it is possible to apply any other thermodynamic notions to one abstract particle.

9.4. Thermodynamic Functions

We now turn to elucidating the probabilistic meaning of some thermodynamic functions. To that end, let us, first, recall a little of thermodynamics.

A thermodynamic state of a macroscopic system is determined by definite values of certain macroscopic properties, such as temperature T, pressure P, specific volume V_0, etc. The equation of state, that is, the relationship between temperature, pressure, and specific volume, which is particular for every thermodynamic system, leads to the independence only of any two of the three variables mentioned. In the simplest cases a thermodynamic state of a macroscopic system is therefore completely determined by any pair of the properties T, P, V_0 possessed by the system. All other thermodynamic properties of the system are functions of any two independent ones and can thus be called state functions.

For a macroscopic system having temperature T, the internal energy U, Helmholtz free energy F, and entropy S are state functions related by

$$U = F + TS. \tag{9.56}$$

The internal energy is that part of the total energy of the system, which depends on the temperature and specific volume. The entropy is defined by the equation

$$\mathrm{d}S = \frac{\mathrm{D}Q}{T} \tag{9.57}$$

valid for reversible processes, where $\mathrm{D}Q$ is an infinitesimal amount of heat Q absorbed by the system at temperature T. The Helmholtz free energy is defined by Eqn. (56) and is that part of the internal energy which can be utilized for producing some work. In the simplest case this work is a purely mechanical one and can be represented as

$$\mathrm{D}A = P\,\mathrm{d}V, \tag{9.58}$$

where $\mathrm{D}A$ is an infinitesimal amount of work A, and $\mathrm{d}V$ is an infinitesimal increase of the total volume V occupied by the system. Both heat and work are not state functions. When an infinitesimal amount of heat is absorbed by a simple system in a reversible process, one part of it goes for increasing the internal energy of the system and another one produces mechanical work. In view of (57) and (58), this can be represented as

$$T\,\mathrm{d}S = \mathrm{d}U + P\,\mathrm{d}V, \tag{9.59}$$

whence

$$\mathrm{d}U = T\,\mathrm{d}S - P\,\mathrm{d}V. \tag{9.60}$$

We now obtain from (56)

$$\mathrm{d}U = \mathrm{d}F + T\,\mathrm{d}S + S\,\mathrm{d}T. \tag{9.61}$$

This yields, in view of (59),

$$\mathrm{d}F = -S\,\mathrm{d}T - P\,\mathrm{d}V. \tag{9.62}$$

Equation (62) allows one to formally define entropy and pressure as

$$S = -\left(\frac{\partial F}{\partial T}\right)_V \tag{9.63}$$

and

$$P = -\left(\frac{\partial F}{\partial V}\right)_T \tag{9.64}$$

respectively. The subscripts in (63) and (64) indicate what variables remain constant in calculating the corresponding partial derivatives.

In view of (63) and (56), we can write

$$\frac{U}{T^2} = \frac{F}{T^2} + \frac{S}{T} = \frac{F}{T^2} - \frac{1}{T}\left(\frac{\partial F}{\partial T}\right)_V = -\frac{\partial}{\partial T}\left(\frac{F}{T}\right)_V. \tag{9.65}$$

This equation allows one to elucidate easily the probabilistic (statistical) meaning of the Helmholtz free energy F. Indeed, the internal energy U of a system contacting a thermostat of temperature T should be identified as the mathematical expectation of energy ϵ of the system. This leads, in view of (36) and (37), to the equation

$$U = \bar{\epsilon} = \frac{\int_0^\infty \epsilon\, \mathrm{e}^{-\epsilon/kT}\omega(\epsilon)\,\mathrm{d}\epsilon}{\int_0^\infty \mathrm{e}^{-\epsilon/kT}\omega(\epsilon)\,\mathrm{d}\epsilon} = \frac{\int_\Gamma \epsilon\,\mathrm{e}^{-\epsilon/kT}\,\mathrm{d}\Gamma}{\int_\Gamma \mathrm{e}^{-\epsilon/kT}\,\mathrm{d}\Gamma} \tag{9.66}$$

which can be represented, with due allowance for (37), as

$$U = kT^2\left(\frac{\partial \ln Z}{\partial T}\right)_V. \tag{9.67}$$

By comparing this with Eqn. (65), we find, taking into consideration that also Eqns. (64) and (56) should be satisfied,

$$F = -kT \ln Z. \tag{9.68}$$

Equations (67) and (68) connect thermodynamics with statistical (probabilistic) physics, and we can now turn to the probabilistic meaning of the entropy. By making use of Eqn. (63) we obtain from (68)

$$S = k \ln Z + kT \left(\frac{\partial \ln Z}{\partial T} \right)_V. \tag{9.69}$$

In view of (67), this gives

$$S = k \ln Z + \frac{1}{T} U, \tag{9.70}$$

which conforms to (56). On the other hand it follows from Eqn. (33) that

$$\ln \rho(q, p) = -\frac{\epsilon}{kT} - \ln Z. \tag{9.71}$$

Therefore, the mathematical expectation of $\ln \rho(q, p)$ is

$$\overline{\ln \rho(q, p)} = -\frac{\overline{\epsilon}}{kT} - \ln Z = -\frac{U}{kT} - \ln Z. \tag{9.72}$$

By comparing this with (70), we find

$$S = k \ln W, \tag{9.73}$$

where

$$\ln W = -\overline{\ln (q, p)} = -\int_\Gamma \rho(q, p) \ln \rho(q, p) \, \mathrm{d}\Gamma. \tag{9.74}$$

Equation (73) is a probabilistic definition of entropy.

Recall that each of the concrete systems, elements of the set corresponding to the canonical distribution under consideration, has definite values of all the random variables, in particular, coordinates and momenta, and that different concrete systems differ in the values of at least one of the random variables. Any real function of random variables is, by definition, also a random variable (cf. Sect. 3.1). Therefore, $\rho(q, p)$ and $\ln \rho(q, p)$ are random variables, too, and every concrete system of the set under consideration has definite values of these. Thus, the mathematical expectation of the random variable $\ln \rho(q, p)$ is at the same time its mean value on the set of concrete systems corresponding to the canonical distribution under consideration (cf. Sect. 3.4).

Equation (66) represents U as the mathematical expectation of the random variable ϵ. Equation (73) represents S as the mathematical expectation of the random variable $-k \ln \rho(q, p)$. Now, it follows from (56) that also F can be represented as the mathematical expectation of a certain random variable, namely,

$$F = \overline{\epsilon + kT \ln \rho(q, p)}. \tag{9.75}$$

Equations (73) and (74) show that the entropy, S, is a functional of $\rho(q, p)$. It can be demonstrated that S has a maximum when $\rho(q, p)$ has the form of Eqn. (33). Indeed, we can rewrite Eqn. (74) as

$$\ln W = -\lim_{\Delta\Gamma_l \to 0} \sum_{l=1}^{M} [\rho(q, p)]_l [\ln \rho(q, p)]_l \Delta\Gamma_l, \tag{9.76}$$

where $\Delta\Gamma_l$ can be supposed to be an infinitesimal cell volume of the corresponding phase space adequate to the set of concrete systems under consideration, at the point of the space indicated for simplicity by the subscript 'l'; M is the number of cells in the phase space. Introducing

$$[\rho(q, p)]_l \Delta\Gamma_l = p_l, \tag{9.77}$$

where p_l is the probability for the respective abstract system to have the values of coordinates and momenta within $\Delta\Gamma_l$ and choosing all the $\Delta\Gamma_l$ of one and the same low volume $\Delta\Gamma$, we obtain from (76) with a sufficiently great accuracy

$$\ln W \cong -\sum_{l=1}^{M} p_l \ln p_l + \sum_{l=1}^{M} p_l \ln \Delta\Gamma$$

$$= -\sum_{l=1}^{M} p_l \ln p_l + \ln \Delta\Gamma \tag{9.78}$$

(here $\Sigma_{l=1}^{M} p_l = 1$ has been taken into account).

On the right-hand side of Eqn. (78), the first term is positive, the second one is negative; their magnitudes increase without limit, while their sum remains approximately constant, tending to the exact value of $\ln W$, when $\Delta\Gamma \to 0$, that is, $M \to \infty$. For every value of $\Delta\Gamma$, the extremum of the right-hand side of Eqn. (78) is obviously determined by that of its first term. One can find the extremum from the conditions

$$-\delta \ln W = \delta \sum_{l=1}^{M} p_l \ln p_l = \sum_{l=1}^{M} (\ln p_l + 1) \delta p_l = 0, \tag{9.79a}$$

$$\delta \sum_{l=1}^{M} p_l = \sum_{l=1}^{M} \delta p_l = 0, \tag{9.79b}$$

$$\delta U = \delta \sum_{l=1}^{M} p_l \epsilon_l = \sum_{l=1}^{M} \epsilon_l \delta p_l = 0. \tag{9.79c}$$

By adding (79a) with (79b) and (79c) multiplied by some constant factors λ and β, respectively, one gets

$$\sum_{l=1}^{M} (\ln p_l + 1 + \lambda + \beta\epsilon_l)\, \delta p_l = 0. \tag{9.80}$$

This equation can be satisfied for any nonzero variation δp_l only if

$$\ln p_l = -(1 + \lambda) - \beta\epsilon_l \tag{9.81}$$

is fulfilled for any l. It follows from (81) that

$$p_l = \text{const}\; \mathrm{e}^{-\beta\epsilon_l}, \tag{9.82}$$

which, in view of (77), coincides with Eqn. (33).

We can make sure that (82) corresponds to the maximum of $\ln W$ by calculating the second variation of it, which gives, in view of the nonnegativity of p_l,

$$\delta^2 \ln W = -\delta \sum_{l=1}^{M} (\ln p_l + 1)\, \delta p_l = -\sum_{l=1}^{M} \frac{1}{p_l} (\delta p_l)^2 \leqslant 0. \tag{9.83}$$

This fact finally confirms that at constant temperature and internal energy the entropy defined by Eqn. (73) has a maximum when $\rho(q, p)$ corresponds to the equilibrium distribution, that is, to the canonical distribution given by Eqn. (33).

Equation (78) allows one to obtain one more representation for the entropy. The set of concrete systems corresponding to the abstract system being in the state described by the canonical distribution with the given value of T is infinite. We may, however, replace this set by a finite set of a very large number N of concrete systems of the same kind, which are formed by the corresponding set of N separate systems contacting a thermostat of the same temperature T, these systems being considered at any certain time instant. Statistical tests carried out on both sets would give approximately the same results; the larger N, the less the difference between the statistical results of testing those two sets.

We can, therefore, replace p_l in Eqns. (77) and (78) by N_l/N, where N_l is the number of concrete systems of the finite set under consideration whose coordinates and momenta have values within the lth cell of the phase space. We obtain thereby

$$[\rho(q, p)]_l\, \Delta\Gamma = \frac{N_l}{N}, \tag{9.84}$$

$$\ln W = -\sum_{l=1}^{M} \frac{N_l}{N} \ln \frac{N_l}{N} + \ln \Delta\Gamma. \tag{9.85}$$

The first term on the right-hand side of (85) can be rewritten as

$$-\frac{1}{N}\sum_{l=1}^{M} N_l(\ln N_l - \ln N) = -\frac{1}{N}\left(\sum_{l=1}^{M} N_l \ln N_l - N\ln N\right). \tag{9.86}$$

By making use of the Stirling formula (cf. Sect. 2.4), we obtain from Eqn. (85)

$$N\ln W = \ln\frac{N!}{\Pi_{l=1}^{M} N_l!} + N\ln\Delta\Gamma = \ln\left(\frac{N!}{\Pi_{l=1}^{M} N_l!}\,\Delta\Gamma^N\right). \tag{9.87}$$

The finite set of N concrete systems under consideration may be regarded as one big concrete system, which is a 'snapshot' of the microstructure of the corresponding concrete big macroscopic system, at a certain time instant, that consists of N noninteracting initial separate systems. At every time instant this set is a finite subset of the infinite set of concrete systems corresponding to the abstract system mentioned above. At different time instants, the separate systems contained in the macroscopic concrete system have different values of coordinates and momenta and yield, therefore, different concrete systems. But the numbers N_l of the concrete systems belonging to the respective cells of the phase space remain approximately unchanged. At any time instant, the distribution of the concrete systems, which represents the instantaneous microstructure of the concrete macroscopic system, over the values of q and p for any of the separate systems, coincides approximately with the canonical distribution for the corresponding abstract system.

Thus we can interpret Eqn. (87) as related to the big macroscopic system consisting of N separate systems contacting a thermostat of temperature T. Since the internal energy of each separate system is U, the internal energy of the big macroscopic system can be written in the form

$$E = NU = N\sum_{l=1}^{M} p_l\epsilon_l = \sum_{l=1}^{M} N_l\epsilon_l. \tag{9.88}$$

It is also obvious that

$$\sum_{l=1}^{M} N_l = N. \tag{9.89}$$

In Eqn. (87), the second factor, $\Delta\Gamma^N$, represents the elementary volume of the phase space for the big system, which is the direct product of the phase spaces of all the separate systems contained in it. The expression $N!/\Pi_{l=1}^{M} N_l!$ on the right-hand side of Eqn. (87) is the number of ways in which N separate systems can be distributed over the corresponding M cells of the phase space of any of them, so that the conditions (88) and (89) be satisfied. In other words, that is the number of the microstructures realizing the given thermodynamic state of the corresponding big macroscopic system under consideration. This number is called the statistical weight of a macroscopic system. It follows from the above, by the way, that the statistical weight has a maximum when

$$N_l = \text{const}\, e^{-\beta\epsilon_l}. \tag{9.90}$$

Each cell of the phase space of the big system conforms to one microstructure of the respective macroscopic system, whose energy is $E = NU$ (cf. Eqn. (88)). Therefore, $(N!/\Pi_{l=1}^{M} N_l!)\, \Delta\Gamma^N$ is approximately the volume of the phase space of the big macroscopic system corresponding to all the microstructures realizing its state with the internal energy E. It is easy to show that this volume is only $(N!)^{-1}$th part of the entire volume of the phase space of the big system, which corresponds to its energy value E.

Indeed, the phase space of the big macroscopic system is the direct product of the phase spaces of all the N separate systems the macroscopic system consists of. But probability, being a relative measure of a corresponding subset of concrete objects, counts the concrete systems, not the separate ones, in the case under consideration. Let the separate systems be indexed by integers from 1 through N. In accordance with the definition (see Sect. 1.1), a concrete system is a separate one having definite values of coordinates and momenta, irrespective of the index assigned to the separate system the concrete one conforms to. On the other hand, a separate system can form, in principle, the entire infinite set of corresponding concrete systems (cf. Sect. 9.1).

This situation is similar to that of throwing a coin. By doing this with any separate coin of a set of indexed coins, one obtains one and the same set of concrete objects (cf. Sect. 1.8). Therefore, when figuring out the probability of appearance of, say, the 'head', no attention should be paid to what is the index of the coin used.

We thus see that the number of concrete microstructures realizing the thermodynamic state of the corresponding big macroscopic system has nothing to do with the indices of the separate systems. At the same time, the coordinate axes in the phase space of this macroscopic system should be marked not only with the indices referring to the coordinates and momenta of a single system, but also with those distinguishing separate systems. If there is a cell in the phase space of the big macroscopic system referring to some approximate values of coordinates and momenta of all the separate systems, then any transposition among the indices of the separate systems leads to some other cell in the phase space, which corresponds, however, to the same energy E. Since the number of the permutations among N indices is $N!$, the volume of the phase space for the big macroscopic system, which corresponds to its energy value E, is $N!$ times larger than the volume of this phase space, which conforms to all the microstructures realizing the thermodynamic state of the big macroscopic system with the same internal energy.

Denoting by $\Omega(E)$ the volume of the big macroscopic system phase space, which corresponds to the energy value E, we can write, in view of Eqns. (73) and (87), and the fact that the entropy S of the big macroscopic system is N times larger than that of the initial system we started with

$$S = k \ln W = k \ln \frac{\Omega(E)}{N!}. \tag{9.91}$$

We note that Eqn. (91) refers to a macroscopic system consisting of N separate noninteracting subsystems of one and the same kind.

It is easy to demonstrate that for any macroscopic system having a large number of degrees of freedom, the energy distribution is very narrw and has a sharp peak at the energy value U, where $U = \overline{\epsilon}$ is the internal energy of the macroscopic system. In order to do this, let us calculate the relative standard deviation (see Sect. 3.4) of the energy distribution for such a case, that is,

$$\frac{\delta\epsilon}{U} = \left(\frac{\overline{\epsilon^2} - U^2}{U^2}\right)^{1/2}. \tag{9.92}$$

The differentiation of Eqn. (66) with respect to T yields

$$\left(\frac{\partial\overline{\epsilon}}{\partial T}\right)_V = \frac{1}{kT^2}\left[\frac{\int_\Gamma \epsilon^2\, e^{-\epsilon/kT}\, d\Gamma}{\int_\Gamma e^{-\epsilon/kT}\, d\Gamma} - \frac{\left(\int_\Gamma \epsilon\, e^{-\epsilon/kT}\, d\Gamma\right)^2}{\left(\int_\Gamma e^{-\epsilon/kT}\, d\Gamma\right)^2}\right]$$

$$= \frac{1}{kT^2}(\overline{\epsilon^2} - U^2), \tag{9.93}$$

whence, in view of (92),

$$\frac{\delta\epsilon}{U} = \left[\frac{kT^2}{U^2}\left(\frac{\partial U}{\partial T}\right)_V\right]^{1/2} = \left[k\left(\frac{\partial U^{-1}}{\partial T^{-1}}\right)_V\right]^{1/2}. \tag{9.94}$$

The internal energy U of any classical system

$$U = N_f\alpha' kT, \tag{9.95}$$

where N_f is the number of degrees of freedom of the system and α' is a number ranging between $\frac{1}{2}$ and 1 (cf. Sect. 9.2). Hence, Eqn. (94) yields

$$\frac{\delta\epsilon}{U} = (N_f\alpha')^{-1/2}, \tag{9.96}$$

which proves the above statement.

If a macroscopic system has, for instance, the number of degrees of freedom of the order of Avogadro's number, that is 10^{23}, then $\delta\epsilon/U$ is of the order of 10^{-12}–10^{-11}, which is a very small number indeed.

From the statement just proven it follows that the probability for a macroscopic system to have an energy value ϵ noticeably differing from the internal energy U

is extremely small. Therefore, for such systems Z can be calculated very accurately from Eqn. (35) by replacing ϵ therein by U and integrating only within the range approximately conforming to the values of $\epsilon \cong U$. This gives

$$Z = \mathrm{e}^{-U/kT} \int \omega(U)\,\mathrm{d}\epsilon = \mathrm{e}^{-U/kT}\,\Omega(U) \quad U \leqslant \epsilon \leqslant U + \Delta U, \tag{9.97}$$

where

$$\Omega(U) = \omega(U)\,\Delta U \tag{9.98}$$

is the volume of the phase space for the macroscopic system, which corresponds to the almost constant energy value U.

Using Eqn. (68), one obtains from Eqn. (97)

$$F = U - kT \ln \Omega(U). \tag{9.99}$$

By comparing this equation with Eqn. (56) it becomes evident that

$$S = -k \ln \Omega(E). \tag{9.100}$$

Equation (100) is often taken as a definition of the entropy of a macroscopic system, the letter E being used instead of U in order to distinguish between the statistical and thermodynamic notations.

It is necessary, however, to emphasize that if a macroscopic system consists of N separate systems of exactly one and the same kind, Eqn. (91) should be used for the calculation of the entropy, instead of Eqn. (100). Equation (91) results immediately from (97) if it is taken into account that, when integrating, only each one of the phase space cells with certain values of coordinates and momenta of all the separate systems contained in the macroscopic system should be considered.

Equation (59) allows one to give the following definition of temperature:

$$\frac{1}{T} = \left(\frac{\partial S}{\partial U}\right)_V. \tag{9.101}$$

This definition is independent of the one above (cf. Sect. 9.2) and can be regarded as a strict definition of the notion of temperature in statistical physics.

All other related quantities, such as Gibbs' free energy, grand partition function, chemical potential, etc., can be easily obtained, proceeding from those already considered. We shall not, however, dwell on these matters here.

9.5. The Ideal (Perfect) Gas

Let us illustrate the above considerations in the example of the ideal gas. The ideal gas may be defined as one which consists of noninteracting molecules having no size. Certainly, no real gas is the ideal one. But any dilute enough gas can, to a certain accuracy, be considered the ideal gas, since the above conditions are approximately satisfied for it.

Let the ideal gas consist of

$$N = \sum_{i=1}^{l} N_i \tag{9.102}$$

molecules, where N_i is the number of molecules of ith species and l is the number of all the different species. Since by definition there is no interaction between molecules, the Hamiltonian function of the ideal gas, in view of Eqn. (102), is

$$H = \sum_{i=1}^{l} N_i H_i, \tag{9.103}$$

where H_i is the Hamiltonian function of any molecule of ith species.

The pressure and internal energy of the ideal gas, which are the mathematical expectations of the corresponding random variables, can be represented as

$$P = \sum_{i=1}^{l} N_i p^{(i)} = \sum_{i=1}^{l} P_i \tag{9.104}$$

and

$$U = \sum_{i=1}^{l} N_i u_i \tag{9.105}$$

respectively. Here $p^{(i)}$ and u_i are the respective quantities related to one molecule of ith species; P_i is the so-called partial pressure exerted by all the molecules of that species.

No matter how complicated a molecule, its translational motion can be completely separated from all the other kinds of motion, that is, one can write

$$H_i = H_i^{(t)}(X_i, Y_i, Z_i, P_{ix}, P_{iy}, P_{iz}) + H_i^{(R)}. \tag{9.106}$$

Here $H_i^{(t)}$ is related to the translational motion, $H_i^{(R)}$ to all the other motions of the ith species of molecule; X_i, Y_i, Z_i are certain Cartesian coordinates of its center of mass, and P_{ix}, P_{iy}, P_{iz} are the respective momentum components. $H_i^{(R)}$ does not depend on X_i, Y_i, Z_i, P_{ix}, P_{iy}, and P_{iz}.

Therefore,

$$u_i = u_i^{(t)} + u_i^{(R)}, \tag{9.107}$$

where $u_i^{(t)}$ is the internal energy of the translational motion of one molecule of the ith species, $u_i^{(R)}$ is its internal energy related to all the other kinds of motion. For the partition function, Z_i, of one molecule of the ith species we have, in view of Eqns. (35) and (106),

$$Z_i = Z_i^{(t)} Z_i^{(R)}. \tag{9.108}$$

Here $Z_i^{(t)}$ and $Z_i^{(R)}$ are obviously related to the translational and all other kinds of motion, respectively.

The partition function, Helmholtz free energy, and entropy of the ideal gas can be represented, respectively, as

$$Z = Z^{(t)} Z^{(R)}, \tag{9.109}$$

$$F = F^{(t)} + F^{(R)}, \tag{9.110}$$

$$S = S^{(t)} + S^{(R)}, \tag{9.111}$$

the upper parenthetic indices (t) and (R) denoting the terms relating to the translational and all the other kinds of motion of gas molecules, respectively.

Since all the motions of a molecule but the translational one need quantum approach, we shall consider here the translational motion of molecules of the ideal gas only, leaving the treatment of the rest of their motions for Chapter 10. But the translational motion of a molecule does not differ from that of an atom. Therefore, we can use here the results of Section 9.3.

From Eqns. (104), (54), and (102) we immediately get

$$P_i = N_i V^{-1} kT; \qquad P = NV^{-1} kT. \tag{9.112}$$

Equations (105), (106), (49), and (102) yield for the translational part, $U^{(t)}$, of internal energy of the ideal gas

$$U^{(t)} = \tfrac{3}{2} NkT. \tag{9.113}$$

But the calculation of $F^{(t)}$ and $S^{(t)}$, proceeding from the results of Section 9.3, requires certain care. We notice, first of all, that it is impossible to use Z as given in Eqn. (40) for calculation of F by Eqn. (68). Indeed, in order for F to have the dimension of energy, as it should, Z is to be dimensionless, whereas in Eqn. (40) it is obviously dimensional, namely, it has the dimension of $p^3 V$. This fact, which does not prevent one from calculating the mathematical expectations of some random variables as well as the differences in Helmholtz free energy and in entropy, nevertheless, does not allow one to calculate the values of these thermodynamic quantities themselves. This trouble is of a general nature and is due to the fact that, in view of Eqn. (35), Z must have the dimension of action (momentum x coordinate) to the power of the number of degrees of freedom.

One can easily improve the situation by dividing in Eqn. (35) $d\Gamma$ by a factor having the same dimension. But the result obtained depends evidently on the choice of the factor. In order for the theoretical results to be in agreement with the results of the experiment, it turned out necessary to choose Planck's constant h to the power needed as the sought-for factor. Hence, we should write the translational partition function for one molecule of mass m_i as

$$Z_i^{(t)} = (2\pi m_i kT)^{3/2} Vh^{-3} \tag{9.114}$$

instead of Eqn. (40).

Furthermore, the considerations which led us to Eqn. (91) show, in view of Eqn. (69), that the translational partition function for N_i molecules of ith species must be of the form

$$(N_i!)^{-1} Z_i^{(t)N_i} = (N_i!)^{-1} [(2\pi m_i kT)^{3/2} Vh^{-3}]^{N_i}. \tag{9.115}$$

Therefore, the translational partition function for the ideal gas under consideration is

$$Z^{(t)} = \prod_{i=1}^{l} (N_i!)^{-1} [(2\pi m_i kT)^{3/2} Vh^{-3}]^{N_i}. \tag{9.116}$$

If all the N_i are large enough, Eqn. (68) yields, in view of Eqn. (2.37),

$$F^{(t)} = -kT \ln Z^{(t)} = \sum_{i=1}^{l} F_i^{(t)}, \tag{9.117}$$

where

$$F_i^{(t)} = -N_i kT \ln [(2\pi m_i kT)^{3/2} VeN_i^{-1} h^{-3}]. \tag{9.118}$$

It follows now from Eqn. (63) that

$$S^{(t)} = -\sum_{i=1}^{l} \left(\frac{\partial F_i^{(t)}}{\partial T} \right)_V = \sum_{i=1}^{l} S_i^{(t)}, \tag{9.119}$$

where

$$S_i^{(t)} = N_i k \ln [(2\pi m_i kT)^{3/2} VeN_i^{-1} h^{-3}] + \tfrac{3}{2} N_i k. \tag{9.120}$$

9.6. Important Notes

In the present chapter, the classical method of the probabilistic treatment of pseudoclosed mechanical systems was set forth in outline. This led, in particular, to the construction of the microcanonical and canonical distributions and to the elucidation of the probabilistic meaning of some thermodynamic quantities. All the results obtained here were based merely on the fact that properties of mechanical systems proved to be Hamiltonian random variables of probabilistics (cf. Sect. 6.2); neither ergodic hypothesis, nor Liouville's equation turned out to be necessary in the course of considerations, and Gibbs's paradox did not appear at all.

In the next chapter we shall consider quantum mechanics, that is, the quantum approach to the probabilistic treatment of mechanical systems, which in some respects is more general than the classical method. In Sections 6.2 and 8.1 it has already been indicated that when both methods are suitable for the consideration of a probabilistic problem, they should lead to the same results. The range, in which the results of the quantum approach coincide with those of the classical method, is that one which actually should be called classical limits of quantum mechanics. This fact immediately leads to the conclusion that classical limits of quantum mechanics must be nothing more than classical statistical mechanics.

Therefore, it should be noted emphatically that it is precisely the distributions determined by the microcanonical one that are the classical limits of the respective quantum mechanical distributions given by the corresponding ψ-function for a conservative mechanical system. Whenever this assertion can be checked it proves to be correct. An illustration of this fact has actually already been given in the example of the one-dimensional harmonic oscillator (see Sects. 8.3, 6.2, and 6.4). We will turn to this point once again elsewhere in Chapters 10 and 14.

Suggested Reference

1. Huang, K.: 1963, *Statistical Mechanics*, Wiley, New York. This book contains enough of thermodynamics and a version of the conventional approach for the construction of statistical mechanics.

Chapter 10

Quantum Mechanics

10.1. Principal Propositions

As a matter of fact, quantum mechanics deals with the quantum approach to the probabilistic treatment of mechanical systems and is, hence, a particular domain of probabilistic physics. Therefore, its principal propositions are those of probabilistic physics, in general, supplemented by statements reflecting the specific features of the quantum approach (see Sects. 8.1 and 5.4). Thus, the principal propositions of quantum mechanics are as follows:

1. All physical quantities (properties) are random variables.
2. A state of an abstract physical system is characterized by definite probability distributions for all the physical quantities possessed by the system. We use the term physical quantity, bearing in mind that it is actually a random variable from the standpoint of probabilistics.
3. A state is described by an appropriate vector or a density operator in some linear space.
4. To every physical quantity there corresponds at least one Hermitian operator defined in that space, whose set of eigenvalues (spectrum) coincides with the totality of values of the physical quantity; each of the eigenvectors of the operator describing the state in which the physical quantity has a definite value, namely, that coinciding with the respective eigenvalue of the operator.

Furthermore, all the results of Section 5.4 are immediately applicable to any abstract physical system. It follows from them, in particular, that in a state of an abstract physical system, the probability density of any physical quantity is proportional to the magnitude squared of the coefficient in the expansion of the state vector describing the state in terms of the eigenvectors of the Hermitian operator corresponding to the physical quantity. The calculation of the mathematical expectations of any functions of physical quantities can be performed proceeding from Eqns. (5.109) or (5.110).

10.2. Operators for Physical Quantities

Since physical quantities possessed by a mechanical system obey the corresponding Lagrangian and Hamiltonian equations, they are Hamiltonian random variables (cf. Sect. 6.2). Therefore, one can use the results of Section 6.4 in order to choose the appropriate Hermitian operators for those physical quantities. We now consider some of them.

10.2.1. *Coordinates and Momenta*

In the coordinate representation, that is in the space of functions of all the coordinates q_j ($j = 1, 2, \ldots, n$) of a mechanical system with n degrees of freedom, the operator $\hat{q}_j$ of multiplication by q_j

$$\hat{q}_j = q_j \hat{I} \tag{10.1}$$

should be taken, in accordance with Eqn. (6.85), as the Hermitian operator for the coordinate q_j.

In accordance with Eqns. (6.86) and (6.89), the operator

$$\hat{p}_j = -i\hbar \left(\frac{\partial}{\partial q_j} + \frac{1}{2} \frac{\mathrm{d} \ln f(q_j)}{\mathrm{d} q_j} \hat{I} \right) \tag{10.2}$$

conforms to the momentum p_j canonically conjugate to the coordinate q_j, in the same representation. Here $f(q_j)$ is the corresponding weight function of Eqn. (6.87) for the case under consideration; $\hbar$ is the Planck constant h divided by 2π, and it is precisely this value of the α-constant in Eqn. (6.86) which has been determined experimentally. Let us consider some examples.

For the Cartesian coordinates x, y, and z of a particle, $f = 1$, and Eqn. (2) yields

$$\hat{p}_\alpha = -i\hbar \frac{\partial}{\partial \alpha} \quad (\alpha = x, y, z). \tag{10.3}$$

For the spherical coordinate r, $f = r^2$, and from Eqn. (2) we get

$$\hat{p}_r = -i\hbar \left(\frac{\partial}{\partial r} + \frac{1}{r} \hat{I} \right). \tag{10.4}$$

For the spherical coordinate θ, $f = \sin\theta$, and hence, as follows from Eqn. (2),

$$\hat{p}_\theta = -i\hbar \left(\frac{\partial}{\partial \theta} + \tfrac{1}{2} \cot \theta \hat{I} \right). \tag{10.5}$$

For the spherical coordinate φ, $f = 1$, and from Eqn. (2) we get

$$\hat{p}_\varphi = -i\hbar \frac{\partial}{\partial \varphi}. \tag{10.6}$$

If q_j takes on values in the interval (a, b), then in order for the operator $\hat{p}_j$ as given in Eqn. (2) to be Hermitian, its domain of definition (hermiticity range) should consist of functions of q_j, such that for any pair $\psi_1(q_j)$ and $\psi_2(q_j)$ of them, the condition

$$\psi_1(a)\bar{\psi}_2(a)f(a) = \psi_1(b)\bar{\psi}_2(b)f(b) \tag{10.7}$$

is satisfied (see Sect. 6.4).

For the Cartesian coordinates x, y, and z, this condition takes the form

$$\psi_1(a)\bar{\psi}_2(a) = \psi_1(b)\bar{\psi}_2(b). \tag{10.8}$$

It is satisfied when the ψ-functions are of the form

$$\psi_\alpha = \text{const} \sum_j \gamma_j \exp\left(im_j \frac{2\pi\alpha}{b-a}\right) \quad (\alpha = x, y, z) \tag{10.9}$$

where m_j are any whole numbers, including zero, and γ_j are arbitrary constants [cf. Eqn. (6.103)], or when $\psi(a) = \psi(b) = 0$. If $a = -\infty$ and $b = \infty$, Eqn. (8) is also satisfied when

$$\lim_{|\alpha| \to \infty} \psi(\alpha) = 0. \tag{10.10}$$

For the spherical coordinate r, for which $a = 0$ and $b = \infty$, Eqn. (7) is satisfied for ψ-functions which are finite at $r = 0$ and vanish at $r = \infty$ faster than r^{-1}, so that

$$\lim_{r \to 0} \psi(r)r = \lim_{r \to \infty} \psi(r)r = 0. \tag{10.11}$$

Since $a = 0$, $b = \pi$, and $f = \sin\theta$ for the spherical coordinate θ, Eqn. (7) is satisfied for ψ-functions finite at $\theta = 0$ and $\theta = \pi$.

For the spherical coordinate φ, for which $a = 0$, $b = 2\pi$, and $f = 1$, Eqn. (7) is satisfied for ψ-functions of the form

$$\psi(\varphi) = \text{const} \sum_j \gamma_j \exp(im_j\varphi) \tag{10.12}$$

only, where m_j are whole numbers, including zero, and γ_j are arbitrary constants.

The above equations determine the respective hermiticity ranges for the corresponding momentum operators. The eigenfunctions of the momentum operator $\hat{p}_j$ [see Eqn. (2)] are of the form given in Eqn. (6.103), namely,

$$\psi_m = (b-a)^{-1/2} \exp\left[im(b-a)^{-1} 2\pi q_j\right] f(q_j), \tag{10.13}$$

where $b - a$ is the range of values of the coordinate q_j, and

$$p_j = m\hbar \frac{2\pi}{b-a} \quad (m = 0, \pm 1, \pm 2, \dots) \tag{10.14}$$

are the respective eigenvalues of the operator $\hat{p}_j$.

10.2.2. *Time and Energy*

Since time, t, is similar to a Cartesian coordinate, the operator for it is

$$\hat{t} = t\hat{I} \tag{10.15}$$

and the operator for energy E canonically conjugate to t with the sign reversed is, in view of Eqn. (2)

$$\hat{E} = i\hbar \frac{\partial}{\partial t}. \tag{10.16}$$

Since t takes on values in the interval $(-\infty, \infty)$, the operator $\hat{E}$ as given in Eqn. (16) is Hermitian, in particular, on functions of the form

$$\psi(t) = \text{const} \exp(i\beta t), \tag{10.17}$$

where β is an arbitrary real constant, as well as on functions vanishing at $|t| = \infty$.

10.2.3. *Hamiltonian*

In classical mechanics the Hamiltonian function, H, is a sum of the kinetic energy, K, and the potential energy, U:

$$H = K + U. \tag{10.18}$$

Hence the operator $\hat{H}$ corresponding to H should be of the form

$$\hat{H} = \hat{K} + \hat{U}, \tag{10.19}$$

where $\hat{K}$ is the operator for K and $\hat{U}$ is that for U; the hermiticity range for $\hat{H}$ having to be the intersection of the hermiticity ranges of the operators $\hat{K}$ and $\hat{U}$.

The construction of the operator $\hat{U}$ offers no difficulty, as a rule. If, for instance, the potential energy U is a certain function of the coordinates of a mechanical system and time, $U = U(q_1, q_2, \ldots, q_n, t)$, then the operator $\hat{U}$ in the coordinate representation is

$$\hat{U} = U(q_1, \ldots, q_n, t)\hat{I}. \tag{10.20}$$

Since the hermiticity range of the operator $\hat{U}$ obviously is the total space of the appropriate functions of all the coordinates involved, the hermiticity range of the operator $\hat{H}$ as given in Eqn. (19) coincides with that of the operator $\hat{K}$.

In the nonrelativistic case, the kinetic energy of a mechanical system consisting of n particles of masses m_i $(i = 1, \ldots, n)$ is

$$K = \frac{1}{2} \sum_{i=1}^{n} m_i(\dot{x}_i^2 + \dot{y}_i^2 + \dot{z}_i^2) = \sum_{i=1}^{n} \frac{p_{ix}^2 + p_{iy}^2 + p_{iz}^2}{2m_i}, \tag{10.21}$$

where $\dot{\alpha}_i$ and $p_{i\alpha}$ are the αth component of the velocity and the momentum canonically conjugate to the αth Cartesian coordinate of ith particle in a Cartesian coordinate system chosen $(\alpha = x, y, z)$.

Making use of Eqn. (3), we get, in view of Eqn. (21),

$$\hat{K} = \sum_{i=1}^{n} \frac{\hat{p}_{ix}^2 + \hat{p}_{iy}^2 + \hat{p}_{iz}^2}{2m_i} = -\frac{\hbar^2}{2} \sum_{i=1}^{n} m_i^{-1} \left(\frac{\partial^2}{\partial x_i^2} + \frac{\partial^2}{\partial y_i^2} + \frac{\partial^2}{\partial z_i^2} \right)$$

$$= -\frac{\hbar^2}{2} \sum_{i=1}^{n} m_i^{-1} \nabla_i^2, \tag{10.22}$$

where $\nabla_i = (\partial/\partial x_i, \partial/\partial y_i, \partial/\partial z_i)$ is the so-called del (or nabla) operator for ith particle. The hermiticity range of $\hat{K}$ in the case under consideration is obviously the intersection of those of all the operators $\hat{p}_{i\alpha}^2$ ($i = 1, \ldots, n$; $\alpha = x, y, z$) $\hat{K}$ consists of, in the space of functions of all the coordinates; those hermiticity ranges being determined by the intervals $(a_{i\alpha}, b_{i\alpha})$ the corresponding Cartesian coordinates take their values in.

When going over to some generalized coordinates q_j ($j = 1, \ldots, 3n$) related to the Cartesian coordinates α_i ($\alpha = x, y, z; i = 1, \ldots, n$) by

$$q_j = q_j(\alpha_i) \quad (j = 1, \ldots, 3n), \tag{10.23}$$

the kinetic energy takes on the form

$$K = \frac{1}{2} \sum_{j,l} g_{jl} p_j p_l. \tag{10.24}$$

Here the p_j are the generalized momenta canonically conjugate to the q_j, and g_{jl} are certain functions of coordinates, namely,

$$g_{jl} = \sum_{i=1}^{n} \sum_{\alpha = x,y,z} m_i^{-1} \frac{\partial q_j}{\partial \alpha_i} \frac{\partial q_l}{\partial \alpha_i}. \tag{10.25}$$

It is obviously impossible to get a Hermitian operator $\hat{K}$ corresponding to K as given in Eqn. (24) by directly replacing the physical quantities in it by the respective Hermitian operators, since the operators for a coordinate and the corresponding momentum do not commute (see Sects. 5.3 and 6.4). Therefore, in order to find the correct Hermitian operator $\hat{K}$ in any coordinates other than the Cartesian ones, it is necessary to transform the expression for $\hat{K}$ as given in Eqn. (22) to the new coordinates. The simplest way to do that is to represent the del operators, ∇_i, in terms of the new coordinates and then to introduce the results obtained into Eqn. (22). The components of ∇_i are $\partial/\partial\alpha_i$ ($\alpha = x, y, z$), and in view of Eqn. (23), we have

$$\frac{\partial}{\partial \alpha_i} = \sum_{j=1}^{3n} \frac{\partial q_j}{\partial \alpha_i} \frac{\partial}{\partial q_j}. \tag{10.26}$$

Hence,

$$\nabla_i^2 = \sum_{\alpha=x,y,z} \sum_{j=1}^{3n} \frac{\partial q_j}{\partial \alpha_i} \frac{\partial}{\partial q_j} \sum_{l=1}^{3n} \frac{\partial q_l}{\partial \alpha_i} \frac{\partial}{\partial q_l}$$

$$= \sum_{\alpha=x,y,z} \sum_{j,l=1}^{3n} \left(\frac{\partial q_j}{\partial \alpha_i} \frac{\partial q_l}{\partial \alpha_i} \frac{\partial^2}{\partial q_j \partial q_l} + \frac{\partial q_j}{\partial \alpha_i} \frac{\partial}{\partial q_j} \left(\frac{\partial q_l}{\partial \alpha_i} \right) \frac{\partial}{\partial q_l} \right). \tag{10.27}$$

By introducing this result into Eqn. (22) we get

$$\hat{K} = -\frac{\hbar^2}{2} \left[\sum_{j,l=1}^{3n} g_{jl} \frac{\partial^2}{\partial q_j \partial q_l} + \sum_{l=1}^{3n} f_l \frac{\partial}{\partial q_l} \right]. \tag{10.28}$$

Here g_{jl} are those as given in Eqn. (25), while

$$f_l = \sum_{i=1}^{n} \sum_{\alpha=x,y,z} \sum_{j=1}^{3n} m_i^{-1} \frac{\partial q_j}{\partial \alpha_i} \frac{\partial}{\partial q_j} \left(\frac{\partial q_l}{\partial \alpha_i} \right). \tag{10.29}$$

In Eqn. (28) the coefficients g_{jl} and f_l are supposed to be expressed in terms of the q_j. Their calculation is rather simple, provided both the functions $q_j(\alpha_i)$ and $\alpha_i(q_j)$ are known. Otherwise certain difficulties arise, which can be overcome, nevertheless, in some particular cases.

We will consider here only the case of spherical (polar) coordinates $q_1 = r$, $q_2 = \theta$, $q_3 = \varphi$ for one particle of mass m, for which no difficulties in calculating the g_{jl} and f_l arise. A method of construction of a general rotation-vibration Hamiltonian for a molecule is given in Appendix 3. We thus have

$$x = r \sin\theta \cos\varphi; \qquad y = r \sin\theta \sin\varphi; \qquad z = \cos\theta \tag{10.30}$$

and, correspondingly,

$$q_1 = r = (x^2 + y^2 + z^2)^{-1/2}; \qquad q_2 = \theta = \arccos z/(x^2 + y^2 + z^2)^{1/2};$$
$$q_3 = \varphi = \arctan y/x. \tag{10.31}$$

Hence

$$\frac{\partial r}{\partial x} = \sin\theta \cos\varphi; \qquad \frac{\partial r}{\partial y} = \sin\theta \sin\varphi; \qquad \frac{\partial r}{\partial z} = \cos\theta;$$

$$\frac{\partial \theta}{\partial x} = \frac{\cos\theta \cos\varphi}{r}; \qquad \frac{\partial \theta}{\partial y} = \frac{\cos\theta \sin\varphi}{r}; \qquad \frac{\partial \theta}{\partial z} = -\frac{\sin\theta}{r}; \tag{10.32}$$

$$\frac{\partial \varphi}{\partial x} = -\frac{\sin\varphi}{r \sin\theta}; \qquad \frac{\partial \varphi}{\partial y} = \frac{\cos\varphi}{r \sin\theta}; \qquad \frac{\partial \varphi}{\partial z} = 0.$$

Simple calculations by Eqns. (25) and (29) give

$$g_{11} = m^{-1}; \qquad g_{22} = (mr^2)^{-1}; \qquad g_{33} = (mr^2 \sin^2 \theta)^{-1}; \qquad g_{12} = g_{13} = g_{23} = 0;$$
$$f_1 = m^{-1} 2r^{-1}; \qquad f_2 = m^{-1} \cot \theta r^{-2}; \qquad f_3 = 0. \tag{10.33}$$

Therefore, Eqn. (28) yields

$$\hat{K} = -\frac{\hbar^2}{2m}\left[\frac{\partial^2}{\partial r^2} + \frac{2}{r}\frac{\partial}{\partial r} + \frac{1}{r^2}\left(\frac{\partial^2}{\partial \theta^2} + \cot\theta \frac{\partial}{\partial \theta}\right) + \frac{1}{r^2 \sin^2\theta}\frac{\partial^2}{\partial \varphi^2}\right]$$
$$= -\frac{\hbar^2}{2m}\left\{\frac{1}{r^2}\frac{\partial}{\partial r}\left(r^2 \frac{\partial}{\partial r}\right) + \frac{1}{r^2}\left[\frac{1}{\sin\theta}\frac{\partial}{\partial\theta}\left(\sin\theta \frac{\partial}{\partial\theta}\right) + \frac{1}{\sin^2\theta}\frac{\partial^2}{\partial\varphi^2}\right]\right\}. \tag{10.34}$$

This is the well-known form of the kinetic part of the Hamiltonian for one particle in spherical coordinates. It is interesting to compare it with the classical kinetic energy, K, which is, in view of Eqns. (24) and (33),

$$K = \frac{1}{2m}\left[p_r^2 + \frac{1}{r^2}\left(p_\theta^2 + \frac{p_\varphi^2}{\sin^2\theta}\right)\right]. \tag{10.35}$$

One might think that it is possible to get $\hat{K}$ from Eqn. (35) by replacing the physical quantities there by the corresponding Hermitian operators. But that is not the case. Indeed, Eqn. (2) yields

$$\hat{p}_j^2 = -\hbar^2\left(\frac{\partial}{\partial q_j} + \frac{1}{2}\frac{f'}{f}\right)\left(\frac{\partial}{\partial q_j} + \frac{f'}{f}\right)$$
$$= -\hbar^2\left[\frac{1}{f}\frac{\partial}{\partial q_j}\left(f \frac{\partial}{\partial q_j}\right) + \frac{1}{2}\frac{f''}{f} - \frac{1}{4}\left(\frac{f'}{f}\right)^2\right]. \tag{10.36}$$

We can easily see that only for $q_1 = r$ and $q_3 = \varphi$, Eqn. (36) gives the operators coinciding with the respective ones in Eqn. (34), whereas for $q_2 = \theta$, the operator given by Eqn. (36) differs from the one in Eqn. (34) by the additional term $-\hbar^2(1 + \sec^2\theta)/4$. This fact should warn one of any attempt to obtain the operator $\hat{K}$ in coordinates other than Cartesian ones by directly replacing the physical quantities in the corresponding classical expression for kinetic energy by the respective operators. It should be emphasized once again the only correct way for obtaining $\hat{K}$ is to transform the right-hand side of Eqn. (22) to the coordinates chosen.

10.2.4. *Angular Momentum*

We shall dwell on the consideration of the angular momentum in more detail, since it possesses some peculiarities, as has already been mentioned in Section 8.4.

(a) *Concrete Mechanical Systems*. Angular momentum is connected to rotational motion of a concrete mechanical system, and hence the specific coordinates describing such motion would be more appropriate for its definition. But, as a rule, it is defined originally in some Cartesian coordinates. For a mechanical system consisting of n particles, the angular momentum with respect to the α-axis of a certain coordinate system, $\mathcal{M}_\alpha$, is defined by

$$\mathcal{M}_\alpha = \sum_{i=1}^{n} (\beta_i p_\gamma^{(i)} - \gamma_i p_\beta^{(i)}) \quad (\alpha \neq \beta \neq \gamma). \tag{10.37}$$

Here α, β, and γ take on any of the values x, y, or z in the indicated succession; β_i and γ_i are the corresponding Cartesian coordinates of ith particle; $p_\beta^{(i)}$ and $p_\gamma^{(i)}$ are the respective momenta.

By making use of the matrices

$$\Sigma_x = \begin{Vmatrix} 0 & 0 & 0 \\ 0 & 0 & 1 \\ 0 & -1 & 0 \end{Vmatrix}; \quad \Sigma_y = \begin{Vmatrix} 0 & 0 & -1 \\ 0 & 0 & 0 \\ 1 & 0 & 0 \end{Vmatrix}; \quad \Sigma_z = \begin{Vmatrix} 0 & 1 & 0 \\ -1 & 0 & 0 \\ 0 & 0 & 0 \end{Vmatrix}, \tag{10.38}$$

Eqn. (37) can be rewritten as

$$\mathcal{M}_\alpha = \sum_{i=1}^{n} \tilde{r}^{(i)} \Sigma_\alpha p^{(i)}, \tag{10.39}$$

where

$$r^{(i)} = (x_i, y_i, z_i) \tag{10.40}$$

and

$$p^{(i)} = (p_x^{(i)}, p_y^{(i)}, p_z^{(i)}) = m_i \dot{r}^{(i)} \tag{10.41}$$

are the corresponding column matrices.

The $\mathcal{M}_\alpha$ obviously depend on the choice of the origin of the frame of reference. To avoid the ambiguity caused by this fact the origin is supposed to coincide with the center of mass of the mechanical system. Then, any infinitesimal motion of the mechanical system can generally be described, at any time instant, by the changes in the three Euler's angles φ, θ, and χ, which determine the spatial instantaneous orientation of the rotating Cartesian coordinate system attached to the mechanical system relative to the originally chosen inertial rest coordinate system (laboratory system), and by some internal coordinates, which determine mutual positions of the particles relative to each other.

The radius vectors $\mathbf{r}^{(i)}$ and $\mathbf{r}^{(i)'}$ of ith particle in the rest and rotating coordinate systems, respectively, are connected by

$$\mathbf{r}^{(i)'} = A_0^{(\Omega)}\mathbf{r}^{(i)}, \qquad \mathbf{r}^{(i)} = \tilde{A}_0^{(\Omega)}\mathbf{r}^{(i)'}, \tag{10.42}$$

where $A_0^{(\Omega)}$ is a corresponding 3 by 3 orthogonal matrix. The same transformations are valid, of course, for the respective column matrices $r^{(i)}$ and $r^{(i)'}$. For a certain choice of the Euler's angles this matrix is

$$A_0^{(\Omega)} = \begin{Vmatrix} \cos\varphi\cos\chi - \sin\varphi\cos\theta\sin\chi & \sin\varphi\cos\chi + \cos\varphi\cos\theta\sin\chi & \sin\theta\sin\chi \\ -(\cos\varphi\sin\chi + \sin\varphi\cos\theta\cos\chi) & -\sin\varphi\sin\chi + \cos\varphi\cos\theta\cos\chi & \sin\theta\cos\chi \\ \sin\varphi\sin\theta & -\cos\varphi\sin\theta & \cos\theta \end{Vmatrix} \tag{10.43}$$

(see Ref. [1]).

Equation (42) yields

$$\dot{r}^{(i)} = \dot{\tilde{A}}_0^{(\Omega)} r^{(i)'} + \tilde{A}_0^{(\Omega)}\dot{r}^{(i)'}. \tag{10.44}$$

Since we are interested in rotations only, the second term in Eqn. (44) will be omitted. We thus have

$$\dot{r}^{(i)} = \frac{\partial \tilde{A}_0^{(\Omega)}}{\partial \tilde{\Omega}} r^{(i)'}\dot{\Omega}, \tag{10.45}$$

where

$$\Omega = (\varphi, \theta, \chi); \qquad \dot{\Omega} = (\dot{\varphi}, \dot{\theta}, \dot{\chi}) \tag{10.46}$$

and

$$\frac{\partial \tilde{A}_0^{(\Omega)}}{\partial \tilde{\Omega}} r^{(i)'} = \frac{\partial r^{(i)}}{\partial \tilde{\Omega}} = \left\Vert \frac{\partial \tilde{A}_0^{(\Omega)}}{\partial \varphi} r^{(i)'} \,\vdots\, \frac{\partial \tilde{A}_0^{(\Omega)}}{\partial \theta} r^{(i)'} \,\vdots\, \frac{\partial \tilde{A}_0^{(\Omega)}}{\partial \chi} r^{(i)'} \right\Vert \tag{10.47}$$

obviously is a composite 3 by 3 matrix. By making use of Eqns. (43) and (38) we get from Eqns. (45) and (47)

$$\dot{r}_i^{(i)} = \tilde{A}_0^{(\Omega)} A_i^{(r)'} C^{(\Omega)} \dot{\Omega} \tag{10.48}$$

where

$$A_i^{(r)'} = -\left\Vert \Sigma_x r^{(i)'} \,\vdots\, \Sigma_y r^{(i)'} \,\vdots\, \Sigma_z r^{(i)'} \right\Vert = \begin{Vmatrix} 0 & z_i' & -y_i' \\ -z_i' & 0 & x_i' \\ y_i' & -x_i' & 0 \end{Vmatrix} \tag{10.49}$$

and

$$C^{(\Omega)} = \begin{Vmatrix} \sin\theta \sin\chi & \cos\chi & 0 \\ \sin\theta\cos\chi & -\sin\chi & 0 \\ \cos\theta & 0 & 1 \end{Vmatrix}. \tag{10.50}$$

For rotational kinetic energy, $K^{(r)}$, Eqn. (21) now yields

$$\begin{aligned} 2K^{(r)} &= \sum_{i=1}^{n} m_i \tilde{\dot{\Omega}} \tilde{C}^{(\Omega)} \tilde{A}_i^{(r)'} A_i^{(r)'} C^{(\Omega)} \dot{\Omega} \\ &= \tilde{\dot{\Omega}} \tilde{C}^{(\Omega)} I'' C^{(\Omega)} \dot{\Omega}, \end{aligned} \tag{10.51}$$

where

$$I'' = \sum_{i=1}^{n} m_i \tilde{A}_i^{(r)'} A_i^{(r)'} = \begin{Vmatrix} I''_{xx} & -I''_{xy} & -I''_{xz} \\ -I''_{xy} & I''_{yy} & -I''_{yz} \\ -I''_{xz} & -I''_{yz} & I''_{zz} \end{Vmatrix} \tag{10.52}$$

is the instantaneous moment of inertia matrix whose elements are

$$\begin{aligned} I''_{\alpha\alpha} &= \sum_i m_i(\beta_i'^2 + \gamma_i'^2) \quad (\alpha \neq \beta \neq \gamma), \\ I''_{\alpha\beta} &= \sum_i m_i \alpha_i' \beta_i' \quad (\alpha \neq \beta) \end{aligned} \tag{10.53}$$

(α, β, and γ take on any of the values x, y, or z).

For the momentum

$$\pi = (p_\varphi, p_\theta, p_\chi) \tag{10.54}$$

canonically conjugate to Ω we obtain from Eqn. (51)

$$\pi = \frac{\partial K^{(r)}}{\partial \tilde{\dot{\Omega}}} = \tilde{C}^{(\Omega)} I'' C^{(\Omega)} \dot{\Omega}, \tag{10.55}$$

whence

$$\dot{\Omega} = C^{(\Omega)'} G'' \tilde{C}^{(\Omega)'} \pi. \tag{10.56}$$

Here

$$G'' = I''^{-1} \tag{10.57}$$

and

$$C^{(\Omega)'} = C^{(\Omega)^{-1}} = \begin{Vmatrix} \dfrac{\sin\chi}{\sin\theta} & \dfrac{\cos\chi}{\sin\theta} & 0 \\ \cos\chi & -\sin\chi & 0 \\ -\sin\chi\cot\theta & -\cos\chi\cot\theta & 1 \end{Vmatrix}. \tag{10.58}$$

In view of Eqn. (56), we get from Eqn. (51)

$$2K^{(r)} = \tilde{\pi}C^{(\Omega)'}G''\tilde{C}^{(\Omega)'}\pi. \tag{10.59}$$

This is the expression for the instantaneous rotational kinetic energy of a concrete mechanical system in terms of the momenta canonically conjugate to Euler's angles.

In order to express this energy in terms of the angular momenta $\mathcal{M}_\alpha$ ($\alpha = x, y, z$) referred to the axes of the initial rest coordinate system, we calculate, first, the angular momenta $\mathcal{M}'_\alpha$ ($\alpha = x, y, z$) referred to the axes of an instantaneous rotating coordinate system, and then make use of the relation

$$\mathcal{M} = \tilde{A}_0^{(\Omega)}\mathcal{M}' \tag{10.60}$$

connecting column matrices $\mathcal{M} = (\mathcal{M}_x, \mathcal{M}_y, \mathcal{M}_z)$ and $\mathcal{M}' = (\mathcal{M}'_x, \mathcal{M}'_y, \mathcal{M}'_z)$. For $\mathcal{M}'_\alpha$ we have, with due allowance for Eqns. (41) and (42), instead of Eqn. (39),

$$\mathcal{M}'_\alpha = \sum_i m_i \tilde{r}^{(i)'}\Sigma_\alpha A_0^{(\Omega)}\dot{r}^{(i)}. \tag{10.61}$$

But it follows from Eqn. (49), in view of $\tilde{\Sigma}_\alpha = -\Sigma_\alpha$, that $\tilde{r}^{(i)'}\Sigma_\alpha$ is the αth row of the matrix $\tilde{A}_i^{(r)}$. Hence Eqn. (61), in view of Eqns. (48) and (52), yields

$$\mathcal{M}' = I''C^{(\Omega)}\dot{\Omega} \tag{10.62}$$

whence, in view of Eqn. (60),

$$\mathcal{M} = \tilde{A}_0^{(\Omega)}I''C^{(\Omega)}\dot{\Omega}, \tag{10.63}$$

and, in view of Eqn. (56),

$$\mathcal{M} = \tilde{A}_0^{(\Omega)}\tilde{C}^{(\Omega)'}\pi. \tag{10.64}$$

We can now rewrite Eqn. (59) as

$$2K^{(r)} = \tilde{\mathcal{M}}\tilde{A}_0^{(\Omega)}G''A_0^{(\Omega)}\mathcal{M}. \tag{10.65}$$

This is the expression for the instantaneous rotational kinetic energy in terms of angular momenta referred to the initial coordinate system.

On the other hand, by introducing

$$\omega = (\omega_x, \omega_y, \omega_z) = \tilde{A}_0^{(\Omega)}C^{(\Omega)}\dot{\Omega}, \tag{10.66}$$

we get from Eqns. (51) and (63)

$$2K^{(r)} = \tilde{\omega}\tilde{A}_0^{(\Omega)}I''A_0^{(\Omega)}\omega \tag{10.67}$$

and

$$\mathcal{M} = \tilde{A}_0^{(\Omega)} I'' A_0^{(\Omega)} \omega. \tag{10.68}$$

It is seen from these two equations that

$$\mathcal{M} = \frac{\partial K^{(r)}}{\partial \tilde{\omega}} \tag{10.69}$$

that is

$$\mathcal{M}_x = \frac{\partial K^{(r)}}{\partial \omega_x}; \qquad \mathcal{M}_y = \frac{\partial K^{(r)}}{\partial \omega_y}; \qquad \mathcal{M}_z = \frac{\partial K^{(r)}}{\partial \omega_z}. \tag{10.70}$$

What do Eqns. (70) mean? The column matrix $\dot{\Omega} = (\dot{\varphi}, \dot{\theta}, \dot{\chi})$ represents the vector of instantaneous angular velocity $\dot{\Omega}$; its elements being the velocity components along the z-axis of the original rest coordinate system, along the x-axis of the first intermediate coordinate system, and along the z-axis of the final coordinate system, respectively. From Eqn. (66) it follows that ω_x, ω_y, and ω_z are the components of the same vector $\dot{\Omega}$ along some other axes. One can make sure that these are the respective axes of the initial rest coordinate system (cf. also Ref. [1]). Thus, Eqns. (70) determine the angular momenta $\mathcal{M}_\alpha$ ($\alpha = x, y, z$) referred to the axes of the initial rest coordinate system as quantities canonically conjugate to the angles φ_x, φ_y, and φ_z of rotation around these axes.

It is necessary to stress, however, that it is impossible to determine uniquely the spatial orientation of a rotating coordinate system with respect to the initial rest coordinate system by the three angles of rotation around the respective axes of the latter, since the final orientation depends on the sequence in which the rotations have been performed. Hence, all three angles φ_x, φ_y, and φ_z cannot serve simultaneously as the generalized coordinates of a mechanical system. If one of the angles has already been chosen as a generalized coordinate, none of the others can simultaneously be chosen as such. Therefore, the angular momenta $\mathcal{M}_x$, $\mathcal{M}_y$, and $\mathcal{M}_z$ determined by Eqns. (70) cannot be simultaneously generalized momenta of a mechanical system, either. It is for this reason that different angular momenta $\mathcal{M}_\alpha$ and $\mathcal{M}_\beta$ do not obey the conditions given by Eqn. (6.18), as would have to be the case if they were generalized momenta simultaneously.

One can easily make sure, by using Eqns. (37) and (6.15), that instead of Eqn. (6.18) they satisfy the relations

$$[\mathcal{M}_\alpha, \mathcal{M}_\beta] = \pm \mathcal{M}_\gamma, \tag{10.71}$$

the sign depending on whether the permutation $\alpha\beta\gamma$ is even or odd.

Thus, if one of the angular momenta $\mathcal{M}_x$, $\mathcal{M}_y$, or $\mathcal{M}_z$ has already been taken as a generalized momentum of a mechanical system, none of the other two can simultaneously be chosen as such. Certainly, the choice of the angular momentum should correspond to that of the angle of rotation.

The results obtained are very important and in agreement with what was discussed on this matter in Section 8.4. Indeed, it is easy to check, by making use of Eqns. (21), (37), and (6.15), that

$$[K, \mathscr{M}_\alpha] = 0 \quad (\alpha = x, y, z). \tag{10.72}$$

This means that if the potential energy of the mechanical system does not depend on the angles $\varphi_x, \varphi_y, \varphi_z$, then any one of the angular momenta $\mathscr{M}_x, \mathscr{M}_y, \mathscr{M}_z$ is conserved separately. But Eqn. (72) cannot give the conservation law for all three angular momenta in common, since the choice of one of them inevitably excludes any possibility of the consideration of the rest of them.

At the same time, as follows from Eqn. (72), the quantity

$$\mathscr{M}^2 = \mathscr{M}_x^2 + \mathscr{M}_y^2 + \mathscr{M}_z^2 \tag{10.73}$$

is conserved as well (under the above condition). This quantity is known to be called the magnitude squared of the angular momentum. But it is more expedient to think of it as of the square of the maximum value of the angular momentum relative to any axis (see Sect. 8.4).

By making use of Eqn. (71), one can check, too, that for any $\mathscr{M}_\alpha$ $(\alpha = x, y, z)$

$$[\mathscr{M}^2, \mathscr{M}_\alpha] = 0. \tag{10.74}$$

This does not mean, however, that $\mathscr{M}^2$ is a generalized momentum. Indeed, the kinetic energy, K, must be, in view of Eqn. (24), quadratic in generalized momenta, which condition obviously cannot be satisfied for $\mathscr{M}^2$. On the other hand, $\mathscr{M}^2$ coincides with $K^{(r)}$ as given in Eqn. (65), when G is a unit matrix.

Since any direction in the three-dimensional physical space can be chosen as z-axis of a Cartesian coordinate system, we can always take $\mathscr{M}_z$ as the only generalized momentum canonically conjugate to the only generalized coordinate φ_z which obviously coincides with Euler's angle φ. From Eqns. (64), (43), and (58) we get

$$\begin{aligned} \mathscr{M}_x &= -\sin\varphi \cot\theta p_\varphi + \cos\varphi p_\theta + \frac{\sin\varphi}{\sin\theta} p_\chi, \\ \mathscr{M}_y &= \cos\varphi \cot\theta p_\varphi + \sin\varphi p_\theta - \frac{\cos\varphi}{\sin\theta} p_\chi, \\ \mathscr{M}_z &= p_\varphi. \end{aligned} \tag{10.75}$$

Hence, $\mathscr{M}_z$ coincides with the generalized momentum p_φ canonically conjugate to Euler's angle φ as should be the case, indeed. For $\mathscr{M}^2$ we obtain from Eqns. (73) and (75)

$$\mathscr{M}^2 = \frac{1}{\sin^2\theta}(p_\varphi^2 + p_\chi^2 - 2\cos\theta p_\varphi p_\chi) + p_\theta^2. \tag{10.76}$$

For $\chi = 0$, Eqn. (76) goes over into

$$\mathscr{M}^2 = \frac{p_\varphi^2}{\sin^2 \theta} + p_\theta^2 \tag{10.77}$$

and Eqn. (35) can be rewritten as

$$K = \frac{1}{2m}\left(p_r^2 + \frac{\mathscr{M}^2}{r^2}\right). \tag{10.78}$$

(b) *Abstract Mechanical Systems* (*Quantum Case*). It follows from the above that the Hermitian operator for $\mathscr{M}_z$ is that given in Eqn. (6), that is,

$$\hat{\mathscr{M}}_z = \hat{p}_\varphi = -i\hbar \frac{\partial}{\partial \varphi} \tag{10.79}$$

whose hermiticity range consists of ψ-functions of the form given in Eqn. (12). The eigenfunctions of $\mathscr{M}_z$ are, in view of Eqn. (13),

$$\psi_m = (2\pi)^{-1/2} \exp (im\varphi), \tag{10.80}$$

the respective eigenvalues being

$$\mathscr{M}_z = m\hbar \quad (m = 0, \pm 1, \pm 2, \ldots) \tag{10.81}$$

We would not need to form operators for $\mathscr{M}_x$ and $\mathscr{M}_y$, since these quantities can no longer be used as generalized momenta if $\mathscr{M}_z$ has already been chosen as such, as was shown above. But we have to construct them, in order to find the Hermitian operator for $\mathscr{M}^2$. It would be imprudent to try to form $\hat{\mathscr{M}}_x$ and $\hat{\mathscr{M}}_y$ by the substitution of the corresponding operators of the form as given in Eqn. (2) for p_φ, p_θ, and p_χ in Eqns. (75), because there is no confidence in the correctness of such a way of constructing operators, as was shown in the particular example of the operator $\hat{K}$ in spherical coordinates. Hence, we have to proceed from the expression for the column matrix of operators for the angular momenta in the Cartesian rest coordinate system, which, in view of Eqn. (60), is

$$\hat{\mathscr{M}} = \tilde{A}_0^{(\Omega)} \hat{\mathscr{M}}', \tag{10.82}$$

where $\hat{\mathscr{M}}'$ is the column matrix of the operators for the angular momenta $\mathscr{M}'_\alpha$ in a rotating coordinate system. In accordance with Eqn. (39), we have

$$\hat{\mathscr{M}}'_\alpha = \sum_i \tilde{r}^{(i)\prime} \Sigma_\alpha A_0^{(\Omega)} \hat{p}^{(i)} \quad (\alpha = x, y, z) \tag{10.83}$$

where

$$\hat{p}^{(i)} = -i\hbar \nabla_i = -i\hbar \frac{\partial}{\partial r^{(i)}}. \tag{10.84}$$

Since we are interested in rotations only, we can write

$$\nabla_i = \nabla_i \tilde{\Omega} \frac{\partial}{\partial \Omega}. \tag{10.85}$$

Then Eqn. (84) yields

$$\hat{p}^{(i)} = \nabla_i \Omega \hat{\Pi}', \tag{10.86}$$

where

$$\hat{\Pi}' = (\hat{\pi}_\varphi, \hat{\pi}_\theta, \hat{\pi}_\chi), \tag{10.87}$$

is the column matrix of the operators

$$\hat{\pi}_\varphi = -i\hbar \frac{\partial}{\partial \varphi}; \qquad \hat{\pi}_\theta = -i\hbar \frac{\partial}{\partial \theta}; \qquad \hat{\pi}_\chi = -i\hbar \frac{\partial}{\partial \chi}. \tag{10.88}$$

For the same reasons as those used in obtaining Eqn. (62), Eqns. (83) and (86) yield

$$\hat{\mathcal{M}}' = \sum_i \tilde{A}_i^{(r)} A_0^{(\Omega)} \nabla_i \tilde{\Omega} \hat{\Pi}'. \tag{10.89}$$

To find $\nabla_i \tilde{\Omega}$ let us consider the identity

$$\frac{\mathrm{d}\Omega}{\mathrm{d}\tilde{\Omega}} = \sum_i \frac{\partial \Omega}{\partial \tilde{r}^{(i)}} \frac{\partial r^{(i)}}{\partial \tilde{\Omega}} = \sum_i \tilde{\nabla}_i \Omega \frac{\partial r^{(i)}}{\partial \tilde{\Omega}} = E_3. \tag{10.90}$$

From Eqns. (45) and (48) we see that

$$\frac{\partial r^{(i)}}{\partial \tilde{\Omega}} = \tilde{A}_0^{(\Omega)} A_i^{(r)} C^{(\Omega)}. \tag{10.91}$$

It is now easy to see that in order to satisfy Eqn. (90), $\nabla_i \tilde{\Omega}$ should be of the form

$$\nabla_i \tilde{\Omega} = m_i \tilde{A}_0^{(\Omega)} A_i^{(r)} G'' \tilde{C}^{(\Omega)\prime} + \tilde{A}_0^{(\Omega)} \tilde{B}_i C, \tag{10.92}$$

where C is an arbitrary 3 by 3 matrix, the B_i are any matrices obeying

$$\sum_i B_i A_i^{(r)} = 0, \tag{10.93}$$

and all the other quantities are the same as before. Hence, Eqn. (89) yields, in view of Eqns. (92) and (93),

$$\hat{\mathcal{M}}' = \tilde{C}^{(\Omega)\prime} \hat{\Pi}', \tag{10.94}$$

whence we get, in view of Eqn. (82),

$$\hat{\mathscr{M}} = \tilde{A}_0^{(\Omega)} \tilde{C}^{(\Omega)'} \hat{\Pi}', \tag{10.95}$$

that is

$$\begin{aligned} \hat{\mathscr{M}}_x &= -\sin\varphi \cot\theta \hat{\pi}_\varphi + \cos\varphi \hat{\pi}_\theta + \frac{\sin\varphi}{\sin\theta} \hat{\pi}_\chi, \\ \hat{\mathscr{M}}_y &= \cos\varphi \cot\theta \hat{\pi}_\varphi + \sin\varphi \hat{\pi}_\theta - \frac{\cos\varphi}{\sin\theta} \hat{\pi}_\chi, \\ \hat{\mathscr{M}}_z &= \hat{p}_\varphi. \end{aligned} \tag{10.96}$$

We see that to obtain Eqn. (96) from Eqn. (75), it is necessary to replace p_θ therein by the operator $\hat{\pi}_\theta$ as given in Eqn. (88), not by the operator $\hat{p}_\theta$ as given in Eqn. (5). The operators $\hat{\pi}_\varphi$ and $\hat{\pi}_\chi$ obviously coincide with $\hat{p}_\varphi$ and $\hat{p}_\chi$, respectively.

Simple calculations based on Eqns. (96) and (88) give

$$\hat{\mathscr{M}}_x^2 + \hat{\mathscr{M}}_y^2 + \hat{\mathscr{M}}_z^2 \equiv \hat{\mathscr{M}}^2 = -\frac{\hbar^2}{\sin^2\theta}\left(\frac{\partial^2}{\partial\varphi^2} + \frac{\partial^2}{\partial\chi^2} - 2\cos\theta \frac{\partial^2}{\partial\varphi\,\partial\chi}\right) - \\ -\hbar^2 \frac{1}{\sin\theta}\frac{\partial}{\partial\theta}\left(\sin\theta \frac{\partial}{\partial\theta}\right). \tag{10.97}$$

Thus, $\hat{\mathscr{M}}^2$ is obtained from Eqn. (76) by replacing p_θ^2 therein by the operator of the same form as that replacing p_θ^2 in Eqn. (35), while operators replacing p_φ and p_χ are $\hat{p}_\varphi$ and $\hat{p}_\chi$, respectively. Note that the operator replacing p_θ^2 in Eqns. (35) and (76) is neither $\hat{p}_\theta^2$ nor $\hat{\pi}_\theta^2$. This fact illustrates once again that one should be careful in constructing a Hermitian operator for a physical quantity in the case when the operators corresponding to the summands, of which the latter is made up, do not commute.

For $\chi = 0$, Eqn. (97) goes over into

$$\hat{\mathscr{M}}^2 = -\hbar^2 \left[\frac{1}{\sin\theta}\frac{\partial}{\partial\theta}\left(\sin\theta\frac{\partial}{\partial\theta}\right) + \frac{1}{\sin^2\theta}\frac{\partial^2}{\partial\varphi^2}\right]. \tag{10.98}$$

This allows one to rewrite Eqn. (34) as

$$\hat{K} = \frac{1}{2m}\left(\hat{p}_r^2 + \frac{\hat{\mathscr{M}}^2}{r^2}\right). \tag{10.99}$$

Let us notice, by the way, that the operators $\hat{\mathscr{M}}_x$, $\hat{\mathscr{M}}_y$, and $\hat{\mathscr{M}}_z$ do not commute. Indeed, from Eqns. (96) and (6.70) one easily gets, in view of

$$[\hat{p}_\varphi \hat{\varphi}] = -i\hbar, \tag{10.100}$$

the relation

$$[\hat{\mathscr{M}}_z \hat{\mathscr{M}}_x] = \hat{\mathscr{M}}_z \hat{\mathscr{M}}_x - \hat{\mathscr{M}}_x \hat{\mathscr{M}}_z = i\hbar \hat{\mathscr{M}}_y. \tag{10.101a}$$

Since $\hat{\mathscr{M}}_x$, $\hat{\mathscr{M}}_y$, and $\hat{\mathscr{M}}_z$ can be obtained from each other by a cyclic permutation of the indices x, y, and z [see Eqn. (37)], Eqn. (101a) yields

$$[\hat{\mathscr{M}}_x \hat{\mathscr{M}}_y] = i\hbar \hat{\mathscr{M}}_z, \tag{10.101b}$$

$$[\hat{\mathscr{M}}_y \hat{\mathscr{M}}_z] = i\hbar \hat{\mathscr{M}}_x. \tag{10.101c}$$

At the same time the operator $\hat{\mathscr{M}}^2$ commutes with any of the operators $\hat{\mathscr{M}}_\alpha$ ($\alpha = x, y, z$). Indeed, since $\hat{\mathscr{M}}_\alpha^2$ commutes with $\hat{\mathscr{M}}_\alpha$, we have for the operator $\hat{\mathscr{M}}^2$, in view of Eqn. (97),

$$[\hat{\mathscr{M}}^2 \hat{\mathscr{M}}_\alpha] = [\hat{\mathscr{M}}_\beta^2 \hat{\mathscr{M}}_\alpha] + [\hat{\mathscr{M}}_\gamma^2 \hat{\mathscr{M}}_\alpha]. \tag{10.102}$$

But

$$[\hat{\mathscr{M}}_\beta^2 \hat{\mathscr{M}}_\alpha] = \hat{\mathscr{M}}_\beta [\hat{\mathscr{M}}_\beta \hat{\mathscr{M}}_\alpha] + [\hat{\mathscr{M}}_\beta \hat{\mathscr{M}}_\alpha] \hat{\mathscr{M}}_\beta. \tag{10.103}$$

Hence, Eqn. (102), in view of Eqns. (101), yields

$$[\hat{\mathscr{M}}^2 \hat{\mathscr{M}}_\alpha] = i\hbar(\pm \hat{\mathscr{M}}_\beta \hat{\mathscr{M}}_\gamma \pm \hat{\mathscr{M}}_\gamma \hat{\mathscr{M}}_\beta \mp \hat{\mathscr{M}}_\gamma \hat{\mathscr{M}}_\beta \mp \hat{\mathscr{M}}_\beta \hat{\mathscr{M}}_\gamma) = 0. \tag{10.104}$$

Equations (101) and (104) are in agreement with Eqns. (71) and (74), as should be the case (cf. Sect. 6.3).

It is easy to make sure, too, by making use of Eqn. (22) and the expression for $\hat{\mathscr{M}}_\alpha$ following from Eqn. (37), namely,

$$\hat{\mathscr{M}}_\alpha = \sum_i (\beta_i \hat{p}_\gamma^{(i)} - \gamma_i \hat{p}_\beta^{(i)}) \quad (\alpha \neq \beta \neq \gamma; \alpha, \beta, \gamma = x, y, \text{ or } z) \tag{10.105}$$

that

$$[\hat{K} \hat{\mathscr{M}}_\alpha] = 0 \quad (\alpha = x, y, z) \tag{10.106}$$

whence it follows that

$$[\hat{K} \hat{\mathscr{M}}^2] = 0. \tag{10.107}$$

Equation (106) conforms to Eqn. (72). If the potential energy, U, does not depend on Euler's angles, then, in view of Eqn. (19) and $[\hat{U} \hat{\mathscr{M}}_\alpha] = 0$,

$$[\hat{H} \hat{\mathscr{M}}_\alpha] = 0, \tag{10.108}$$

$$[\hat{H} \hat{\mathscr{M}}^2] = 0 \tag{10.109}$$

are valid. These equations, in view of Eqn. (6.95), yield

$$\dot{\hat{\mathscr{M}}}_\alpha = 0 \quad (\alpha = x, y, z) \tag{10.110}$$

and

$$\dot{\hat{\mathscr{M}}}^2 = 0 \tag{10.111}$$

respectively, which reflects the corresponding conservations laws. It is necessary to remember, however, that any two different $\hat{\mathscr{M}}_\alpha$ ($\alpha = x, y, z$) do not commute. Therefore, as was shown in Section 5.3, they have no common eigenfunctions describing a state of the corresponding abstract mechanical system and hence cannot have definite eigenvalues in any state simultaneously. The commutativity of the operators $\hat{H}$, $\hat{\mathscr{M}}^2$, and any of $\hat{\mathscr{M}}_\alpha$ ($\alpha = x, y, z$) leads thus to the existence of common eigenfunctions of the former two operators and only one of the latter ones.

Hence, only those states of an abstract mechanical system can exist, which are determined by the definite eigenvalues of $\hat{H}$, $\hat{\mathscr{M}}^2$, and, say, $\hat{\mathscr{M}}_z$. This means, in view of Eqn. (8.4), that an abstract mechanical system, whose potential energy does not depend on Euler's angles, has constant values of energy, angular momentum relative to z-axis, and angular momentum magnitude squared (that is, definite maximum value of the angular momentum with respect to any axis).

Hence all the concrete mechanical systems corresponding to this abstract mechanical system have the same constant values of the physical quantities indicated. The conservation laws for the angular momentum are thus the same both for concrete and abstract mechanical systems, as should be the case. This assertion is correct, irrespective of whether the mechanical systems are macroscopic or microscopic. It might seem that there is an essential difference between macroscopic and microscopic systems, since the values of $\mathscr{M}_z$ for the former are continuous, whereas for the latter they are discrete. But that is only a seeming difference which results from the real fact that however small may be the measurable changes in macroscopic values of angular momenta, they always are, nevertheless, enourmously big in comparison with $\hbar$.

Let us now find the set of values of $\mathscr{M}^2$, that is, the eigenvalues of $\hat{\mathscr{M}}^2$. Since $\hat{\mathscr{M}}^2$ and $\hat{\mathscr{M}}_z$ commute, so do $\hat{\mathscr{M}}_x^2 + \hat{\mathscr{M}}_y^2$ and $\hat{\mathscr{M}}^2$, as well as $\hat{\mathscr{M}}_x^2 + \hat{\mathscr{M}}_y^2$ and $\hat{\mathscr{M}}_z$. Hence, $\hat{\mathscr{M}}_z$, $\hat{\mathscr{M}}_x^2 + \hat{\mathscr{M}}_y^2$, and $\hat{\mathscr{M}}^2$ have common eigenfunctions. The eigenvalues of these operators corresponding to a common eigenfunction of them are connected, in view of Eqn. (97), by

$$(\mathscr{M}^2) = (\lambda^2) + (\mathscr{M}_z)^2, \tag{10.112}$$

where $(\mathscr{M}^2)$, (λ^2), and $(\mathscr{M}_z)$ are the corresponding eigenvalues of $\hat{\mathscr{M}}^2$, $\hat{\mathscr{M}}_x^2 + \hat{\mathscr{M}}_y^2$, and $\hat{\mathscr{M}}_z$, respectively. It follows from Eqn. (112) that for a given eigenvalue $(\mathscr{M}^2)$ of the operator $\hat{\mathscr{M}}^2$ the condition

$$(\mathscr{M}^2)^{1/2} \geqslant (\mathscr{M}_z) \geqslant -(\mathscr{M}^2)^{1/2} \tag{10.113}$$

should be satisfied. Hence, in view of Eqn. (81), $(\mathscr{M}_z)$ has in this case a finite set of $2l+1$ valus $m\hbar$ ($m = 0, \pm 1, \pm 2, \ldots, \pm l$), where l is an integer closest to $(\mathscr{M}^2)^{1/2}/\hbar$.

For a given $(\mathcal{M}^2)$ the operator $\hat{\mathcal{M}}_z$ is thus represented on the basis of its eigenfunctions by a diagonal matrix $\mathcal{M}_z$ of the form

$$\mathcal{M}_z = \hbar \|\delta_{kj} m_j\| \quad (k, j = 1, 2, \ldots, n = 2l + 1), \tag{10.114}$$

where

$$m_j = l + 1 - j. \tag{10.115}$$

If we know the matrices $\mathcal{M}_x$ and $\mathcal{M}_y$ which represent operators $\hat{\mathcal{M}}_x$ and $\hat{\mathcal{M}}_y$, respectively, on that basis, we will obtain the diagonal matrix $\mathcal{M}^2$ representing $\hat{\mathcal{M}}^2$ on the same basis, by making use of the equation

$$\mathcal{M}^2 = \mathcal{M}_x^2 + \mathcal{M}_y^2 + \mathcal{M}_z^2 \tag{10.116}$$

following from Eqn. (97). So, we will, first, find the matrices $\mathcal{M}_x$ and $\mathcal{M}_y$.

Let us write

$$\mathcal{M}_x = \hbar \|a_{kj}\|; \qquad \mathcal{M}_y = \hbar \|b_{kj}\|; \quad (k, j = 1, 2, \ldots, n). \tag{10.117}$$

Then, in view of Eqn. (114), Eqns. (101a) and (101c) yield, respectively,

$$m_k a_{kj} - m_j a_{kj} = i b_{kj}, \tag{10.118}$$

$$m_k b_{kj} - m_j b_{kj} = -i a_{kj}. \tag{10.119}$$

Introducing also

$$a_{kj} + i b_{kj} = 2 c_{kj}, \tag{10.120}$$

$$a_{kj} - i b_{kj} = 2 d_{kj}, \tag{10.121}$$

we get from Eqns. (118) and (119)

$$m_k c_{kj} - m_j c_{kj} = c_{kj}, \tag{10.122}$$

$$m_k d_{kj} - m_j d_{kj} = -d_{kj}. \tag{10.123}$$

It follows from Eqns. (122) and (123), in view of Eqn. (115), that

$$c_{kj} = 0 \quad (j \neq k + 1); \qquad d_{kj} = 0 \quad (j \neq k - 1). \tag{10.124}$$

Since equations (120) and (121) yield

$$a_{kj} = c_{kj} + d_{kj}, \tag{10.125}$$

$$b_{kj} = i(d_{kj} - c_{kj}), \tag{10.126}$$

the a_{kj} and b_{kj} can differ from zero only if $|k - j| = 1$. Furthermore, considering $\mathcal{M}_x$ a symmetric matrix, we get from Eqn. (125)

$$a_{k,k-1} = a_{k-1,k} = d_{k,k-1} = c_{k-1,k} \equiv a_{k-1}. \tag{10.127}$$

Then, Eqn. (126) gives

$$b_{k,k-1} = ia_{k-1}; \qquad b_{k-1,k} = -ia_{k-1}. \tag{10.128}$$

We thus get for the elements of matrices $\mathcal{M}_x$ and $\mathcal{M}_y$

$$(\mathcal{M}_x)_{kj} = \hbar(\delta_{k,j+1}a_j + \delta_{k,j-1}a_{j-1}), \tag{10.129}$$

$$(\mathcal{M}_y)_{kj} = i\hbar(\delta_{k,j+1}a_j - \delta_{k,j-1}a_{j-1}). \tag{10.130}$$

Equation (101b) yields, in view of Eqn. (114),

$$\sum_j (\delta_{k,j+1}a_j + \delta_{k,j-1}a_{j-1})(\delta_{j,l+1}a_l - \delta_{j,l-1}a_{l-1}) -$$

$$- \sum_j (\delta_{k,j+1}a_j - \delta_{k,j-1}a_{j-1})(\delta_{j,l+1}a_l + \delta_{j,l-1}a_{l-1}) = \delta_{kl}m_l. \tag{10.131}$$

We get from this, after simple calculation,

$$2(a_k^2 - a_{k-1}^2) = m_k, \tag{10.132}$$

whence we obtain, in view of Eqn. (115),

$$2a_k^2 = \sum_{j=1}^{k} m_j \tag{10.133}$$

and, hence,

$$a_k = 2^{-1}[(n-k)k]^{1/2}. \tag{10.134}$$

Equations (129) and (130) can thus be rewritten as

$$(\mathcal{M}_x)_{kj} = 2^{-1}\hbar\,\{\delta_{k,j+1}[j(n-j)]^{1/2} + \delta_{k,j-1}[(j-1)(n-j+1)]^{1/2}\}. \tag{10.135}$$

$$(\mathcal{M}_y)_{kj} = 2^{-1}i\hbar\,\{\delta_{k,j+1}[j(n-j)]^{1/2} - \delta_{k,j-1}[(j-1)(n-j+1)]^{1/2}\}. \tag{10.136}$$

In terms of l and m_j they take on the conventional form

$$(\mathscr{M}_x)_{kj} = \frac{\hbar}{2} \{\delta_{k,j+1} [(l + m_j)(l + 1 - m_j)]^{1/2} +$$

$$+ \delta_{k,j-1} [(l - m_j)(l + 1 + m_j)]^{1/2}\}, \tag{10.137}$$

$$(\mathscr{M}_y)_{kj} = \frac{i\hbar}{2} \{\delta_{k,j+1} [(l + m_j)(l + 1 - m_j)]^{1/2} -$$

$$- \delta_{k,j-1} [(l - m_j)(l + 1 + m_j)]^{1/2}\}. \tag{10.138}$$

Let us finally calculate the matrix $\mathscr{M}^2$. We get from Eqn. (116), in view of Eqns. (129), (130), and (114),

$$\mathscr{M}^2_{kj} = \hbar^2 \sum_l [(\delta_{k,l+1}a_l + \delta_{k,l-1}a_{l-1})(\delta_{l,j+1}a_j + \delta_{l,j-1}a_{j-1}) -$$

$$- (\delta_{k,l+1}a_l - \delta_{k,l-1}a_{l-1})(\delta_{l,j+1}a_j - \delta_{l,j-1}a_{j-1})] +$$

$$+ \hbar^2 \delta_{kj}m_j^2 = \hbar^2 \delta_{jk} [2(a_k^2 + a_{k-1}^2) + m_k^2]. \tag{10.139}$$

By making use of Eqns. (133) and (115), we obtain

$$2(a_k^2 + a_{k-1}^2) + m_k^2 = \frac{n+1}{2} \frac{n-1}{2} = l(l+1). \tag{10.140}$$

Hence, the final result is

$$\mathscr{M}^2_{kj} = \hbar^2 \delta_{kj} l(l+1). \tag{10.141}$$

This means that the operator $\hat{\mathscr{M}}^2$ is represented on the basis of the eigenfunctions of $\hat{\mathscr{M}}_z$ by a scalar matrix

$$\mathscr{M}^2 = l(l+1)\hbar^2 E, \tag{10.142}$$

where E is a pertinent unit matrix, as should be the case, the eigenvalues of $\hat{\mathscr{M}}^2$ being

$$(\mathscr{M}^2) = l(l+1) \quad (l = 0, 1, 2, \ldots). \tag{10.143}$$

The eigenfunctions of $\hat{\mathscr{M}}^2$ (of the form as given in Eqn. (98)) are well known. They are the so-called spherical harmonic functions

$$Y_{lm}(\theta, \varphi) = N_{lm} P_l^m(\cos\theta)\, e^{im\varphi}, \tag{10.144}$$

where $P_l^m(\cos\theta)$ are known to be associated Legendre polynomials and N_{lm} are the pertinent normalization factors.

10.2.5. *Derivatives*

Operators for derivatives of physical quantities with respect to time, coordinates, and momenta are defined by Eqns. (6.95), (6.96), and (6.97). In those equations α should be replaced by $\hbar$, $\hat{t}_k$ by $\hat{q}_k$, and $\hat{t}_{n+k}$ by $\hat{p}_k$. Since $t_n \equiv t$ (time) and $-t_{2n} \equiv E$ (energy), Eqn. (6.95), in view of Eqn. (16), yields

$$\hat{\dot{F}} = \frac{i}{\hbar}[\hat{H}'\hat{F}] = \frac{i}{\hbar}[-\hat{E} + \hat{H}, \hat{F}]$$

$$= \frac{\partial \hat{F}}{\partial t} + \frac{i}{\hbar}[\hat{K}\hat{F}] + \frac{i}{\hbar}[\hat{U}\hat{F}]. \tag{10.145}$$

If F does not depend on t explicitly, Eqn. (145) takes the form

$$\hat{\dot{F}} = \frac{i}{\hbar}[\hat{H}\hat{F}] = \frac{i}{\hbar}[\hat{K}\hat{F}] + \frac{i}{\hbar}[\hat{U}\hat{F}]. \tag{10.146}$$

For $F = \beta_j$ ($\beta = x, y, z$), in particular, that is, for the βth Cartesian coordinate of the jth particle of a mechanical system, Eqns. (146) and (22) (nonrelativistic case) yield, in view of Eqn. (6.68),

$$\hat{\dot{\beta}}_j = \frac{i}{\hbar}[\hat{K}\hat{\beta}_j] = \frac{\hat{p}_{j\beta}}{m_j}. \tag{10.147}$$

This equation defines the operator for the βth Cartesian component of the velocity of the jth particle through the operator for the corresponding momentum.

For $F = p_{j\beta}$ ($\beta = x, y, z$), Eqn. (146), in view of Eqn. (22), yields

$$\hat{\dot{p}}_{j\beta} \equiv \hat{f}_{j\beta} = \frac{i}{\hbar}[\hat{U}\hat{p}_{j\beta}] = -\frac{\partial U}{\partial \beta_j}\hat{I}. \tag{10.148}$$

Equation (148) defines the operator for the βth Cartesian component of the force acting on the jth particle as the operator for the corresponding Cartesian component of $\nabla_j U$. Both Eqns. (147) and (148) follow immediately from Eqns. (6.94).

10.2.6. *Quantities having Finite Number of Values*

The intrinsic angular momentum (spin), intrinsic magnetic moment, and charge of a particle are examples of physical quantities having a finite number of values. Since the spectrum of an operator for a physical quantity should coincide with the set of its values, a pertinent n by n Hermitian matrix can be chosen as an operator for a physical quantity having n discrete values.

It follows from the experiment that spin is similar to the angular momentum in all respects. In particular, every concrete particle possesses a definite value of its intrinsic angular momentum with respect to only one, say, z-axis. A concrete particle cannot have definite values of its intrinsic angular momentum with respect to any two different axes simultaneously. The maximum value of the intrinsic angular momentum of a particle with respect to any axis, in terms of $\hbar$, is called the spin of the particle. The spin can be, as the experiment shows, an integer or a half-integer.

If the spin of a particle is s, then the intrinsic angular momentum can have any of $n = 2s + 1$ values

$$\sigma_j = s,\ s-1,\ \ldots,\ -(s-1),\ -s \tag{10.149}$$

with respect to any axis. Because of the similarity of the intrinsic angular momentum and angular momentum of a mechanical system, the operator for the former with respect to the z-axis can be chosen as

$$s_z = \|\delta_{jk}\sigma_j\| \quad (k, j = 1, \ldots, n = 2s+1), \tag{10.150}$$

where

$$\sigma_j = s + 1 - j \quad (j = 1, 2, \ldots, n = 2s+1). \tag{10.151}$$

Although intrinsic angular momenta of a particle with respect to the x- and y-axes do not really exist simultaneously with that with respect to the z-axis, it can be useful, nevertheless, to deal with them as some auxiliary quantities. It is clear that, since Eqn. (150) is similar to Eqn. (114), the operators s_x and s_y should be determined by equations similar to Eqns. (135) and (136), respectively. We thus have in terms of n and j:

$$(s_x)_{kj} = \tfrac{1}{2}\{\delta_{k,j+1}[j(n-j)]^{1/2} + \delta_{k,j-1}[(j-1)(n-j+1)]^{1/2}\}, \tag{10.152}$$

$$(s_y)_{kj} = \frac{i}{2}\{\delta_{k,j+1}[j(n-j)]^{1/2} - \delta_{k,j-1}[(j-1)(n-j+1)]^{1/2}\}. \tag{10.153}$$

In terms of s and σ_j, these equations take on, in view of Eqn. (151), the forms similar to those as given in Eqns. (137) and (138), respectively:

$$(s_x)_{kj} = \tfrac{1}{2}\{\delta_{k,j+1}[(s+\sigma_j)(s-\sigma_j+1)]^{1/2} + \\ + \delta_{k,j-1}[(s-\sigma_j)(s+\sigma_j+1)]^{1/2}\}, \tag{10.154}$$

$$(s_y)_{kj} = \frac{i}{2}\{\delta_{k,j+1}[(s+\sigma_j)(s-\sigma_j+1)]^{1/2} - \\ - \delta_{k,j-1}[(s-\sigma_j)(s+\sigma_j+1)]^{1/2}\}. \tag{10.155}$$

The magnitude squared of spin can be obtained from Eqn. (143) by replacing therein l by s. Hence, we have

$$(s^2) = s(s+1). \tag{10.156}$$

Many particles, in particular, the electron and position, have spin $\frac{1}{2}$. For these particles $n = 2s + 1 = 2$, and Eqns. (152)–(153) and (150) yield

$$s_x = \frac{1}{2}\begin{Vmatrix} 0 & 1 \\ 1 & 0 \end{Vmatrix}; \qquad s_y = \frac{1}{2}\begin{Vmatrix} 0 & -i \\ i & 0 \end{Vmatrix}; \qquad s_z = \frac{1}{2}\begin{Vmatrix} 1 & 0 \\ 0 & -1 \end{Vmatrix}. \tag{10.157}$$

The matrices

$$\sigma_x = \begin{Vmatrix} 0 & 1 \\ 1 & 0 \end{Vmatrix}; \qquad \sigma_y = \begin{Vmatrix} 0 & -i \\ i & 0 \end{Vmatrix}; \qquad \sigma_z = \begin{Vmatrix} 1 & 0 \\ 0 & -1 \end{Vmatrix} \tag{10.158}$$

are known as Pauli matrices. Simple calculations give them the well-known relations

$$\sigma_\alpha \sigma_\beta + \sigma_\beta \sigma_\alpha = 0 \qquad (\alpha \neq \beta; \alpha, \beta = x, y, \text{ or } z) \tag{10.159}$$

and

$$\sigma_\alpha^2 = \begin{Vmatrix} 1 & 0 \\ 0 & 1 \end{Vmatrix}. \tag{10.160}$$

For these particles Eqn. (156) yields $\langle s^2 \rangle = \frac{3}{4}$.

10.3. Schrödinger Equation

One of the main problems of quantum physics is finding the probability distributions for various physical quantities in feasible (real) states of abstract physical systems and calculation of the respective mathematical expectations in those states. The probability distributions are given, in particular, by the corresponding ψ-functions describing the states.

Since, for a mechanical system, the physical quantities possessed are Hamiltonian random variables, all the results of Section 6.4 are immediately applicable. In the case under consideration, the random variable t_n is time, t, while the random variable $-t_{2n}$ is energy E, as was mentioned in the preceding section. Therefore, Eqn. (6.91) takes on the form

$$\hat{E}\psi = \hat{H}\psi, \tag{10.161}$$

where $\hat{E}$ is the operator for energy of the form given in Eqn. (16) and $\hat{H}$ is the operator for Hamiltonian function of the form given in Eqn. (19).

Equation (161) is the famous Schrödinger equation. In each particular case, the ψ-functions obeying it describe the states of the respective abstract mechanical system which correspond to real motion of the concrete mechanical systems that are elements of the set, whose image is the abstract mechanical system.

If the potential energy U does not depend on t explicitly (a conservative system), $\hat{E}$ and $\hat{H}$ commute and have common eigenfunctions. The eigenfunctions of $\hat{E}$ are always of the form

$$\psi_E(x, t) = \psi(x) \exp(iEt/\hbar) \tag{10.162}$$

and satisfy the equality

$$\hat{E}\psi_E(x, t) = E\psi(x)\exp(iEt/\hbar) \tag{10.163}$$

(x denotes the totality of all spatial coordinates of the system).

The eigenfunctions of $\hat{H}$ for a conservative system are of the form

$$\psi_H(x, t) = \psi_H(x)f(t), \tag{10.164}$$

where $\psi_H(x)$ obey

$$\hat{H}\psi_H(x) = H\psi_H(x) \tag{10.165}$$

and H is, hence, the corresponding eigenvalue of $\hat{H}$.

The common eigenfunctions of $\hat{E}$ and $\hat{H}$ have, therefore, the form

$$\psi(x, t) = \psi_H(x)\exp(iEt/\hbar). \tag{10.166}$$

In a state of an abstract mechanical system described by a function of this form, energy has a definite value E and the Hamiltonian has a definite value H. Schrödinger equation (161) is satisfied by functions of the form as given in Eqn. (166) if

$$H = E, \tag{10.167}$$

that is, if the values of energy and Hamiltonian function coincide. But Eqn. (167) coincides with Eqn. (8.4), which is the condition for the motion of any concrete mechanical system to be real. Hence, ψ-functions satisfying Schrödinger's equation describe real states of abstract mechanical systems, that is, states which conform to real motion of the corresponding concrete mechanical systems.

The probability density of the spatial coordinates x in these states is proportional, in view of Eqn. (166), to

$$|\psi(x, t)|^2 = |\psi_E(x)|^2 \tag{10.168}$$

and does not depend on time, becuase of which such states are called stationary.

To find real states for a conservative system one should solve Eqn. (165), which is often called a Schrödinger equation without time.

The general solution to the Schrödinger equation (161) for a conservative system is obviously

$$\psi(x, t) = \int a(E)\psi_E(x)\exp(-iEt/\hbar)\,dE, \tag{10.169}$$

where the $a(E)$ are constant coefficients to be determined from some conditions. In a state described by a function of this form, energy of the corresponding abstract mechanical system is indefinite – it has different values with respective probabilities – and $|\psi(x, t)|^2$ depends both on x and t. The latter fact can be interpreted as the dependence of the probability distribution for coordinates on time, because of which, states described by such functions are called nonstationary.

For a conservative system, however, the only physically reasonable condition is the one requiring that energy of the abstract system should have quite a definite value, say E'. This condition can be written as

$$a(E) = \delta(E - E'), \tag{10.170}$$

which reduces Eqn. (169) to Eqn. (166):

$$\psi(x, t) = \psi_{E'}(x) \exp(-iE't/\hbar). \tag{10.171}$$

Indeed, an abstract conservative system is the image of the corresponding set of concrete ones. But for any concrete conservative system the energy conservator law is valid. Hence, it should be valid for the respective abstract system as well. Note, by the way, that this reasoning is true with respect to any conservation law. Therefore, if several conservation laws are fulfilled in common for each concrete system of a set of them, then they are valid for the corresponding abstract system too. This means that such an abstract system can only be in states with definite values of all the physical quantities for which conservation laws are fulfilled in common. In particular, an abstract conservative system can be solely in states with definite values of energy, that is, in stationary states. It is very important to stress this fact, because the conventional opinion that a conservative system can be in nonstationary states described by wave functions of the type given in Eqn. (169) leads to numerous misunderstandings, unnecessary debates, and unceasing flow of needless works (cf. Sect. 14.5).

As was mentioned before, Eqn. (168) immediately determines the probability distribution for coordinates in the corresponding stationary state of an abstract conservative system. By the utilization of Eqn. (5.115), the probability distributions for all the other physical quantities, in particular, momenta in that state can also been obtained from the same function $\psi(x, t)$. This fact is in agreement with the results obtained in Section 6.2 when considering Hamiltonian random variables by the classical method. It has been demonstrated there that probability distributions for any functions of the Hamiltonian random variables $t_1, t_2, \ldots, t_{2n}$ are fully determined by the probability distribution for the first n ones, that is, $t_1, t_2, \ldots, t_n$. But these correspond to coordinates and time (t_n), while $t_{n+1}, t_{n+2}, \ldots, t_{2n}$ correspond to the canonically conjugate momenta and energy ($-t_{2n}$).

Hence, in the classical case it is sufficient to know the probability distribution for coordinates of a conservative mechanical system in order to calculate distributions for both momenta and any functions of coordinates and momenta. The same is true in the quantum case (that is, when utilizing the quantum approach), provided the respective Hermitian operators are known (see also Sect. 6.4).

Moreover, quantum probability distributions determined by a ψ-function describing a stationary state of a mechanical system go over into the respective classical probability distributions related to a constant value of energy of the system when the quantum numbers characterizing the state become very large (are approaching infinity). Whenever this assertion can be checked it proves to be correct. For closed systems, such probability distributions result from the corresponding microcanonical distribution of the equilibrium classical statistical mechanics. Therefore, one may say that the quantum mechanical probability distributions determined by ψ-functions, describing stationary states, go over into the respective classical statistical distributions resulting from the corresponding microcanonical ones, when proceeding to the classical limit, that is, for large enough quantum numbers.

The conventional belief that the classical limit of quantum mechanics in classical mechanics thus turns out to be wrong. As a matter of fact, the classical limit of quantum mechanics is nothing other than classical statistical mechanics (see also Sect. 14.2).

Let us now consider some particular examples.

10.4. A Free Particle

(a) *Nonrelativistic Case*. For a free nonrelativistic particle of mass m moving along X-axis of an inertial coordinate system, the operators for the momentum and Hamiltonian function are

$$\hat{p} = -i\hbar \frac{\partial}{\partial x} \tag{10.172}$$

and

$$\hat{H} = \frac{\hat{p}^2}{2m} = -\frac{\hbar^2}{2m}\frac{\partial^2}{\partial x^2} \tag{10.173}$$

respectively. Therefore, in this particular case, the Schrödinger equation (161) takes on the form

$$i\hbar \frac{\partial \psi}{\partial t} = -\frac{\hbar^2}{2m}\frac{\partial^2 \psi}{\partial x^2}. \tag{10.174}$$

The eigenfunctions of $\hat{p}$ corresponding to its eigenvalues p are

$$\psi_p = \text{const} \exp(ipx/\hbar). \tag{10.175}$$

These functions are also the eigenfunctions of $\hat{H}$ conforming to its eigenvalues

$$H = p^2/2m. \tag{10.176}$$

Therefore, solutions to Eqn. (174) are

$$\psi_{E,p}(x, t) = \psi_p \exp(-iEt/\hbar) = \text{cost} \exp[i(px - Et)/\hbar], \tag{10.177}$$

where $E = H = p^2/2m$.

A function of the form as given in Eqn. (177) describes a state of an abstract nonrelativistic free particle with definite values of energy, E, and momentum, p. The probability density $\rho(x) = |\psi_E(x)|^2$ of the coordinate x, in view of Eqn. (177), is obviously constant.

The results obtained fully coincide with those we got in Section 8.3 when considering a free nonrelativistic particle by the classical method. Both methods give the same probability distribution for the coordinate x, while energy and momentum of the particle have quite definite values.

To have the explicit expression for $\rho(x)$ we should normalize the function (177) by the condition

$$\int \rho(x)\,\mathrm{d}x = 1, \tag{10.178}$$

which gives

$$\int |\psi_{E,p}(x, t)|^2\,\mathrm{d}x = \int |\psi_p|^2\,\mathrm{d}x = |\mathrm{const}|^2 \int \mathrm{d}x = 1. \tag{10.179}$$

But a free particle considered a closed system should move in the interval $(-\infty, \infty)$. Then, the integral in Eqn. (179) is infinite, and the constant in Eqn. (177) is zero. Hence, for a free particle regarded as a closed system we have $\rho(x) = 0$. This fact does not allow one to calculate mathematical expectations of any quantities in a stationary state of a free particle.

In order to avoid this difficulty, we can approximate the closed system by a pseudoclosed one, that is, by a particle moving freely in an finite interval $(-b/2, b/2)$, where b is any large number. It is clear that the greater b, the more accurate approximation is being achieved. In this approximation, as follows from Eqn. (179), the constant in Eqn. (177) is $b^{-1/2}$.

By making use of the function $\delta_\epsilon(x)$ [see Eqn. (3.17)], we can rewrite the approximate ψ-function as

$$\psi_{E,p}(x, t) \cong \delta_b^{1/2}(x) \exp\left[i(px - Et)/\hbar\right]. \tag{10.180a}$$

Then the exact ψ-function can be represented in the symbolic form

$$\psi_{E,p}(x, t) = \delta_\infty^{1/2}(x) \exp\left[i(px - Et)/\hbar\right], \tag{10.180b}$$

where

$$\delta_\infty^{1/2}(x) = \lim_{b \to \infty} \delta_b^{1/2}(x). \tag{10.181}$$

When using the $\delta_\infty(x)$-symbol, one must proceed to the limit *after* the calculation with $b \neq \infty$ have been accomplished.

The functions

$$\psi_p = \delta_\infty^{1/2}(x) \exp(ipx/\hbar) \tag{10.182}$$

are the eigenfunctions of the operator $\hat{p}$ normalized to the continuous Kroneker symbol $\delta_{pp'}$. Indeed,

$$(\psi_p, \psi_{p'}) = \lim_{b\to\infty} b^{-1} \int_{-b/2}^{b/2} e^{i(p-p')x/\hbar}\, dx$$

$$= \lim_{b\to\infty} \frac{\sin (p-p')b/2\hbar}{(p-p')b/2\hbar}. \tag{10.183}$$

No matter how small $|p - p'|$, the right-hand side of Eqn. (183) is zero, since $|\sin \alpha| \leqslant 1$ for any real α. For $p = p'$, however, the right-hand side of Eqn. (183) is unity. Hence,

$$(\psi_p, \psi_{p'}) = \delta_{pp'} = \begin{cases} 0 & (p \neq p'), \\ 1 & (p = p'). \end{cases} \tag{10.184}$$

The probability density of x takes on the form

$$\rho(x) = \delta_\infty(x), \tag{10.185}$$

the condition $\int \rho(x)\, dx = \int \delta_\infty(x)\, dx = 1$ being satisfied [see Eqn. (3.18)]. It is now possible to calculate mathematical expectations of any related quantities. For instance,

$$Mx = \int \delta_\infty(x)x = 0, \tag{10.186}$$

$$Mx^2 = \int \delta_\infty(x)x^2\, dx = \lim_{b\to\infty} b^2/12 = \infty. \tag{10.187}$$

In view of Eqn. (184),

$$Mp = (\hat{p}\psi_p, \psi_p) = p, \tag{10.188}$$

$$Mp^2 = p^2. \tag{10.189}$$

For the variance of p Eqns. (188) and (189) yield

$$Dp = 0. \tag{10.190}$$

From Eqns. (186) and (187) it follows that

$$Dx = \lim_{b\to\infty} \frac{b^2}{12} = \infty. \tag{10.191}$$

From Eqns. (190) and (191) we now have

$$\mathrm{D}x\ \mathrm{D}p = 0 \tag{10.192}$$

for any value of b, and hence, for $b = \infty$, too. This is in agreement with the results of Section 6.5, since the ψ-function under consideration is also an eigenfunction of $\hat{p}$.

Let us now calculate the velocity $\hat{v} = \hat{\dot{x}}$ of a nonrelativistic free particle. By making use of Eqns. (147), we get for the operator $\hat{\dot{x}}$

$$\hat{\dot{x}} = \hat{p}/m, \tag{10.193}$$

therefore, the velocity in a stationary state of an abstract free particle with a definite value p of the momentum also has a definite value $v = p/m$.

(b) *Relativistic Case*. The Hamiltonian function for a concrete free relativistic particle of rest mass m_0 moving along the X-axis of an inertial coordinate system is

$$H = c(p^2 + m_0^2 c^2)^{1/2}. \tag{10.194}$$

Here c is the ultimate velocity of real motion, which is supposed to be that of light in vacuum, and p is the particle momentum.

Hence, the Hamiltonian operator for the abstract particle is

$$\hat{H} = c(\hat{p}^2 + m_0^2 c^2 \hat{I})^{1/2}, \tag{10.195}$$

where $\hat{I}$ is the identity operator and $\hat{p}$ is the momentum operator of the form given in Eqn. (172). It follows from definition of a function of an operator (see Sect. 5.3) that the eigenfunctions of $\hat{H}$ in the case under consideration coincide with the eigenfunctions of the Hamiltonian operator for a free nonrelativistic particle, and describe stationary states of an abstract free relativistic particle, which are determined by definite values of energy and momentum; the respective eigenvalues being those as given in Eqn. (194). The Schrödinger equation

$$\hat{E}\psi = c(m_0^2 c^2 \hat{I} + \hat{p}^2)^{1/2}\psi \tag{10.196}$$

is satisfied when

$$E = H = (m_0^2 c^4 + c^2 p^2)^{1/2}, \tag{10.197}$$

as should be the case for real motion.

The probability density of the coordinate x, and hence, all the mathematical expectations are the same as in the case of a free nonrelativistic particle. But for the velocity operator $\hat{\dot{x}}$ we now get, in view of Eqns. (195) and (5.67),

$$\hat{\dot{x}} = \frac{i}{h}[\hat{H}\hat{x}] = c\hat{p}\hat{H}^{-1}. \tag{10.198}$$

The value of the velocity in a stationary state of an abstract free relativistic particle with a definite value p of the momentum also is definite, namely,

$$v = cpH^{-1}. \qquad (10.199)$$

Thus, both for nonrelativistic and relativistic free particles, the results of the quantum approach coincide with those of the classical method.

10.5. A Particle in a One-Dimensional Potential Box

This is an idealized example: a concrete nonrelativistic particle of mass m is moving along, say, x-axis of an inertial coordinate system in the range $(0, a)$, provided the potential field is of the form

$$U(x) = \begin{cases} 0 & (0 < x < a), \\ \infty & (x < 0; x > a). \end{cases} \qquad (10.200)$$

The particle can never be outside the interval $(0, a)$ (therefore, it is called 'a particle in a box'). It is not a free particle, since there are forces acting on it, but it is a conservative system, because the forces do not depend on time. Hence, its energy is conserved, whereas its momentum is not.

Let us now apply the methods of probabilistics to the case under consideration.

(a) *Classical Method.* In a state of the corresponding abstract particle with a definite energy value E, the probability density $\rho(x)$ of the coordinate is

$$\rho(x) = \delta_a\left(x - \frac{a}{2}\right) \qquad (10.201)$$

[see Eqn. (5.10)]. In this state, the momentum p of the abstract particle can have two values: $p = p_E$ or $p = -p_E$, where

$$p_E = (2mE)^{1/2} \qquad (10.202)$$

with equal probabilities 1/2. Therefore, by making use of Eqn. (3.15), we can write

$$\rho(p) = \tfrac{1}{2}\,\delta(p - p_E) + \tfrac{1}{2}\,\delta(p + p_E). \qquad (10.203)$$

To simplify the further considerations, we introduce the dimensionless quantities

$$\eta = x/a; \qquad E_d = E/E_0; \qquad \epsilon = p/(2mE_0)^{1/2}, \qquad (10.204)$$

where E_0 is some energy unit.

Equations (201) and (203) yield, then, respectively

$$\rho(\eta) = \delta_1(\eta - \tfrac{1}{2}) \qquad (10.201a)$$

and

$$\rho(\epsilon) = \tfrac{1}{2}\,\delta(\epsilon - E_d^{1/2}) + \tfrac{1}{2}\,\delta(\epsilon + E_d^{1/2}). \tag{10.203a}$$

It will prove even more convenient to introduce the quantity

$$\chi = \frac{E_d^{1/2} - \epsilon}{E_d^{1/2}} \tag{10.205}$$

in further calculations. Instead of Eqn. (203a), we now get

$$\rho(\chi) = \tfrac{1}{2}\,\delta(\chi) + \tfrac{1}{2}\,\delta(2 - \chi). \tag{10.206}$$

(b) *Quantum Approach*. The Schrödinger equation (without time) for the case under consideration is

$$\hat{H}\psi = -\frac{\hbar^2}{2m}\frac{d^2\psi}{dx^2} = E\psi. \tag{10.207}$$

It is easy to verify that the normalized eigenfunctions of $\hat{H}$ of the form

$$\psi_n(x) = \delta_a\left(x - \frac{a}{2}\right)(2a)^{1/2}\sin\frac{n\pi}{a}x \quad (n = 1, 2, \ldots) \tag{10.208}$$

are solutions to Eqn. (207); the conditions $\psi_n(0) = \psi_n(a) = 0$ ensuring the continuity of the probability density $\rho(x) = |\psi_n(x)|^2$ at both $x = 0$ and $x = a$.

The energy values in stationary states described by the $\psi_n(x)$ are

$$E_n = \frac{\hbar^2}{2m}\frac{n^2\pi^2}{a^2} \quad (n = 1, 2, \ldots). \tag{10.209}$$

In view of Eqn. (204), we get from Eqn. (208)

$$\psi_n(\eta) = \delta_1(\eta - \tfrac{1}{2})\,2^{1/2}\sin n\pi\eta \quad (n = 1, 2, \ldots). \tag{10.210}$$

Hence,

$$\rho_n(\eta) = |\psi_n(\eta)|^2 = \delta_1(\eta - \tfrac{1}{2})\,2\sin^2 n\pi\eta$$
$$= \delta_1(\eta - \tfrac{1}{2})(1 - \cos 2\pi n\eta). \tag{10.211}$$

To get the probability density of χ we should find both the operator for ϵ and its eigenfunctions, and then we have to obtain the expansion of $\psi_n(x)$ as given in Eqn. (210) in those eigenfunctions. By making use of Eqn. (209), we can choose

$$E_0 = E_1 = \frac{\hbar^2}{2m}\frac{\pi^2}{a^2}. \tag{10.212}$$

Then it follows from Eqn. (204) that

$$\epsilon = \frac{pa}{\pi\hbar}, \tag{10.213}$$

therefore,

$$\hat{\epsilon} = -\frac{i}{\pi}\frac{d}{d\eta}. \tag{10.214}$$

The eigenfunctions of $\hat{\epsilon}$ normalized to the δ-function, which correspond to its eigenvalues ϵ, are

$$\psi_\epsilon(\eta) = 2^{-1/2}\exp(i\pi\epsilon\eta). \tag{10.215}$$

They obey the conditions

$$(\psi_\epsilon(\eta), \psi_{\epsilon'}(\eta)) = \delta(\epsilon - \epsilon'). \tag{10.216}$$

The expansion of $\psi_n(\eta)$ in these functions is of the form

$$\psi_n(\eta) = \int_{-\infty}^{\infty} \alpha_n(\epsilon)\psi_\epsilon(\eta)\,d\epsilon, \tag{10.217}$$

where $\alpha_n(\epsilon)$ is determined, in view of Eqns. (210), (215), and (216), by

$$\alpha_n(\epsilon) = (\psi_n(\eta), \psi_\epsilon(\eta)) = \int_0^1 \sin n\pi\eta \exp(-i\pi\epsilon\eta)\,d\eta. \tag{10.218}$$

Simple calculations yield

$$\alpha_n(\epsilon) = \frac{n}{\pi(n^2 - \epsilon^2)}[1 + \exp(i\pi\epsilon)] \quad (n \text{ odd}), \tag{10.219a}$$

$$\alpha_n(\epsilon) = \frac{n}{\pi(n^2 - \epsilon^2)}[1 - \exp(-i\pi\epsilon)] \quad (n \text{ even}), \tag{10.219b}$$

whence follows

$$\rho_n(\epsilon) = |\alpha_n(\epsilon)|^2 = \frac{4n^2}{\pi^2(n^2 - \epsilon^2)^2}\cos^2\frac{\pi\epsilon}{2} \quad (n \text{ odd}), \tag{10.220a}$$

$$\rho_n(\epsilon) = \frac{4n^2}{\pi^2(n^2 - \epsilon^2)^2}\sin^2\frac{\pi\epsilon}{2} \quad (n \text{ even}). \tag{10.220b}$$

In view of Eqn. (205) and

$$E_{dn} = n^2 \tag{10.221}$$

resulting from Eqns. (204) and (212), we get from Eqns. (220) for both n odd and even

$$\rho_n(\chi) = \frac{4}{n\pi^2\chi^2(2-\chi)^2} \sin^2 \frac{\pi n\chi}{2}. \tag{10.222}$$

Equations (211) and (222) give the probability densities for the dimensionless coordinate, η, and the relative difference, χ, between the momentum and energy root squared, respectively, in the nth stationary state of the abstract particle under consideration.

Let us now examine what these probability densities are in the limiting case $n \to \infty$. One may guess that when $n \to \infty$, the term $\cos 2\pi n\eta$ in Eqn. (211) contributes nothing to the probability distribution of η. This guess can be affirmed by the calculation of the distribution function (see Sect. 3.2)

$$F_n(\eta') = \int_{-\infty}^{\eta'} \rho_n(\eta)\, \mathrm{d}\eta. \tag{10.223}$$

Equation (211) yields for $0 < \eta' < 1$

$$F_n(\eta') = \int_0^{\eta'} (1 - \cos 2\pi n\eta)\, \mathrm{d}\eta = \eta' - \frac{1}{2\pi n} \sin 2\pi n\eta'. \tag{10.224}$$

When $n \to \infty$, the term containing $\sin 2\pi n\eta'$ vanishes, which proves that

$$\lim_{n\to\infty} \rho_n(\eta) = \delta_1(\eta - \tfrac{1}{2}). \tag{10.225}$$

When proceeding to the limit in Eqn. (222), we have

$$\lim_{n\to\infty} \rho_n(\chi) = 0 \qquad (\chi \neq 0, 2). \tag{10.226}$$

Then, since $\rho_n(0) = \rho_n(2) = n/4$,

$$\lim_{n\to\infty} \rho_n(\chi) = \infty \qquad (\chi = 0, 2). \tag{10.227}$$

On the other hand, in view of Eqn. (205) and

$$\int_{-\infty}^{\infty} \rho_n(\chi)\, \mathrm{d}\chi = 1, \tag{10.228}$$

we get

$$\int_{-\infty}^{1} \rho_n(\chi)\, \mathrm{d}\chi = \int_1^{\infty} \rho_n(\chi)\, \mathrm{d}\chi = \tfrac{1}{2} \tag{10.229}$$

for any n and, hence, for $n \to \infty$.

Thus, $\lim_{n\to\infty} \rho_n(\chi)$ possesses all the properties of δ-function, and we can finally write

$$\lim_{n\to\infty} \rho_n(\chi) = \tfrac{1}{2}\,\delta(\chi) + \tfrac{1}{2}\,\delta(2-\chi). \tag{10.230}$$

The right-hand sides of Eqns. (225) and (230) coincide with the right-hand sides of Eqns. (201a) and (206), respectively. This means that for large enough quantum numbers n, that is, for sufficiently large energy values, E_d, the quantum mechanical probability distributions coincide with the classical ones for the example under consideration too, as should be the case.

10.6. One-Dimensional Harmonic Oscillator

This example has actually been considered in Sections 8.3, 6.2, 6.4, and 3.3. The classical method presented in Sections 8.3 and 6.2 gives for the dimensionless coordinate η and momentum ϵ [see Eqns. (8.7) and (8.8)] the probability densities of the form

$$\rho(\eta) = \pi^{-1}(E_d' - \eta^2)^{-1/2}, \tag{a}$$

$$\rho(\epsilon) = \pi^{-1}(E_d' - \epsilon^2)^{-1/2} \tag{b}$$

[see Eqn. (8.14)], where E_d' is the value of the dimensionless energy E_d [see Eqn. (8.11)], which characterizes the state of the abstract one-dimensional harmonic oscillator under consideration.

The quantum approach of Section 6.4 gives, in this case, for $E_d' = 2n+1$,

$$\rho(\eta) = \pi^{-1/2}(2^n n!)^{-1} \exp(-\eta^2) H_n^2(\eta), \tag{c}$$

$$\rho(\epsilon) = \pi^{-1/2}(2^n n!)^{-1} \exp(-\epsilon^2) H_n^2(\epsilon) \tag{d}$$

[see Eqn. (6.110)]. As has been shown there, if the quantum number n is large enough, the quantum mechanical probability densities of η and ϵ practically coincide with the respective probability densities obtained when the classical method has been used [see Eqns. (6.111) and (6.112)].

10.7. A Particle in a Central Field

We consider a concrete nonrelativistic particle of mass m moving in a central field $U = U(r)$, where r is the distance between the particle and the origin of a frame of reference. It is convenient here to use the polar (spherical) coordinates r, θ, and φ. The classical Hamiltonian function of the particle is, in view of Eqns. (78) and (18),

$$H = \frac{1}{2m}\left(p_r^2 + \frac{\mathcal{M}^2}{r^2}\right) + U(r), \tag{10.231}$$

where p_r and $\mathcal{M}^2$ are the momentum canonically conjugate to r and the angular momentum squared, respectively. Let us now apply the same two methods of probabilistics to the corresponding abstract particle.

(a) *Classical Method.* The microcanonical distribution for the case under consideration is, in view of Eqns. (9.1), (9.2), and (231),

$$\rho(r, \theta, \varphi, p_r, p_\theta, p_\varphi) = c\,\delta\left(\frac{p_r^2}{2m} + \frac{\mathcal{M}^2}{2mr^2} + U(r) - E\right), \tag{10.232}$$

where $\mathcal{M}^2$ is that given in Eqn. (77). A state of the abstract particle is characterized by a definite constant value E of energy, as should be the case for the microcanonical distribution. At the same time, since U does not depend on θ and φ, both $\mathcal{M}_z = p_\varphi$ and $\mathcal{M}^2$ also have definite constant values, which give an additional characterization of a state.

The probability density of φ is constant:

$$\rho(\varphi) = (2\pi)^{-1}. \tag{10.233}$$

In the state of the abstract particle characterized by the value μ of $\mathcal{M}_z$ and the value l^2 of $\mathcal{M}^2$, the probability densities of θ and p_θ are determined, in view of Eqn. (77), by

$$\rho(\theta, p_\theta) = \text{const}\;\delta\left(p_\theta^2 + \frac{\mu^2}{\sin^2\theta} - l^2\right). \tag{10.234}$$

It follows from Eqn. (77), too, that

$$p_\theta = \pm\left(l^2 - \frac{\mu^2}{\sin^2\theta}\right)^{1/2}. \tag{10.235}$$

We now get from Eqn. (234), with due allowance for the considerations of Section 6.2 and for the fact that spherical coordinates have been used,

$$\rho(\theta) = \text{const}\left(l^2 - \frac{\mu^2}{\sin^2\theta}\right)^{-1/2}\sin\theta. \tag{10.236}$$

By introducing

$$x = \cos\theta \tag{10.237}$$

we obtain from Eqn. (236)

$$\rho(x) = \text{const}\left(l^2 - \frac{\mu^2}{1 - x^2}\right)^{-1/2}. \tag{10.238}$$

Let us note, by the way, that, since $\mu \leqslant l$, the condition

$$|x| \leqslant \left(1 - \frac{\mu^2}{l^2}\right)^{1/2} \tag{10.239}$$

should be satisfied.

We will introduce, too, the dimensionless quantities:

$$\eta = r(2mE_0)^{1/2}\mathcal{M}_0^{-1}; \qquad \epsilon = p_r(2mE_0)^{-1/2};$$
$$U_d(\eta) = U(r)/E_0; \qquad E_d = E/E_0; \qquad H_d = H/E_0. \tag{10.240}$$

In terms of these quantities we can write

$$\epsilon^2 + l'^2/\eta^2 + U_d(\eta) - E_d = 0 \tag{10.241}$$

($l' = l/\mu_0$ is the value of the dimensionless angular momentum), whence the probability densities of η and ϵ are determined by

$$\rho(\eta, \epsilon) = \text{const}\ \delta(\epsilon^2 + l'^2/\eta^2 + U_d(\eta) - E_d). \tag{10.242}$$

Equation (241) yields

$$\epsilon = \pm(E_d - U_d(\eta) - l'^2/\eta^2)^{1/2}. \tag{10.243}$$

This leads, in view of the fact that we use spherical coordinates, to

$$\rho(\eta) = \text{const}\ (E_d - U_d(\eta) - l'^2/\eta^2)^{-1/2}\eta^2. \tag{10.244}$$

The constants in Eqns. (238) and (244) can be determined from the conditions of the type given in Eqn. (6.43). On the strength of Eqns. (235) and (243), p_θ is a function of x, and ϵ is a function of η, respectively. Therefore, the probability densities of p_θ and ϵ can be found, if desirable, on the basis of the considerations of Section 3.3.

For the important particular case, when

$$U(r) = -\alpha/r, \tag{10.245}$$

we choose

$$E_0 = m\alpha^2/2\mathcal{M}_0^2. \tag{10.246}$$

Then

$$r = \eta\mathcal{M}_0^2/m\alpha, \tag{10.247}$$

$$p_r = \epsilon m\alpha/\mathcal{M}_0, \tag{10.248}$$

$$U_d(\eta) = -2/\eta, \tag{10.249}$$

and Eqn. (241) takes on the form

$$\epsilon^2 + l'^2/\eta^2 - 2/\eta - E_d = 0. \tag{10.250}$$

Since $U_d(\eta)$ is negative, so is E_d. It is easy to prove that

$$-E_d \leqslant l'^{-2}, \tag{10.251}$$

$$\eta_1^{-1} \leqslant \eta^{-1} \leqslant \eta_2^{-1}, \tag{10.252}$$

$$\epsilon_1 \leqslant \epsilon \leqslant \epsilon_2, \tag{10.253}$$

$$\eta_1^{-1} = l'^{-1}(l'^{-1} + \epsilon_1), \qquad \eta_2^{-1} = l'^{-1}(l'^{-1} + \epsilon_2), \tag{10.254}$$

where

$$\epsilon_1 = -(l'^{-2} + E_d)^{1/2}, \qquad \epsilon_2 = +(l'^{-2} + E_d)^{1/2}. \tag{10.255}$$

Equation (244) now takes the form

$$\rho(\eta) = \text{const}\ (2/\eta - l'^2/\eta^2 + E_d)^{-1/2}\, \eta^2. \tag{10.256}$$

(b) *Quantum Approach*. In view of Eqn. (99), Schrödinger equation (without time) for a particle in a central field is

$$\left[\frac{1}{2m}\left(\hat{p}_r^2 + \frac{\hat{\mathcal{M}}^2}{r^2}\right) + (U(r) - E)\hat{I}\,\right]\psi = 0, \tag{10.257}$$

where

$$\hat{p}_r^2 = -\frac{\hbar^2}{r^2}\frac{\partial}{\partial r}\left(r^2\frac{\partial}{\partial r}\right) \tag{10.258}$$

and $\hat{\mathcal{M}}^2$ is that as given in Eqn. (98). The operators $\hat{H}$, $\hat{\mathcal{M}}^2$, and $\hat{\mathcal{M}}_z$ commute and, hence, have common eigenfunctions of the form

$$\psi = \psi_{E,l}(r)\,\psi_{l,m}(\theta)\,\psi_m(\varphi), \tag{10.259}$$

where $\psi_m(\varphi)$, $\psi_{l,m}(\theta)$ and $\psi_{E,l}(r)$ obey the equations

$$\hat{\mathcal{M}}_z\psi_m(\varphi) = \hbar m\,\psi_m(\varphi), \tag{10.260}$$

$$\left\{\frac{1}{\sin\theta}\frac{\mathrm{d}}{\mathrm{d}\theta}\left(\sin\theta\frac{\mathrm{d}}{\mathrm{d}\theta}\right) - \left[\frac{m^2}{\sin^2\theta} - l(l+1)\right]\hat{I}\right\}\psi_{l,m}(\theta) = 0, \tag{10.261}$$

and

$$\left\{\frac{\hbar^2}{2mr^2}\frac{\mathrm{d}}{\mathrm{d}r}\left(r^2\frac{\mathrm{d}}{\mathrm{d}r}\right) + \left[E - U(r) - \frac{l(l+1)}{r^2}\right]\hat{I}\right\}\psi_{E,l}(r) = 0 \tag{10.262}$$

respectively.

A state of the corresponding abstract particle is characterized by a definite value m of $\mathcal{M}_z$, a definite value l related to $\mathcal{M}^2$, and a definite value E_0 of H and energy. The probability density of φ obviously is constant and precisely of the form as given in Eqn. (233). The probability density of $x = \cos\theta$ is given, in view of Eqn. (144), by

$$\rho_{l,m}(x) = (l + \tfrac{1}{2})\,\frac{(l-m)!}{(l+m)!}\,[P_l^m(x)]^2 \quad (m = 0, 1, 2, \ldots, l; l = 0, 1, \ldots). \quad (10.263)$$

The probability density of r depends on the form of $U(r)$. We consider here only the case represented by Eqn. (245). By choosing

$$\mathcal{M}_0 = \hbar \quad (10.264)$$

we get from Eqn. (262) the corresponding dimensionless equation

$$\frac{d^2R(\eta)}{d\eta^2} + \frac{2}{\eta}\frac{dR(\eta)}{d\eta} + \left(\frac{2}{\eta} - \frac{l(l+1)}{\eta^2} + E_d\right)R(\eta) = 0. \quad (10.265)$$

The solution of this equation gives discrete negative values for E_d:

$$E_d = -n^{-2} \quad (n = 1, 2, \ldots) \quad (10.266)$$

and for $R(\eta)$ the corresponding functions:

$$R_{nl} = -\frac{2}{n^2}\left\{\frac{(n-l-1)!}{[(n+l)!]^3}\right\}^{1/2}\left(\frac{2\eta}{n}\right)^l \exp\left(-\frac{\eta}{n}\right) L_{n+l}^{2l+1}\left(\frac{2\eta}{n}\right), \quad (10.267)$$

where L_{n+l}^{2l+1} are the so-called associated Laguerre polynomials.

The probability density of η is

$$\rho(\eta) = \frac{4}{n^4}\frac{(n-l-1)!}{[(n+l)!]^3}\left(\frac{2\eta}{n}\right)^{2l}\exp\left(-\frac{2\eta}{n}\right)\left[L_{n+l}^{2l+1}\left(\frac{2\eta}{n}\right)\right]^2\eta^2. \quad (10.268)$$

In the case under consideration, the coincidence of, some results of the classical method with the corresponding ones of the quantum approach is evident, in particular, for the probability density $\rho(\varphi)$ and the magnitude of the E_d-values. To compare all the results of the classical method with those of the quantum approach in the classical limit, it is necessary to know, first of all, the asymptotic expressions for the associated Laguerre and Legendre polynomials at the corresponding large quantum numbers. If they were known, one would be able to verify the coincidence of both results. But they are not known at present, and one might draw some conclusions about the asymptotic behavior of those polynomials, proceeding from the fact that the results of the classical method must coincide with the results of the quantum approach in the classical limit.

10.8. Equilibrium Quantum Statistical Mechanics

'Quantum statistical mechanics' seems to be the proper name for statistical mechanics of any systems whose quantum properties should be taken into consideration. As a matter of fact, however, it is mainly applied to the topics related to the probability distributions which correspond to the canonical one in the classical case. Therefore, quantum statistical mechanics is supposed here to deal with systems contacting thermostats, and we will confine ourselves mainly to obtaining some basic equations.

We consider, first, a concrete closed mechanical system consisting of two pseudoclosed subsystems I and II (see Sect. 8.2). Let us denote the totalities of the spatial coordinates of these subsystems by x' and x'', respectively. We will regard time in both subsystems as formally different and denote it, respectively, by t' and t''. Since each of the concrete subsystems is pseudoclosed, so are the corresponding abstract subsystems. They have, therefore, their own Hamiltonians, $\hat{H}'$ and $\hat{H}''$, and energy operators, $\hat{E}'$ and $\hat{E}''$, respectively. The latter are of the form

$$\hat{E}' = i\hbar \frac{\partial}{\partial t'}; \qquad \hat{E}'' = i\hbar \frac{\partial}{\partial t''}. \tag{10.269}$$

The energy values of each entire concrete closed system equals the sum of the energy values of the concrete pseudoclosed subsystems it consists of. Therefore, the energy operator $\hat{E}$ for the entire abstract closed system is

$$\hat{E} = \hat{E}' + \hat{E}''. \tag{10.270}$$

For the same reason, the Hamiltonian $\hat{H}$ for the whole system is

$$\hat{H} = \hat{H}' + \hat{H}''. \tag{10.271}$$

The abstract subsystems I and II by themselves can be, in any of their closedness intervals, only in stationary states with certain definite energy values E' and E'', respectively, described by the corresponding functions $\psi_{E'}(x', t')$ and $\psi_{E''}(x'', t'')$, which satisfy the respective Schrödinger equations

$$\hat{E}'\psi_{E'} = \hat{H}'\psi_{E'}, \tag{10.272}$$

$$\hat{E}''\psi_{E''} = \hat{H}''\psi_{E''} \tag{10.273}$$

in those time intervals. The entire abstract system is thereby in a stationary state with the energy value

$$E = E' + E'', \tag{10.274}$$

the state being described by a function $\psi_E(x', t'; x'', t'')$ satisfying the equation

$$\hat{E}\psi_E = i\hbar\left(\frac{\partial \psi_E}{\partial t'} + \frac{\partial \psi_E}{\partial t''}\right) = H\psi_E. \tag{10.275}$$

Beyond the closedness intervals, the subsystems interact, changing thereby their energy values, the energy value of the whole system being kept constant. Therefore, when considering the entire closed system, we have to think that in the stationary state of the system with the energy value E, the energy values E' and E'' of the pseudoclosed subsystems I and II can vary, but must obey Eqn. (274). Then the function describing the state, normalized to unity, should be of the form

$$\psi_E(x', t'; x'', t'') = \int a(E')\psi_{E'}(x', t')\psi_{E-E'}(x'', t'')\,\mathrm{d}E'. \tag{10.276}$$

Indeed, this function satisfies Eqns. (272) and (273), and hence, Eqn. (275) as well. It obviously is a common eigenfunction of the operators $\hat{H}$ and $\hat{E}$ corresponding to their common eigenvalue $E = H$.

The coefficients $a(E')$ in Eqn. (276) are connected to the energy distribution in subsystem I and, hence, also to that in subsystem II, as will be shown below.

In a state where the entire closed system has a definite energy value, each of the pseudoclosed subsystems it consists of does not. Therefore, a state of a separate subsystem of a closed system can no longer be described by a function of the form given in Eqn. (171). A question may now be asked: How to describe such a state? Certainly, it is enough to answer the question, say, for subsystem I.

If some physical quantity F is inherent in subsystem I, then the eigenvectors of the corresponding operator $\hat{F}$ in the x-representation are functions of only x' and t'. The coefficients of the expansion of a function of the form as given in Eqn. (276), which describes a stationary state of the entire closed system, in the eigenfunctions of the operator $\hat{F}$ and, hence, the probability density of F, are functions of x'' and t''. Given definite values of x'' and t'', we can obtain the corresponding conditional probability distributions for the physical quantity F. It is possible, however, to characterize a state of subsystem I by the probability distribution of F at any values of x'' and t''. To this end, it is necessary to integrate the squared modulus of the coefficients of the indicated expansion over all values of x'' and t''. A state of subsystem I described in such a way is obviously uniquely determined by the state of the entire closed system and, hence, by its energy value (if the state is nondegenerate).

Let us consider the mathematical expectation of a quantity F inherent in subsystem I in the state of the entire closed system described by a function $\psi = \psi(x', t'; x'', t'')$ normalized to unity. In accordance with Eqn. (5.109), we write this mathematical expectation as

$$MF = (\hat{F}\psi, \psi). \tag{10.277}$$

Since $\hat{F}$ operates only on x' and t', we can integrate over x'' and t'' beforehand and introduce the density matrix

$$W(x', t'; x, t) = \iint \psi(x', t'; x'', t'')\bar{\psi}(x, t; x'', t'')\,\mathrm{d}x''\,\mathrm{d}t'' \tag{10.278}$$

(cf. Sect. 5.4). Equation (277) can now be rewritten, in accordance with Eqn. (5.110), as

$$MF = \mathrm{Tr}\,\hat{F}\hat{W}, \tag{10.279}$$

where $\hat{W}$ is the density operator whose matrix in the x', t'-representation is that as given in Eqn. (278).

For the particular case, when the function $\psi(x', t'; x'', t'')$ is of the form of Eqn. (276), we have, in view of

$$\psi_{E'}(x', t') = \psi_{E'}(x') \exp(-iE't'/\hbar) \tag{10.280}$$

and

$$\psi_{E-E'}(x'', t'') = \psi_{E-E'}(x'') \exp[-i(E-E')t''/\hbar], \tag{10.281}$$

$$W(x', t'; x, t) = \int |a(E')|^2 \psi_{E'}(x')\bar{\psi}_{E'}(x) \exp[-iE'(t'-t)]\, \mathrm{d}E' \tag{10.282}$$

(the intermediate calculation being omitted).

For $\hat{F} = \hat{E}'$, Eqns. (279) and (282) yield

$$ME' = \int |a(E')|^2 E'\, \mathrm{d}E' \tag{10.283}$$

provided the $\psi_{E'}(x')$ are normalized to unity. Equation (283) shows that $|a(E')|^2$ is the probability density for subsystem I to be in a state with a definite energy value E', which is described by the function $\psi_{E'}(x', t')$. At the same time, it is the probability density of the energy $E - E'$ of subsystem II, since the entire closed system is in a state with a definite energy value E.

In view of the indicated fact, we can rewrite Eqn. (282) for the density matrix at $t' = t$ as

$$W(x', t; x, t) = W(x'; x) = \int |a(E')|^2 \psi_{E'}(x')\bar{\psi}_{E'}(x)\, \mathrm{d}E'. \tag{10.284}$$

This shows, by the way, that the probability density of x', which is

$$\rho(x') = W(x', x') = \int |a(E')|^2 |\psi_{E'}(x')|^2\, \mathrm{d}E' \tag{10.285}$$

does not depend on time.

Equation (285) has the most usual probabilistic meaning (cf. Sect. 2.3), since the integrand is the product of the conditional probability density of x', when subsystem I is in a state described by the function $\psi_{E'}(x')$, and the probability density for this subsystem to be in that state.

The probability distributions and mathematical expectations of any physical quantities inherent in subsystem I can be obtained, if related to x', by making use of Eqn. (285) or (284) and (279), respectively.

Let us now suppose subsystem II to be a macroscopic one and very large relative to subsystem I, and let E' be very small relative to $E - E'$. In this case, subsystem II can be considered a thermostat, while $|a(E')|^2$ is the probability density of its energy $E - E'$. Since the classical method is applicable to finding the energy distribution for a big macroscopic system (this can be even a monatomic ideal gas), we may use for the calculation of $|a(E')|^2$ the considerations of Section 9.2. Thereby we obtain

$$|a(E')|^2 = Z^{-1} \exp(-E'/kT), \tag{10.286}$$

where T is the absolute temperature of the thermostat and Z^{-1} is the normalization factor to be determined from the condition

$$\int |a(E')|^2 \, \mathrm{d}E' = Z^{-1} \int \exp(-E'/kT) \, \mathrm{d}E' = 1. \tag{10.287}$$

This gives

$$Z = \int \exp(-E'/kT) \, \mathrm{d}E'. \tag{10.288}$$

If the energy values are discrete, Eqn. (288) goes over into

$$Z = \sum_i \exp(-E'/kT). \tag{10.289}$$

In view of Eqn. (286), Eqn. (284) can be rewritten as

$$W(x';x) = Z^{-1} \int \exp(-E'/kT) \psi_{E'}(x') \bar{\psi}_{E'}(x) \, \mathrm{d}E'. \tag{10.290}$$

Since $E' = H'$ is the eigenvalue of $\hat{H}'$ corresponding to its eigenfunction $\psi_{E'}(x')$, we can finally write, by making use of the mathematics of Section 5.3,

$$\hat{W} = Z^{-1} \exp(-\hat{H}/kT), \tag{10.291}$$

while Z can obviously be represented in a form valid for both continuous and discrete energy values, namely,

$$Z = \mathrm{Tr} \exp(-\hat{H}/kT). \tag{10.292}$$

Equations (291) and (292) are widely used in conventional quantum statistical physics, in particular, for the representation of the density operator $\hat{W}$.

Equation (290) corresponds to Eqn. (9.33) of classical statistical mechanics, and when it is possible to proceed to the classical limit, the former practically coincides with the latter, as should be the case. The reservation made about the possibility of proceeding to the classical limit is conditioned by the fact that some quantum numbers, such as those of spin, for instance, are limited by the very nature of the physical quantity and can therefore never be sufficiently large.

The equations of Chapter 9 connecting the thermodynamic functions with Z remain unchanged, but it is necessary to take into consideration, too, that the methods of calculating Z in the quantum case differ, in view of Eqn. (289), from those in the classical case. More about this matter can be found in the conventional literature (see Refs. [2] and [4]).

Problems

10.1. A particle of mass m is in a central field having constant values of the potential energy $U(r)$ in three different intervals, namely, $U(r) = 0$ for $0 \leqslant r < r_1$ (interval 1), $U(r) = U_1$ for $r_1 < r < r_2$ (interval 2), and $U(r) = 0$ for $r_2 < r < \infty$ (interval 3). For this system find the wave functions $\psi_E(r)$ describing its stationary states corresponding to the zero value of the angular momentuam and the energy values $E < U_1$.

10.2. Consider the particular case of the preceding problem: $r_2 = \infty$.

10.3. Find the probability density $\rho(v)$ of the radial velocity of the particle in a stationary state, in problem 1.

10.4. Calculate the vibrational part of the Helmholtz free energy, F, of a sample consisting of N identical diatomic gas molecules, provided the vibrations are harmonic.

Suggested References

1. Goldstein, H.: 1980, *Classical Mechanics*, 2nd edn., Addison-Wesley, Reading. This book contains the detailed consideration of the rotations of a mechanical system and the discussion of some peculiarities of the angular momentum.
2. Huang, K.: 1963, *Statistical Mechanics*, Wiley, New York. This book, among a number of others, contains methods for obtaining thermodynamic functions, proceeding from quantum statistical mechanics.
3. Landau, L. D., and E. M. Lifshitz: 1977, *Quantum Mechanics, Non-Relativistic Theory*, tr. by J. B. Sykes and J. S. Bell, 3rd edn., Pergamon Press, New York. In this book, one can find the detailed analysis of the problem of a particle in the central field.
4. Landau, L. D., and E. M. Lifshitz: 1969, *Statistical Physics*, tr. by J. B. Sykes and M. J. Kearsley, 2nd rev. edn., Pergamon Press, New York. This book also contains methods for obtaining thermodynamic functions, proceeding from quantum statistical mechanics.
5. Mayants, L. S.: 1967, *Quantum Physics and Transformations of Matter* (in Russian), VINITI, Moscow. In this monograph the basic principles of quantum mechanics as resulting directly from the probabilistic treatment of physical systems were formulated for the first time.
6. Mayants, L. S.: 1973, *Found. Phys.* **3**, 413. This paper claims for the first time that quantum physics is a particular domain of probabilistic physics.

7. Mayants, L. S.: 'New Rotation-Vibration Hamiltonian' (to be published). This paper, based on the talk given by the author at the atomic physics seminar at the University of Massachusetts (Amherst) on 21 September 1981, considers for the first time a new approach to the construction of the proper operator for the angular momentum.
8. Schiff, L. I.: 1968, *Quantum Mechanics*, 3rd edn., McGraw-Hill, New York. In this book there is a detailed analysis of the problem of a particle in the central field.

Chapter 11

Kinetics of Physical Transformations

11.1. Processes in Concrete Physical Systems

A concrete physical system may change in time. Any such change will be called a process, provided the change exists in every inertial coordinate system. In this broad sense, not only chemical reactions, positron–electron 'annihilation', nuclear reactions, etc., but also the emission and absorption of photons, and even a change in a purely mechanical motion, are particular cases of processes. Free motion of a concrete physical object is not a process, because there exists an inertial coordinate system in which the object is at rest.

Since free motion of a concrete physical object is possible only when it is not acted upon by other physical objects, the necessary condition for the occurrence of a process is the interaction of concrete physical systems. Hence, for a concrete physical system a process may be due either to its interaction with other concrete physical systems or to interaction among different parts of the system itself. In the second case the system may be considered closed (more exactly pseudoclosed, which does not, however, make any difference for the subsequent considerations; see Sect. 8.2), and then the processes in it are internal. It may turn out that in the first case all the interacting systems may be united to form one closed system. In this case, too, one may regard the processes as internal for the corresponding overall concrete closed physical system.

Thus, a common property of processes in concrete physical systems is their causal conditionality. One may consider this a manifestation of a universal principle of nature – the causality principle – which may be formulated as follows: Any concrete process is conditioned by the interaction of concrete physical objects.

Internal processes in concrete closed physical systems also have such properties, which reveal themselves, in particular, through the fulfillment of the conservations laws of physics. One may regard these properties as a manifestation of another universal principle of nature – the principle of nonannihilatibility of matter (cf. Sect. 8.4). The universality of this principle is reflected in the universality of the way in which any process is written down. Every concrete process may be written down in the form of an equation, like a chemical reaction, the notation of the initial and of the final physical objects being on the left and on the right, respectively; the equality sign corresponds to the principle of nonannihilatibility of matter and shows that all conservation laws are fulfilled.

In the following we shall be interested in processes of transformation (in short, in transformations) of physical systems, that is, in processes, as a result of which a closed physical system is transformed into one or several other physical systems. Decay of nuclei and intramolecular rearrangements are the important examples of such processes. According to the causality principle, transformation of a concrete physical system is conditioned by the specific interaction of different parts of the system. But such specific interaction can arise only when a definite spatial arrangement of these parts is achieved, that is, when an appropriate configuration of the system occurs. This idea was utilized for the construction of the theory of absolute rates of chemical reactions. On that theory this specific configuration was called the 'activated complex' or the 'transition state'. I call it a transitional configuration.

The causality principle thus leads to the principle of transitional configuration, according to which the necessary condition for the possibility of transformation of a concrete physical system is the emergence of an appropriate transitional configuration.

The essential features of transformations are conveniently illuminated by physical systems that can be treated by means of a model of the one-dimensional motion of a particle in a field whose potential energy $U(q)$ is independent of time (q is the coordinate of the particle). A closed physical system represented by such a model consists of the particle and another physical object whose mass is large enough compared to that of the particle. The inertial coordinate system is supposed to be the center-of-mass system of the whole physical system under consideration.

In order that some definite transformation could be picked out of the (generally compound) whole internal motion of a system, an objective criterion should exist, which allows one to distinguish between what was before the transformation and what became after it. Since one and the same closed physical system is considered, only the existence of physically different situations accessible to this system can provide such a criterion. Therefore, the notion of 'transformation' may have a meaning only when a system can be in at least two physically different situations. But in the one-dimensional motion model the existence of these situations may be associated only with the presence of the particle in different regions of the one-dimensional space.

Thus, we see that to be able to speak of the transformation of a concrete physical system it is necessary, first of all, that two regions occur in the one-dimensional space of the model representing this system, the location of a particle in the regions corresponding to at least two physically different possible situations for the system. For example, for a system consisting of an α-particle and some nucleus these regions are determined by the region inside the nucleus and that outside it, and the two corresponding physically different situations are the location of the α-particle inside and outside the nucleus, respectively.

A system that is contained in any one of the physically different situations will be called a variant of the system. Then, any transformation taking place in a concrete system is a transition (transformation) of one variant of the system into another. In the above example one variant is the α-emitting nucleus and the other is the decay products (the new nucleus plus the α-particle), the decay corresponding to the transition of the first variant into the second.

Let us consider the case where in the one-dimensional motion model the displacement of a particle from one space region (I) into another (II) corresponds to the transition of one system variant (I) into another (II). Since any concrete particle trajectory is continuous, the particle cannot move from region I into region II without passing through the transitional point with the coordinate value q_0, which divides the two regions.

Certainly, it would be more correct to speak about a small transitional interval rather than a transitional point. However, the subsequent deductions do not depend on this refinement, and for the sake of simplicity we shall always make use of the notion of a transitional point.

The above considerations are applicable not only to concrete closed systems that can be represented by a one-dimensional motion model, but also to concrete closed systems whose transformations are connected with the change of one internal coordinate q. It is natural to call the q coordinate the transformation path coordinate, and any configuration of the system for which the q coordinate has the value q_0 corresponding to the transitional point the transitional configuration. The fact that any transformation of a concrete closed physical system has to go through a corresponding transitional configuration represents the transitional configuration principle.

A concrete particle in region I at a short enough distance Δq from the transitional point will necessarily move into region II after a time not exceeding Δt if its velocity is directed toward region II (we denote it then by v_+), being connected to Δq by the relation

$$\Delta q < v_+ \Delta t. \tag{11.1}$$

Hence, Eqn. (1) is the necessary and sufficient condition for the transformation of variant I of the corresponding concrete physical system into variant II over a time not exceeding Δt (provided, of course, that Δt and, therefore, Δq are small enough).

11.2. Probabilistic Treatment of Transformations

To sets of concrete variants I and II of a physical system there correspond abstract objects: the respective variants I and II of the corresponding abstract physical system. In the one-dimensional motion model the abstract object 'a particle in region I' stands for the whole set of concrete particles in region I for which the property 'coordinate' has any value in the interval (a, q_0) and the property 'velocity' has any value in the interval $(-c, c)$; a is the least value of q, and c is the maximum particle velocity. All the concrete particles of the class with a definite velocity value v_+ will necessarily move from region I into region II in a given short enough time Δt if they have coordinate values in the interval $(q_0 - v_+\Delta t, q_0)$. Therefore, the probability $\Delta W(v_+, \Delta t)$ that the abstract particle having the velocity value v_+ will go over from region I into region II during a time Δt is

$$\Delta W(v_+, \Delta t) = \int_{q_0 - v_+ \Delta t}^{q_0} \rho(q, v_+)\, dq = \rho(q_0, v_+) v_+ \Delta t, \tag{11.2}$$

where $\rho(q, v)$ is the probability density of the two-dimensional random variable (q, v), which relates to region I and satisfies the condition

$$\int_{-c}^{c} \int_{a}^{q_0} \rho(q, v)\, dq\, dv = 1. \tag{11.3}$$

The probability $\Delta W(\Delta t)$ that the abstract particle, having any velocity value in the interval $(0, c)$, will pass from region I into region II during a time Δt is obtained from Eqn. (2) by integration over all values v_+:

$$\Delta W(\Delta t) = \Delta t \int_{0}^{c} \rho(q_0, v) v\, dv \tag{11.4}$$

(since here all values of v are positive, the plus sign is omitted). Hence, the probability per unit time of the abstract particle making a transition from region I into region II, κ, is determined by the expression

$$\Delta W(\Delta t)/\Delta t \equiv \kappa = \int_{0}^{c} \rho(q_0, v) v\, dv. \tag{11.5}$$

The same expression is obviously valid for the probability per unit time of the transformation of variant I into variant II for the corresponding abstract physical system. This transformation takes place, of course, in a state of the abstract physical system under consideration that is characterized by definite values of all the conserved physical quantities; in particular, the state has to be stationary.

It is convenient to rewrite Eqn. (5) in the form

$$\kappa = \rho(q_0) \bar{v}_+(q_0), \tag{11.6}$$

where

$$\bar{v}_+(q_0) = \int_{0}^{c} \rho_{q_0}(v) v\, dv \tag{11.7}$$

and

$$\rho_{q_0}(v) = \rho(q_0, v)/\rho(q_0) \tag{11.8}$$

is the conditional probability density of the particle velocity at $q = q_0$.

If there is a large enough number N of concrete variants I of a physical system, then on the basis of Axiom 1 of probabilistics (see Sect. 1.6), the decrease $-\Delta N$ of this number during a sufficiently short time Δt, due to transformation into variant II, is determined by the approximate equation

$$-\Delta N \cong N\,\Delta W(\Delta t) = N\kappa\,\Delta t, \tag{11.9}$$

whence

$$\kappa \cong -\Delta N/(N\,\Delta t). \tag{11.10}$$

Thus, κ has also the meaning of the rate constant for the transformation of variant I of a physical system into its variant II. It follows immediately from Eqn. (10) that

$$N \cong N_0 \exp(-\kappa t), \tag{11.11}$$

where N_0 is the number of concrete variants I of the physical system at $t = 0$. Since for closed systems the probability distributions of any physical quantities are independent of time (see Sects. 6.2 and 8.3), the quantity κ determined by Eqn. (6) is a constant. Therefore, the law describing the decrease of the number of concrete variants I of any physical system, due to transformation into variants II, is always approximately exponential (of course, assuming N_0 is large enough and t is not too large). If time t is viewed as a random variable whose values are time instants when the transformation under consideration takes place, then it follows from Eqn. (11) that

$$F(t) = 1 - \exp(-\kappa t) \tag{11.12}$$

is the probability distribution function of this random variable, and the probability that this transformation does not happen prior to the instant t has a strictly exponential form.

It should be emphasized that all of the quantitative reasoning above concerns just variant I, irrespective of what happens further with variant II. Therefore, in particular, the law describing the increase in the number of final concrete physical systems, which are the ultimate result of the transformation, can be quite different from the exponential form.

Note that when we deduced Eqn. (6) no suppositions were made as to whether a physical system is classical or microscopic. Therefore, this equation has to be valid in both cases. But for microsystems this result cannot be utilized at present, because, as indicated in Section 6.4, the probabilistic treatment of microsystems (that is, quantum mechanics) does not permit us to find a joint probability distribution for the coordinate and velocity (except the case of a free particle, which is of no interest here). The following reasoning makes it possible to some extent to avoid this difficulty.

First, a number of conclusions which follow from the above considerations do not depend on whether or not we can make use of Eqn. (6) to calculate κ. Among these is the conclusion, in particular, that for any physical system the probability of preservation of variant I until a time instant t is strictly exponential. This means that the law of decrease of a large enough number of concrete variants I is approximately exponential.

Second, in many cases one may replace the exact Eqn. (6) by the approximate one

$$\kappa_0 = \rho(q_0)\overline{v}_+, \tag{11.13}$$

where

$$\overline{v}_+ = \int_0^c \rho(v)v \, dv \tag{11.14}$$

and where $\rho(v)$ is the probability density of the change rate of the q-coordinate in the corresponding stationary state of a system, which satisfies the condition

$$\int_{-\infty}^{\infty} \rho(v) \, dv = 1. \tag{11.15}$$

In a number of cases Eqn. (13) enables one to calculate approximate values of transformation rate constants for physical systems, by using quantum mechanical techniques and, what is more, to reveal important regularities concerning the decay of physical systems in particular. Of course, generally $\overline{v}_+ \neq \overline{v}_+(q_0)$. However, in many cases both quantities are of the same order of magnitude, since they are both certain mean rates of the q-coordinate change. In these cases the orders of magnitude of κ and of κ_0 coincide, and a quantum mechanical calculation of κ_0 allows one to judge the order of magnitude of κ. So, for a one-dimensional harmonic oscillator in the nth stationary state, for instance, $\overline{\eta}_+ \cong \omega(2n)^{1/2}\pi^{-1}$, if $n \geqslant 5$ [ω is the circular vibration frequency of the oscillator, η is the dimensionless coordinate defined by Eqn. (8.7)]. For the classical oscillator, which corresponds to sufficiently large values of n, $\overline{\eta}_+(\eta_0) = 2^{-1}\omega(2n+1-\eta_0^2)^{1/2} \cong \omega(2n)^{1/2}2^{-1}$.

It is always possible to write the relation of $\overline{v}_+$ to $\overline{v}_+(q_0)$ in the form $\overline{v}_+(q_0) = G\overline{v}_+$, where G is an indeterminable, for the present, positive factor having in many cases a magnitude of order unity. Therefore, one may also write, making use of Eqns. (6) and (13),

$$\kappa = G\kappa_0. \tag{11.16}$$

11.3. Basic Principles of the Transitional Configuration Theory

As noted in Section 1, any process in which a finite number of concrete physical systems takes part may be considered internal for the corresponding concrete closed system. Furthermore, any concrete physical objects, including the so-called elementary particles, necessarily possess internal structure, because they cannot be either points or perfect solid bodies. Finally, the transitional configuration principle following from the universal causality principle may also be considered universal and extended to any physical systems, including those whose internal structure is so far unknown. Therefore, generalizing the results of the foregoing sections, one may formulate the basic principles of the universal theory of transformations of physical systems – of the transitional configuration theory – as follows:

1. Any transformation, being an internal process for a certain concrete closed physical system, takes place at definite values of all physical quantities which have to be conserved.
2. Every transformation becomes possible only when a specific configuration (transitional configuration) arises.
3. The probability per unit time (the rate constant), κ, of transformation of variant I of a closed physical system into variant II is determined by Eqn. (6). In this equation $\rho(q_0)$ is the probability density of the transitional configuration (the latter being characterized by the value q_0 of the transformation path coordinate q) calculated for variant I, and $\overline{v}_{+}(q_0)$ is the conditional average rate of the q-coordinate change in the transformation direction at $q = q_0$ in the stationary state under consideration.

If variant II is a stable final transformation product, then the transformation is a one-step process and the rate constant for it is fully determined by Eqn. (6). However, transformations of physical systems are, as a rule, many-stage processes and their rates (that is, the rates of generation of the final products of transformations) depend generally on the rates of the separate stages in a complicated manner. In case the rate of a certain stage is much less than the rates of all the other stages, the rate of the whole process is, in the main, determined just by the rate of this slowest stage.

If a one-step transformation can be realized in several independent ways, then the probability per unit time of this process is equal to the sum (or to the integral) of probabilities per unit time over all the possible routes for its realization. As a result of a transformation, a concrete physical system can either remain preserved as a whole or decay to separate subsystems which do not interact subsequently. All the above considerations are obviously valid for both cases equally. In the second case each of the subsystems becomes pseudoclosed and can be considered independently with all the processes feasible in it.

Let us now illustrate the transitional configuration theory with the aid of some important examples.

11.4. Leaving a Potential Well Through a Potential Barrier

Many transformations, including any decay, may be treated by using the model of a particle leaving a potential well through a potential barrier. In case a potential barrier has a maximum it is natural to assume that the transitional configuration is determined by the coordinate value corresponding to this maximum. If a barrier is square-topped, the transitional configuration is determined by the coordinate value of the barrier wall which is farthest from the center of the well.

First, we consider the well-known problem of a particle of mass m in a state with orbital quantum number $l = 0$ leaving a spherically symmetric 'square' potential well through a spherically symmetric 'square-topped' potential barrier. For the sake of simplicity we assume that the potential energy $U(r)$ is

$$U(r) = \begin{cases} 0 & 0 \leqslant r < r_1 \\ U_1 & r_1 < r < r_2 \\ 0 & r_2 < r < \infty \end{cases} \tag{11.17}$$

(cf. Problem 10.1).

The Schrödinger equation in this case can be solved exactly. The energy values are positive and belong to a continuous spectrum. The normalized wave function corresponding to the energy value E is

$$\psi_E(r) = r^{-1} \chi_E(r), \tag{11.18}$$

where

$$\chi_E(r) = \begin{cases} A \sin kr, & 0 \leqslant r \leqslant r_1 \\ AB \exp[k'(r - r_1)] + AC \exp[-k'(r - r_1)], & r_1 \leqslant r \leqslant r_2 \\ AD \exp[ik(r - r_2)] + A\bar{D} \exp[-ik(r - r_2)], & r_2 \leqslant r < \infty \end{cases} \tag{11.19}$$

$$k = \hbar^{-1}(2mE)^{1/2}; \qquad k' = \hbar^{-1}[2m(U_1 - E)]^{1/2},$$

$$\begin{aligned} B &= (2k')^{-1}(k' \sin kr_1 + k \cos kr_1), \\ C &= (2k')^{-1}(k' \sin kr_1 - k \cos kr_1), \\ D &= 2^{-1}[(1 - ik^{-1}k')B \exp(k'L) + (1 + k^{-1}k')C \exp(-k'L), \end{aligned} \tag{11.20}$$

$L = r_2 - r_1$, A is the normalization factor.

Let us consider the case $E < U_1$, for which the k' values are real. According to the above, the transitional configuration is determined by the value $r = r_2$. Then the normalization factor A is obtained from the condition

$$4\pi \int_0^{r_2} |\chi_E(r)|^2 \, dr = 1. \tag{11.21}$$

The probability density of the transitional configuration is

$$\rho(r_2) = 4\pi |\chi_E(r_2)|^2 = 4\pi A^2 [B \exp(k'L) + C \exp(-k'L)]^2. \tag{11.22}$$

It now follows from Eqn. (16) that

$$\kappa = 4\pi G A^2 [B \exp(k'L) + C \exp(-k'L)]^2 \overline{v}_+. \tag{11.23}$$

If $\exp(k'L) \gg 1$ (only this case is usually considered) and B is noticeably different from zero, then the rate constant for a particle leaving the well is great. This means that a particle is practically (that is, for the overwhelming part of a sufficiently large number of concrete physical objects) outside the well in this instance. The same conclusion follows also from the form of $\chi_E(r)$ [Eqn. (19)] when $\exp(k'L) \gg 1$ and $B \neq 0$. But if $\exp(k'L) \gg 1$ and $B = 0$, then κ, being determined by

$$\kappa = 4\pi G A^2 C^2 [\exp(-2k'L)] \overline{v}_+ \tag{11.24}$$

is small. In this case a particle is practically inside the well and leaves it only with a small probability per unit time; this is confirmed by the form of $\chi_E(r)$ as well.

The condition $B = 0$ leads to the familiar equation

$$k' \tan kr_1 = -k \tag{11.25}$$

(cf. Refs. [2] and [13], and Problem 10.2), which defines the discrete energy values E_j when $r_2 = \infty$. Calculating A_j^2 to the necessary accuracy for each such value, which is

$$E_j = (2m)^{-1}\hbar^2 k_j^2 \tag{11.26}$$

we find

$$A_j^2 = k_j'[2\pi(1 + k_j' r_1)]^{-1}. \tag{11.27}$$

In addition, as follows from Eqns. (25) and (20),

$$C_j^2 = k_j^2(k_j^2 + k_j'^2)^{-1} = U_1^{-1} E_j. \tag{11.28}$$

The operator of the change rate of r, in view of Eqn. (10.4), is

$$\hat{\dot{r}} = \frac{\hat{p}_r}{m} = -\frac{i\hbar}{m}\left(\frac{\partial}{\partial r} + \frac{1}{r}\hat{I}\right). \tag{11.29}$$

Its eigenfunctions corresponding to the eigenvalues v are

$$\varphi(v, r) = \left(\sin\frac{mvr}{\hbar}\right) r^{-1}. \tag{11.30}$$

Performing the necessary calculations, we find that

$$\rho(v') = 2^{-1}[\delta(v' - v) + \delta(v' + v) \tag{11.31}$$

(cf. Problem 10.3), and hence,

$$\overline{v}_+ = 2^{-1} v = \hbar k/2m. \tag{11.32}$$

Therefore, for the stationary state with the energy value E_j, Eqn. (24) yields

$$\kappa(E_j) = G_j \kappa_0(E_j), \tag{11.33}$$

where

$$\kappa_0(E_j) = \hbar k_j^3 k_j' [m(k_j^2 + k_j'^2)(1 + k_j' r_1)]^{-1} \exp(-2k_j' L). \tag{11.34}$$

Thus, the condition $B = 0$ defines the discrete set of stationary states, in which the probability per unit time of a particle leaving the well is small and determined by Eqn. (33). However, as seen from Eqn. (23), κ values have the same order of magnitude as the corresponding $\kappa(E_j)$ for all values of B for which $|B|$ has the order of magnitude of $\exp(-2k'L)$. Hence, the $\kappa(E)$ values have the same order of magnitude for the continuum of stationary states, for which the energy takes values in the interval of 'relative stability'. We define the 'width' Δ_j of this interval as $\Delta_j = \delta_j^+ - \delta_j^-$, where δ_j^+ and δ_j^- are, respectively, the positive and negative values of δ_j satisfying the condition

$$\kappa(E_j + \delta_j) = 2\kappa(E_j). \tag{11.35}$$

To calculate δ_j we make use of the relation

$$\delta_j = \left[\left(\frac{dB}{dE}\right)_{E=E_j}\right]^{-1} \delta B_j = 2\hbar^2 k_j k_j'^2 [m(k_j^2 + k_j'^2)^{1/2}(1 + k_j' r_1)]^{-1} \delta B_j \tag{11.36}$$

and write δB_j as

$$\delta B_j = b C_j \exp(-2k_j' L). \tag{11.37}$$

Substituting δB_j for B into Eqn. (23) and using Eqn. (24), one obtains, to the necessary accuracy, $(b + 1)^2 = 2$. Taking into consideration Eqns. (36) and (34), one finally gets

$$\Delta_j = 4 \times 2^{1/2} k_j^{-1} k_j' \hbar \kappa_0(E_j). \tag{11.38}$$

Thus, in many cases Δ_j may turn out to be of the same order of magnitude as $\hbar\kappa(E_j)$.

The example considered allows one to understand of what form can be the solution to the general problem of a particle leaving a potential well through a potential barrier (and departing to infinity). If the minimum energy of a particle when it still can leave the well is E_0, then the stationary states corresponding to energy values belonging to a continuous spectrum $E_0 < E < U_{max}$ (U_{max} is the height of the potential barrier) are roughly divided into two groups: the states belonging to a particle inside the well with sufficiently small values of $\kappa(E)$ and the states for which values of $\kappa(E)$ are so large that the particle is practically outside the well. In the continuous energy spectrum there are narrow intervals of 'relative stability around E_j values whose 'width' Δ_j can be of the same order of magnitude as $\hbar\kappa(E_j)$. These intervals correspond to the states belonging to the first group; E_j is the energy corresponding to the least value of $\kappa(E)$ for the jth interval. The ψ_E functions normalized over the region of the well and describing these states differ only slightly from the corresponding normalized ψ_{E_j} functions, the difference increasing as does $|E - E_j|$; the magnitudes of these functions outside the well are small in comparison with the magnitudes inside.

For the states belonging to the second group, which relate to a particle that is practically outside the well, the magnitudes of the ψ_E functions inside the well are small in comparison with those outside it.

In view of the above considerations one may suppose that a system whose decay can be represented by a model involving a particle leaving a potential well through a potential barrier is relatively stable, that is, decays with small probability per unit time, only in the stationary states whose energy takes values on an enumerable set of narrow intervals, the 'width' of these intervals being the smaller the less the probability of decay per unit time. In states with intermediate energy values such a system is practically unstable. This result is, in the main, connected with the fact that the probability densities of a transitional configuration are different for different stationary states. Since the transitional configuration principle is universal, such situations can generally arise for any physical systems capable of decaying.

In particular, the emission of a photon by a system is also a decay process. Therefore, one may assert that there should exist such stationary states of the system in which it is relatively stable with respect to this decay, whereas in states with other energy values the system is fully unstable with respect to such a decay. The states of relative stability of a system capable of photon emission are just the stationary states corresponding to discrete energy levels obtained as a result of solving the Schrödinger equation. It is in these and in adjacent (in narrow energy intervals) stationary states where the photon emission as a rule takes place with small probability per unit time. Consequently, in these states the system is relatively stable with respect to photon emission. In all other states it is apparently fully unstable with respect to photon emission.

A system that in some stationary states is relatively unstable with respect to a certain decay process can be practically quite stable in these same states against all the other possible decay processes. However, strictly speaking, no physical objects exist which could be fully stable forever, and therefore in stationary states with sufficiently large energy values any physical object may become more or less unstable with respect to some possible decay process.

11.5. Nonexponential Decay Law

As seen from Eqn. (6), for any one-step process κ has a quite definite value independent of time in every stationary state determined by definite values of all the quantities conserved. Therefore, in view of Eqn. (11), the decay law for an abstract physical system being in a certain stationary state must be strictly exponential if the decay is a one-step process. But a totality of a large number of concrete one-species systems capable of decaying is distributed over a continuous set of energy values determining different stationary states in a certain relative stability interval and, hence, over the corresponding continuous set of κ values. Therefore, the real decay law for a totality of a large number of concrete systems is nonexponential in principle.

Indeed, let the least value of κ in such an interval be κ^0, and let $W_0(\kappa - \kappa^0)$ be the initial statistical density of concrete systems that have not decayed to the initial time instant $t = 0$. Then the total number of these systems that have not decayed to time t is

$$N = \int_{\kappa^0}^{\kappa_{max}} N(\kappa, t)\,d\kappa = N_0 \int_{\kappa^0}^{\kappa_{max}} W_0(\kappa - \kappa_0) \exp(-\kappa t)\,d\kappa, \qquad (11.39)$$

where N_0 is the initial total number of nondecayed systems at $t = 0$, $N(\kappa, t)$ is the number of systems that did not decay to the time instant t which are characterized by the value κ, and κ_{max} is the maximum value of κ.

By introducing $x = \kappa - \kappa^0$, from Eqn. (39) we get

$$N = N_0 \exp(-\kappa^0 t) \int_0^{x_{max}} W_0(x) \exp(-xt)\,dx. \qquad (11.40)$$

The particular form of the decay law for a totality of concrete systems obviously depends on the specific form of the initial statistical distribution $W_0(x)$, but it is evidently nonexponential in essence. At the same time, the statistical average of κ, $\bar{\kappa}(t)$, is always positive and tends to κ^0 when $t \to \infty$, as can be seen from its expression

$$\bar{\kappa}(t) = -\frac{dN}{N\,dt} = \kappa^0 + \frac{\int_0^{x_{max}} x W_0(x) \exp(-xt)\,dx}{\int_0^{x_{max}} W_0(x) \exp(-xt)\,dx}$$

$$= \kappa^0 + \overline{x \exp(-xt)}. \qquad (11.41)$$

The result obtained is quite reasonable from the standpoint of physics (cf. Sect. 14.6).

If $W_0(x)$ is, for instance, the normal distribution of the form

$$W_0(x) = (2/\pi\gamma)^{1/2} \kappa^{0\,-1} \exp(-x^2/2\gamma\kappa^{0\,2}), \qquad (11.42)$$

where γ is a constant, then

$$N = N_0 \exp[-\kappa^0 t(1 - 2^{-1}\gamma\kappa^0 t)]\,\{1 - \mathrm{erf}\,[(\gamma/2)^{1/2}\kappa^0 t]\}. \qquad (11.43)$$

For small values of $(\gamma/2)^{1/2}\kappa^0 t$, Eqn. (43) gives approximately

$$N = N_0 \exp\left[-\kappa^0 t(1 - 2^{-1}\gamma\kappa^0 t)\right] \left[1 - (2\gamma/\pi)^{1/2}\kappa^0 t\right], \tag{11.44}$$

whereas for large enough values of $(\gamma/2)^{1/2}\kappa^0 t$, it yields

$$N = N_0(2/\pi\gamma)^{1/2}(\kappa^0 t)^{-1} \exp(-\kappa^0 t). \tag{11.45}$$

11.6. Intramolecular Rearrangements

Every set of atomic nuclei and electrons with one equilibrium configuration, that is, with one least value of the potential energy, will be considered a single molecule. This means that in case a set of atomic nuclei and electrons has several equilibrium configurations of the atomic cores, a different molecule corresponds to each equilibrium configuration. Not only different molecules in the common chemical sense, but also various conformations of molecules, chemical radicals, various chemical complexes of poor stability, etc., fall under such a definition of a molecule. Transformation of a molecule into one or several other molecules is called an intramolecular rearrangement.

Since every concrete molecule capable of undergoing a definite intramolecular rearrangement is considered to be a closed physical system, it has definite values of all the conserved physical quantities, in particular of the energy, and corresponds to an abstract molecule which is in a definite stationary state. Let N concrete initial molecules, of which N_α correspond to the αth stationary state of the abstract molecule, be in the reaction volume at time t. If N_α is large enough, then

$$N_\alpha \cong \beta_\alpha(t)N, \tag{11.46}$$

where $\beta_\alpha(t)$ is the probability that the initial abstract molecule is in the αth stationary state, this probability depending generally on time. Furthermore, let κ_α be the rate constant of the intramolecular rearrangement under consideration in the αth stationary state. Then, on the basis of Eqn. (10), one gets

$$-\Delta N_\alpha/\Delta t \cong \kappa_\alpha N_\alpha \cong \kappa_\alpha \beta_\alpha(t)N \tag{11.47}$$

and from

$$N = \sum_\alpha N_\alpha \tag{11.48}$$

it follows that

$$-\Delta N/\Delta t \cong \sum_\alpha \kappa_\alpha \beta_\alpha(t)N = \kappa(t)N, \tag{11.49}$$

where

$$\kappa(t) = \sum_{\alpha} \kappa_\alpha \beta_\alpha(t) \tag{11.50}$$

is the rate constant characterizing the decline of the number of initial molecules at time t. In experiments one usually measures the average of this quantity for a certain time interval, that is, the quantity

$$\overline{\kappa} = \sum_{\alpha} \kappa_\alpha \overline{\beta}_\alpha, \tag{11.51}$$

where

$$\overline{\beta}_\alpha = (t_2 - t_1)^{-1} \int_{t_1}^{t_2} \beta_\alpha(t)\, \mathrm{d}t. \tag{11.52}$$

Equation (51) is valid for both nonstationary and stationary processes. In the second case $\overline{\beta}_\alpha = \beta_\alpha(t) = \text{const}$, $\overline{\kappa} = \kappa(t) = \text{const}$, and Eqn. (51) goes over into

$$\kappa = \sum_{\alpha} \kappa_\alpha \beta_\alpha. \tag{11.53}$$

In Eqns. (51) and (53) the sum should be taken over all the states liable to contribute. Since summands in each sum can be of very different orders of magnitude, the magnitudes of $\overline{\kappa}$ and κ are practically determined only by the summands that have leading magnitudes and differ not more than in three to four orders (depending on the calculation accuracy).

The appearance of a transitional configuration in the case of an intramolecular rearrangement may be due to intramolecular vibrations only. Therefore, to calculate the rate constant of an intramolecular rearrangement one has to start from an analysis of the vibrations of the corresponding molecule. We confine ourselves to the case of stationary processes and assume that a molecule is in thermal equilibrium with its environment, its properties depending on the environment. The last assumption allows one in principle to treat intramolecular rearrangements also in case the environment is a real physical system rather than an ideal gas. In addition we assume that $G = 1$ in Eqn. (16), that is, we utilize Eqn. (13).

Even with these restrictions, a calculation of the rate constant of an intramolecular rearrangement through Eqn. (53) in general encounters huge difficulties, which are connected with the necessity of taking account of various electronic and rotational states as well as of anharmonicity of the vibrations responsible for the rearrangement. In some cases one may confine oneself to the ground electronic state (this is possible if energy values of other electronic states differ greatly from that of the ground state) and disregard both rotational states (to some extent this is possible if the probability distributions of coordinates in vibrational states depend only weakly on rotational quantum numbers) and the anharmonicity of vibrations. However, even then calculation of $\kappa_{0\alpha}$ in vibrational stationary states with the aid of Eqn. (13) requires, for polyatomic molecules, that one finds the probability distribution of the internal coordinate that is the process path coordinate (cf. Sect. 2), and this leads to additional, though purely computational, difficulties. Therefore, we discuss the simplest case – an intramolecular rearrangement for which the process path coordinate q coincides with a certain normal coordinate.

In this case an intramolecular rearrangement may be represented in the one-dimensional model of motion by means of a particle leaving a potential well through a potential barrier, the transitional configuration being determined by the coordinate value q_0 corresponding to the potential barrier maximum. Accurate solution, even of this problem, requires a knowledge of the potential energy $U(q)$ in the whole of space and exact solution of the Schrödinger equation. In the 'harmonic approximation' we confine ourselves to a 'harmonic' potential well, defined by a potential energy curve which obeys

$$U(q) = \tfrac{1}{2}Kq^2 \tag{11.54}$$

inside the interval $(-\infty, q_0)$ and monotonically declines to zero outside, and make use of the solutions of the Schrödinger equation for a one-dimensional harmonic oscillator. This may be done because the true wave functions for the stationary states that are of interest differ from those of the harmonic oscillator only when $q > q_0$, but this is practically inessential if q_0 is sufficiently large, as is usually the case.

Let us introduce the dimensionless coordinate η and momentum ϵ connected with q and $\dot{q}$, respectively, by the relations

$$q = 2^{1/2}\sigma\eta; \qquad \dot{q} = 2^{1/2}\sigma'\epsilon, \tag{11.55}$$

where

$$\sigma = (\hbar\, \partial\omega/\partial K)^{1/2}; \qquad \sigma' = \omega\sigma. \tag{11.56}$$

Taking into account that $\rho(q) = 2^{-1/2}\sigma^{-1}\rho(\eta)$ and $\rho(\dot{q}) = 2^{-1/2}\sigma'^{-1}\rho(\epsilon)$, we obtain from Eqn. (13) for the lth stationary state

$$\kappa_{0l} = \sigma^{-1}\sigma'\rho_l(\eta_0)\overline{\epsilon}_+ = \omega\rho_l(\eta_0)\overline{\epsilon}_+, \tag{11.57}$$

where

$$\overline{\epsilon}_+ = \int_0^\infty \rho_l(\epsilon)\epsilon\, d\epsilon. \tag{11.58}$$

When calculating $\rho(\eta_0)$ and $\overline{\epsilon}_+$, we make use of the wave functions for the stationary states of a harmonic oscillator normalized over the interval $(-\infty, \infty)$. Then we obtain finally

$$\overline{\epsilon}_+ = (2\pi)^{-1/2}I_l \tag{11.59}$$

and

$$\kappa_l = \nu(2^l l!)^{-1} \exp(-\eta_0^2) H_l^2(\eta_0) I_l, \tag{11.60}$$

where $\nu = (2\pi)^{-1}\omega$ is the vibrational frequency of the oscillator; I_l are numbers defined by the equations

$$I_{2l} = \frac{(2l-1)!!}{(2l)!!}(4l+1); \qquad I_{2l+1} = \frac{(2l-1)!!}{(2l)!!}(4l+2) \quad (l = 0, 1, 2, \ldots) \tag{11.61}$$

(the subscript zero on κ is omitted, for the sake of simplicity, here and in what follows). When l is sufficiently large, Eqn. (61) gives

$$I_l \cong 2(2l)^{1/2}\pi^{-1/2} \cong 1.6 l^{1/2}. \tag{11.62}$$

This expression for I_l is obtained, as indeed it should, for a classical one-dimensional harmonic oscillator. Equation (62) leads already for I_5 to a value which differs from the true one by not more than by 5%.

We now consider, in the harmonic approximation, the case when the transformation path coordinate is a certain internal coordinate q_j. The q_j coordinate is connected with the properly normalized dimensionless normal coordinates η_l $(l = 1, 2, \ldots, n)$ by the relation

$$q_j = \sum_{l=1}^{n} 2^{1/2} \sigma_j^{(l)} \eta_l, \tag{11.63}$$

and $\dot{q}_j$ is connected with the corresponding dimensionless momenta ϵ_l by the relation

$$\dot{q}_j = \sum_{l=1}^{n} 2^{1/2} \sigma_j^{\prime(l)} \epsilon_l, \tag{11.64}$$

where

$$\sigma_j^{(l)} = (\hbar\, \partial\omega_l/\partial u_{jj})^{1/2}; \qquad \sigma_j^{\prime(l)} = \omega_l \sigma_j^{(l)}, \tag{11.65}$$

ω_l is the circular frequency of the lth normal vibration and u_{jj} is the diagonal element of the matrix of force coefficients that corresponds to the q_j coordinate. To calculate the rate constant κ_α of the transformation under consideration in the αth vibrational stationary state of the molecule [this state being characterized by the vibrational quantum numbers m_l $(l = 1, 2, \ldots, n)$], one has to find the probability distributions of q_j and of $\dot{q}_j$ in this state. These can be represented in the form

$$\rho_\alpha(q_j) = \sum_{s=0}^{M} a_s \rho_s^{(\sigma)}(q_j); \qquad \rho_\alpha(\dot{q}_j) = \sum_{s=0}^{M} a_s' \rho_s^{(\sigma')}(\dot{q}_j), \tag{11.66}$$

where

$$\rho_s^{(\sigma)}(q_j) = [(2\pi)^{1/2}\,\sigma 2^s s!]^{-1}\,\exp\,(-q_j^2/2\sigma^2)H_s^2(q_j/2^{1/2}\sigma), \tag{11.67}$$

and

$$\rho_s^{(\sigma')}(\dot{q}_j) = [(2\pi)^{1/2}\sigma' 2^s s!]^{-1}\,\exp\,(-\dot{q}_j^2/2\sigma'^2)H_s^2(\dot{q}_j/2^{1/2}\sigma'), \tag{11.68}$$

$M = \Sigma_{l=1}^{n}\, m_l$ is the summed quantum number of the αth vibrational level; σ and σ' are defined as

$$\sigma = \left(\sum_{l=1}^{n} \sigma_j^{(l)\,2}\right)^{1/2}; \qquad \sigma' = \left(\sum_{l=1}^{n} \sigma_j^{\prime(l)\,2}\right)^{1/2}, \tag{11.69}$$

a_s and a_s' are positive numbers which depend on totalities m_l and $\sigma_j^{(l)}$ $(l = 1, 2, \ldots, n)$ and satisfy the relations

$$\sum_{s=0}^{M} a_s = 1; \qquad \sum_{s=0}^{M} a_s' = 1. \tag{11.70}$$

Determination of a_s and a_s' involves a laborious calculation which practically prevents one from making an exact computation of κ_α by the equation

$$\kappa_\alpha = \rho_\alpha(q_0)\overline{v}_{\alpha+} \tag{11.71}$$

(here q_0 is the value of q_j corresponding to the transitional configuration and $\overline{v}_{\alpha+} = \overline{\dot{q}}_{j+(\alpha)}$). However, such an exact computation is hardly necessary because a number of approximations have already been made anyway. At the same time, an estimation of the right-hand side of Eqn. (53) can be made without knowledge of the values of a_s and of a_s', on the basis of the following reasoning.

First, note that the β_α, which are determined by

$$\beta_\alpha = Q \prod_{l=1}^{n} \beta_l^{m_l}, \tag{11.72}$$

where

$$\beta_l = \exp\,(-h\omega_l/kT); \qquad Q = \prod_{l=1}^{n}\,(1-\beta_l), \tag{11.73}$$

T being the absolute temperature, are positive. Since the κ_α are also positive, the following relation holds:

$$\sum_{\alpha} \kappa_\alpha\beta_\alpha \leqslant \sum_{M=0}^{\infty}\;\sum_{\alpha(M)} \kappa_\alpha\beta_\alpha \equiv \kappa_\infty, \tag{11.74}$$

where $\Sigma_{\alpha(M)}$ signifies summation over all the α states corresponding to one summed quantum number M. Let us represent κ_∞ in the form

$$\kappa_\infty = \kappa_{N-1} + \delta_N \kappa, \tag{11.75}$$

where

$$\kappa_{N-1} = \sum_{M=0}^{N-1} \sum_{\alpha(M)} \kappa_\alpha \beta_\alpha \tag{11.76}$$

and

$$\delta_N \kappa = \sum_{M=N}^{\infty} \sum_{\alpha(M)} \kappa_\alpha \beta_\alpha . \tag{11.77}$$

Taking a number N such that the inequality $\delta_N \kappa / \kappa_\infty < \gamma$ is satisfied (γ is any given small number), one may replace κ_∞ by κ_{N-1}. Indeed, calculating

$$\bar{v}_{\alpha+} = \int_0^\infty \rho_\alpha(\dot{q}_j) \dot{q}_j \, d\dot{q}_j, \tag{11.78}$$

we find, in view of Eqns. (66), (68), and (59),

$$\bar{v}_{\alpha+} = (2\pi)^{-1/2} \sigma' \sum_{s=0}^{M} a_s' I_s = (2\pi)^{-1/2} \sigma' \bar{I}_\alpha, \tag{11.79}$$

where, due to Eqn. (70),

$$1 \leqslant \bar{I}_\alpha \leqslant I_M . \tag{11.80}$$

Since the relation

$$(2^s s!)^{-1} [\exp(-x^2) H_s^2(x) \leqslant k_0, \; k_0 \cong 1 \tag{11.81}$$

is valid, one gets, making use of Eqns. (66) and (70),

$$\rho_\alpha(q_0) \leqslant (2\pi)^{-1/2} \sigma^{-1} . \tag{11.82}$$

Utilizing in addition Eqns. (72) and (79), we obtain from Eqn. (77)

$$\delta_N \kappa < (2\pi\sigma)^{-1} \sigma' Q \sum_{M=N}^{\infty} n_M \beta^M I_M , \tag{11.83}$$

where

$$n_M = \binom{M+n-1}{n-1}$$

is the number of different vibrational states corresponding to the quantum number M; $\beta = \exp(-\hbar\omega_{\min}/kT)$ is the largest of the β_l. In view of Eqn. (62), and provided $\beta_N < 1$, we find

$$\delta_N\kappa < \frac{\sigma'}{2\pi\sigma} Q\beta^N I_N n_N \frac{1}{1-\beta_N}\left[1 + \frac{\beta_N}{2N(1-\beta_N)}\right] \equiv \Delta_N\kappa, \tag{11.84}$$

with $\beta_N = \beta(N + n)(N + 1)^{-1}$. Furthermore, in view of Eqn. (80), it follows from Eqns. (74), (71), and (79) that

$$\kappa_\infty > (2\pi)^{-1/2}\sigma' \sum_{M=0}^{\infty} \sum_{\alpha(M)} \rho_\alpha(q_0)\beta_\alpha = (2\pi)^{-1/2}\sigma'\rho(q_0) \equiv \kappa_{(0)}, \tag{11.85}$$

where

$$\rho(q_0) = (2\pi)^{-1/2}\sigma_j^{-1} \exp(-q_0^2/2\sigma_j^2) \tag{11.86}$$

is the probability density of the transitional configuration in thermal equilibrium and

$$\sigma_j = \left(\sum_{l=1}^{n} \sigma_j^{(l)2} \coth \frac{\hbar\omega_l}{2kT}\right)^{1/2} \tag{11.87}$$

is the mean-square amplitude of the coordinate q_j (in a number of cases σ_j can be found from the electronographic experiment). We conclude from Eqns. (84) and (85) that

$$\delta_N\kappa/\kappa_\infty < \Delta_N\kappa/\kappa_{(0)}. \tag{11.88}$$

Therefore, if a value $N = N_0$ is taken such that $\Delta_{N_0}\kappa/\kappa_{(0)} < \gamma$, then, with a relative error which is certainly less than γ, one may replace κ_∞ by κ_{N_0-1}. On the other hand, it follows from Eqns. (71), (75), (76), (79), (80), and (85) that

$$\kappa_\infty \cong \kappa_{N_0-1} < \kappa_{(0)} I_{N_0-1}. \tag{11.89}$$

Hence, if a quantity κ_e defined as

$$\kappa_e = I'\kappa_{(0)} \tag{11.90}$$

with $I' = I_{N_0-1}^{1/2} \cong 1.27(N_0 - 1)^{1/4}$, is introduced to estimate κ_∞, then from Eqns. (85) and (89) it follows that $I'^{-1} < \kappa_\infty/\kappa_e < I'$. When $N_0 < 82$, $I' < 3.8$; when $N_0 < 257$, $I' < 5.1$.

The above considerations have actually concerned the conversion of variant I of a system (that is, of the initial molecule) into variant II. But this may happen to be only the first step of the process, since initial and final molecules generally have different sets of energy levels. The second step of the process is, therefore, 'stabilization' of variant II in the form of the final molecule, the stabilization being either spontaneous or due to interaction with the environment. Thus, rearrangements of molecules are schematically two-step rather than one-step processes.

If the energy value of the stationary state in which the process takes place differs greatly from that off any stationary state of the final molecule, then the rate of the second step is relatively small and the rearrangement rate is small as well. By contrast, if the energy value of the stationary state in which the process is taking place is very near that of a certain stationary state of the final molecule, the rate of the second step is great in comparison with the rate of the first step, and the rearrangement rate is determined by the corresponding κ_α value. Therefore, the rate of a proper intramolecular rearrangement is actually determined only by the stationary states of the initial molecule whose energy values coincide approximately with the energy values of some stationary states of the final molecule, and the sum in Eqn. (53) must be taken over these states only.

Let such coincidences occur for the energy values E_l ($l = 1, 2, \ldots, p$) of the initial-molecule vibrational levels. Then Eqn. (53), in view of Eqn. (72), goes over into

$$\kappa = Q \sum_{l=1}^{p} \kappa_l \exp\left(-\frac{E_l}{kT}\right), \tag{11.91}$$

where

$$E_l = \sum_{l'=1}^{n} m_{l'} \hbar \omega_{l'}$$

is the energy of the αth vibrational level, and

$$\kappa_l = \sum_{\alpha} \kappa_\alpha \quad (E_\alpha = E_l). \tag{11.92}$$

If, in a certain temperature interval, separate summands in Eqn. (91) are greatly different (as is usually the case) and the largest of them corresponds to $l = j$, then Eqn. (91) yields

$$\kappa = Q\kappa_j \exp(-E_j/kT), \tag{11.93}$$

which may coincide with the Arrhenius formula to greater accuracy because of the slight dependence of Q on T.

11.7. A Criterion for the Possibility of an Intramolecular Rearrangement

To find such a criterion let us first give another approximate estimation of the rate constant κ for an intramolecular rearrangement. If we assume that, in the initial Eqn. (13), the quantity $\rho(q_0)$ is given by Eqn. (86) and $\overline{v}_+$ is the mean rate of change of the q_j coordinate in the transformation direction in thermal equilibrium, then we obtain

$$\kappa_s = (2\pi)^{-1} \sigma_j^{-1} \sigma_j' \exp(-q_0^2/2\sigma_j^2), \tag{11.94}$$

where

$$\sigma_j' = \left[\sum_{l=1}^{n} \sigma_j'^{(l)2} \coth\left(\frac{\hbar\omega_l}{2kT}\right) \right]^{1/2}. \tag{11.95}$$

Equation (94) was first derived by Slater (see Ref. [14]) and later, independently, by the present author (see Ref. [5]). It is easily seen that κ_e and κ_s are related by

$$\kappa_e/\kappa_s = I'\sigma'/\sigma_j'. \tag{11.96}$$

Hence, κ_s is of the same order of magnitude as κ_∞ and may serve to estimate it. Note also that, for $T = 0$, $\kappa_s = \kappa_\infty = \kappa_{(0)}$.

Now we can obtain the criterion sought for. Indeed, κ is of about the same order of magnitude as κ_s. Since in Eqn. (94) the pre-exponential factor is a certain mean frequency of molecular vibrations [this follows from Eqns. (95), (87), and (65)], it cannot exceed 10^4 sec^{-1}. If one assumes that the least value of κ capable of being experimentally observed is of the order of magnitude of 10^{-8} sec^{-1}, then, as is readily seen, the relation

$$q_0/\sigma_j \cong 10 \tag{11.97}$$

defines the approximate boundary between the possibility and impossibility of running a corresponding process. Although this criterion is of very approximate character, in some cases it can prove to be helpful in order to reject some unrealizable theoretical schemes of chemical processes.

Problems

11.1. Inversion of ammonia (NH_3) is a process, which results in turning out a concrete molecule of ammonia in such a way that its new equilibrium configuration is a mirror reflection of the old one in the plane that passes through its center of mass parallel to the plane of the three hydrogen atoms. What are the variants of the molecule, the transformation path coordinate, the value q_0 of the transitional point, and the transitional configurations, in this process?

11.2. Estimate the rate constant of the ammonia inversion at $T = 0° K$, if it is given that $l = 0.36$ Å, and the calculation of the normal vibrations of ammonia yields: $\sigma_q = 0.0824$, $\kappa' = (2\pi)^{-1}\sigma_q^{-1} \times \sigma_{\dot{q}} = 3.088 \times 10^{13}$ sec^{-1} (in the calculation dimensionless coordinates are used and the equilibrium length of C—H bond in methane, that is, 1.09 Å is taken as a length unity).

Suggested References

1. Bateman, H., and A. Erdelyi: 1953, *Higher Transcendental Functions*, Vol. 2, McGraw-Hill, New York. Relation (91) can be found in this book.
2. Blokhintsev, D. I.: 1964, *Quantum Mechanics*, tr. of original Russian edn. (1961) by J. B. Sykes and M. J. Kearsley, Reidel, Dordrecht. The conventional discussion of the problem of a particle in a spherically symmetric square-well potential can be found in particular in this book.
3. Evans, M. G., and H. Polanyi: 1935, *Trans. Farad. Soc.* **31**, 875. In this paper the absolute rate theory of chemical reactions is based on the notion of transition state.
4. Eyring, H.: 1935, *J. Chem. Phys.* **3**, 107. This paper presents the absolute rate theory based on the notion of activated complex.
5. Mayants, L. S.: 1963, *DAN SSSR* **151**, 624.
6. Mayants, L. S.: 1964, *Zh. Fiz. Khim.* **38**, 623.
   These two papers consider intramolecular rearrangements as due to molecular vibrations and also contain some examples and the criterion for the possibility of the corresponding processes.
7. Mayants, L. S.: 1964, *Zh. Vsesouzn. Khim. Obshch. im. Mendel.* **9**, 705.
8. Mayants, L. S.: 1966, 1967, *Zh. Teor. i Eksper. Khim.* **2**, 563 and **3**, 425.
   In these two papers the transitional configuration theory was formulated for the first time.
9. Mayants, L. S.: 1965, *DAN SSSR* **164**, 852. The derivation of Eqns. (66) is given in this paper.
10. Mayants, L. S.: 1967, *Quantum Physics and Transformations of Matter* (in Russian), VINITI, Moscow. The monograph contains a more detailed exposition of the transitional configuration theory and its applications.
11. Mayants, L. S.: 1969, *Izv. Akad. Nauk SSSR* (*ser. khim.*) **5**, 1026. An estimation of the rate constant and consideration of the compatibility of the theory with the Arrhenius formula are given in this paper.
12. Mayants, L. S.: 1976, *Found. Phys.* **6**, 485. This paper presents transitional configuration theory with some applications, proceeding from the fundamentals of probabilistic physics.
13. Schiff, L. I.: 1968, *Quantum Mechanics*, 3rd edn., McGraw-Hill, New York. This book also contains the conventional discussion of the problem mentioned in Ref. [2].
14. Slater, N. B.: 1959, *Theory of Intramolecular Reactions*, Cornell, U.P., Ithaca. Equation (94) appeared first in this book and further has been rejected in the same author's article:
15. Slater, N. B.: 1961, *J. Chem. Soc.* **1961**, 606.

Chapter 12

Electromagnetic Field and Photons

12.1. Preliminaries

The probabilistic treatment of the electromagnetic field in vacuum would have to be based on the results of Chapter 7 dealing with random fields. But the definition of a random field there proceeds, as should be the case in probabilistics, from the notion of a continuous concrete object regarded as the carrier of a field. This notion is, however, irrelevant to the case under consideration. Hence, it would be illogical to apply the results of Chapter 7 to the electromagnetic field, even though it obeys Maxwell equations, which allow one to think of it as a Hamiltonian random field. The equations of the electromagnetic field in vacuum can be reduced to the form given in Eqn. (7.46). This leads, when making use of the quantum approach, to the expressions of the form given in Eqn. (7.77) for the field energy values, which, in view of Eqn. (7.78), $\alpha = \hbar$, and $\lambda_j = \omega_j^2$ (ω_j is the circular frequency of the corresponding plane electromagnetic wave), yields

$$E = \sum_j (n_j + \tfrac{1}{2})\hbar\omega_j. \tag{12.1}$$

Here $\hbar\omega_j = h\nu_j$ is the energy of one quantum of the electromagnetic field (of one photon) and n_j is the number of photons of the frequency ν_j; the additional term, $2^{-1}\hbar\omega_j$, which leads to an infinite zero state energy, is usually omitted.

So the formal quantization of the electromagnetic field seemingly explains the real existence of photons, which is well verified experimentally. A photon should have the same velocity as an electromagnetic wave in vacuum is supposed to have, namely c, which is the ultimate velocity of motion of any physical object. Because of this circumstance, as follows from special relativity, a photon must have zero rest mass.

There is thus the following situation. On the one hand, the only realistic way of introducing the concept of random field provided by probabilistics does not apply to the electromagnetic field. On the other hand, the really existing concrete photons, which do not differ from any other concrete particles in their mechanical behavior, must have zero rest mass provided they are regarded as the outcome of the electromagnetic field. Since from the standpoint of probabilistics the electromagnetic field cannot be considered a random field, the photons are not its outcome and, hence, do not necessarily have zero rest mass.

The problem of the real relationship between the electromagnetic field and photons will be discussed, after we have considered the related problem whether there can in general exist particles of zero rest mass. Could, say, the neutrino have a vanishing rest mass? It is usually believed that the neutrino rest mass is very small or zero. And one might even think that it practically makes no difference whether a particle has a very small rest mass or one that equals zero exactly. But as a matter of fact this is a question of principle, which deserves the most careful examination.

Rest mass is a property inherent in every concrete particle, and all the particles of one kind have the same rest mass. An abstract particle corresponding to a set of concrete particles of one kind, being the image of the set of particles, has the same rest mass. But to avoid any misunderstanding, we have to deal with concrete particles and, hence, to pose the problem as follows. Can a concrete particle have zero rest mass? When solving this problem it is necessary to remember the obvious principle of the concreteness of experiment, too, which reads: any experiment is concrete and is carried out on concrete objects and under concrete conditions. Certainly, also, any mental experiment should be regarded as obeying this principle.

Since we shall have to deal with special relativity, it might be helpful to recall some of its postulates and results. One postulate of this theory states that all the inertial coordinate systems (ICS) are equivalent. There is no absolute coordinate system, and all the ICS are moving relative to each other with constant velocities. One of the principal results of special relativity theory is the statement that there exists the ultimate velocity c for the real motion of any concrete particle, which is believed to be that of light in vacuum, and that velocity is one and the same in all the ICS.

Any event, that is, the existence of a concrete particle in space and time, is described in different ICS not only by different sets of the Cartesian spatial coordinates x, y, and z, but also by different instants of time t. This means that equal clocks run at different rates and show, therefore, different instants of time for one and the same event in different ICS. An event is described in every ICS by the corresponding four-vector

$$R = (x_1, x_2, x_3, x_4), \tag{12.2}$$

where $x_1 = x$, $x_2 = y$, $x_3 = z$, $x_4 = ct$. The vectors R and R' referring to one event in different ICS, K and K', respectively, are related to each other by a linear transformation which is known as the Lorentz transformation. In the particular case, when K' is moving along the x-axis of K with the constant velocity V, the respective axes of both ICS being parallel, the vector $R' = (x_1', x_2', x_3', x_4')$ is related to R by the equations

$$\begin{aligned} x_1' &= (x_1 - bx_4)/(1 - b^2)^{1/2}; \qquad x_2' = x_2; \qquad x_3' = x_3; \\ x_4' &= (x_4 - bx_1)/(1 - b^2)^{1/2}, \end{aligned} \tag{12.3}$$

where $b = V/c$.

The components of any other four-vector, when referring to one event in different ICS, are also transformed like R. The Lorentz transformation keeps, however, the magnitude squared of any four-vector $f = (f_1, f_2, f_3, f_4)$, which is defined as

$$|f|^2 = f_4^2 - f_1^2 - f_2^2 - f_3^2 \tag{12.4}$$

unchanged. This means that in any ICS the magnitude squared of any four-vector has one and the same value. For a free particle moving with the velocity v and having started its motion at the zero point $0 = (0, 0, 0, 0)$ of space-time, Eqn. (4) gives, in view of Eqn. (2) and $x_1 = v_x t$, $x_2 = v_y t$, and $x_3 = v_z t$,

$$R^2/c^2 \equiv \tau^2 = (1 - \beta^2)t^2 \tag{12.5}$$

in any ICS, where

$$\beta^2 = \beta_x^2 + \beta_y^2 + \beta_z^2; \qquad \beta_x = v_x/c, \quad \beta_y = v_y/c, \quad \beta_z = v_z/c \tag{12.6}$$

and

$$v_x = v \cos \overline{\alpha}, \qquad v_y = v \cos \overline{\beta}, \qquad v_z = v \cos \overline{\gamma} \tag{12.7}$$

($\overline{\alpha}$, $\overline{\beta}$, and $\overline{\gamma}$ are the direction angles of the particle motion).

For real motion the velocity of a particle cannot exceed c in any ICS. Therefore, $\tau^2 \geqslant 0$ and

$$x_1^2 + x_2^2 + x_3^2 \leqslant x_4^2. \tag{12.8}$$

This means that the straight lines on which a free particle can really move must lie inside a cone described by Eqn. (8) in every ICS. The cone is usually called the null cone, and those straight lines are known as world lines. The scalar τ is known to be the proper time of the particle, that is, the time which is shown by the clock rigidly attached to this particle. From Eqn. (5) it follows that in this ICS the particle has to be at rest.

By dividing the four-vector R as given in Eqn. (2) by the scalar $c\tau$ one gets for a free particle the four-vector

$$u = (u_1, u_2, u_3, u_4) \tag{12.9}$$

whose components are

$$\begin{aligned} u_1 &= \beta_x/(1 - \beta^2)^{1/2}, & u_2 &= \beta_y/(1 - \beta^2)^{1/2}, \\ u_3 &= \beta_z/(1 - \beta^2)^{1/2}, & u_4 &= 1/(1 - \beta^2)^{1/2}. \end{aligned} \tag{12.10}$$

This vector is called the four-velocity, and its magnitude, in view of Eqn. (10), is unity. The multiplication of the four-vector u by the scalar $m_0 c$, where m_0 is the rest mass of the particle, gives the four-vector

$$M = (M_1, M_2, M_3, M_4) \tag{12.11}$$

which is called the four-momentum. The components of this four-vector can be written as follows

$$M_1 = p_x = mc\beta_x, \qquad M_2 = p_y = mc\beta_y,$$
$$M_3 = p_z = mc\beta_z, \qquad M_4 = E/c = mc, \tag{12.12}$$

where p_x, p_y, and p_z are the components of the three-momentum of the particle, E is its energy, and m is its mass in the ICS chosen. It is evident that the equation

$$m = m_0/(1 - \beta^2)^{1/2} \tag{12.13}$$

is also valid in any ICS. The magnitude squared of the four-momentum is

$$M_4^2 - M_1^2 - M_2^2 - M_3^2 = m_0^2 c^2. \tag{12.14}$$

12.2. The Four-Dimensional Space Treatment

One might try to solve the problem in question by making use of the four-dimensional space formalism. No matter how small the nonzero rest mass of a free particle is, its velocity is less than c in any ICS, and its world line should lie inside the null cone (all world lines are supposed to pass through 0, the vertex of the cone). The world lines which might lie on the null cone would have, hence, to refer to the particles of zero rest mass only. This conclusion is usually made on the basis of Eqn. (13) and its related expressions because if $\beta = 1$, then $m_0 = 0$ must be valid in order that the mass and momentum of the particle be finite, and conversely when $m_0 = 0$, then $\beta = 1$ has to be satisfied in order that these quantities be nonzero. To prove that a particle has zero rest mass it is, hence, sufficient to demonstrate that its world line lies on the null cone.

One might use for this purpose Eqn. (5), trying to find out whether or not for the suspected particle the equation $\beta = 1$ or $\tau = 0$ is valid, as should be the case. To prove that the velocity of the particle is exactly c (that is, β is exactly 1) in the ICS chosen, one has not only to measure it to an absolute accuracy, but also has to know the exact value of c. No measurement, however, can be performed to absolute accuracy, while the exact value of c is not known either. This way of verification thus proves to be unfit.

A few words should be said about the possibility of testing the condition $\tau = 0$. In order to measure the proper time τ, a clock should be attached rigidly to the particle under test. If a particle moves at the rate c, so does the clock, which must therefore have zero rest mass too. But a real clock must possess an internal structure, and this prevents it from having zero rest mass. So, should a clock move at the rate c exactly, it would not be a clock at all and could not indicate any values of τ, including $\tau = 0$. Hence, it is impossible to test whether $\tau = 0$ for a particle, while the supposition that it might have zero rest mass leads to the above internal contradiction. This supposition, thus, proves to be at least doubtful.

Let us now move on to the consideration of other quantities related to a free particle. The four-velocity u as given in Eqn. (9) could be obtained from the four-vector R as given in Eqn. (2), for a free particle, by dividing the latter by $c\tau$ only when $\tau \neq 0$, and u does not exist at all when $\tau = 0$. But the four-quantity

$$B = (\beta_x, \beta_y, \beta_z, 1) \tag{12.15}$$

which is not a four-vector, exists when $\tau = 0$. This four-quantity is related to the four-momentum M, in view of Eqn. (12), by the equation

$$M = mcB, \tag{12.16}$$

when $\tau \neq 0$. Since M also exists when $\tau = 0$, it must be connected with B by a factor of proportionality in this case too, and the relation of M to B has to be of the same form as that given in Eqn. (16) by considerations of continuity. In order that the components of M be finite when $\tau = 0$, it is necessary to believe, in view of Eqn. (13), that the equality $m_0 = 0$ should be satisfied in this case. This result coincides with the one above, which was obtained from the consideration of the world lines which are supposed to be lying on the null cone.

So, when $\tau = 0$, then $m_0 = 0$ and the magnitude of the four-vector M, in view of Eqn. (14), is zero. But what are the components of M themselves in that case? To answer the question it is necessary to examine what m is when $\tau = 0$. And this can be done by proceeding from the physical nature of the problem under consideration.

12.3. Physical Consideration

Let us define a quantity η by

$$\beta^2 = 1 - \eta^2 \quad (0 \leqslant \eta^2 \leqslant 1). \tag{12.17}$$

Then Eqn. (13) takes on the form

$$m = m_0 \eta^{-1} \quad (1 \leqslant \eta^{-1} < \infty). \tag{12.18}$$

Recall now that Eqn. (18) refers to a concrete particle and a concrete ICS, in which its motion is considered. Therefore, when treating this equation there is no reason at all for the application of any limiting process with respect to m_0. Since for every concrete particle with the rest mass m_0, Eqn. (18) describes the corresponding straight line, there is no reason for the application of any limiting process with respect to η^{-1} either. In the preceding section it was shown that when $\tau = 0$, then also $m_0 = 0$. Therefore, in order to get to know what m is when $\tau = 0$ it is necessary only to put $m_0 = 0$ on the right-hand side of Eqn. (18). Then we can obviously obtain $m = 0$ for any value of η^{-1} and, hence, for $\eta^{-1} = \infty$, also. The equation under consideration does not differ from the well-known equation for the x-axis in the xy-plane, which is $y = 0$ and can be rewritten as $y = 0x$. The answer to the question: 'What is y when $x = \infty$?' is, of course, $y = 0$. To avoid any misunderstanding it might be helpful to remember that the record $\eta^{-1} = \infty$ (or $x = \infty$) only means that η^{-1} (or x) exceeds any arbitrarily large positive number, but it does not determine, of course, a definite point on the corresponding axis.

Since $m = 0$, when $m_0 = 0$, in any ICS, it follows from Eqn. (16) that all components of the four-momentum M are also zero in every ICS in this case. Consequently, a concrete particle of zero rest mass, in other words, one having no rest mass, cannot have either energy or momentum in any ICS. But this means that such a particle does not really exist at all.

It has been demonstrated in Section 2 that the assumption of the existence of concrete particles with zero rest masses leads to an internal contradiction which makes that doubtful. Here it has been proved finally that indeed no concrete particle can have zero rest mass. But what does it mean for a photon? If that has zero rest mass, as is generally believed, then it cannot exist as a concrete particle. But if a photon is a concrete particle, then it cannot have zero rest mass. Let us now endeavor to settle this dilemma.

12.4. The 'Emon'

As has been mentioned in Section 1, the photons are usually believed to be the outcome of the electromagnetic field. Therefore, the velocity of photons is supposed to be that of light in vacuo, which in turn, is believed to be the ultimate velocity c of the special relativity theory; all their mechanical properties being expressed in terms of the wave properties. In particular, the energy and momentum of a photon are represented in every ICS as

$$E = h\nu \tag{12.19}$$

and

$$p = h\nu/c, \tag{12.20}$$

where ν is the frequency of the corresponding electromagnetic wave and h is the Planck constant. The four-momentum has the form

$$M = \left(\frac{h\nu}{c}\cos\bar{\alpha}, \frac{h\nu}{c}\cos\bar{\beta}, \frac{h\nu}{c}\cos\bar{\gamma}, \frac{h\nu}{c}\right) \tag{12.21}$$

where $\bar{\alpha}$, $\bar{\beta}$, and $\bar{\gamma}$ are the respective angles between the direction of the photon velocity and the Cartesian coordinate axes in the ICS chosen. The four-quantity $B = (\beta_x, \beta_y, \beta_z, 1)$ for a photon is

$$B = (\cos\bar{\alpha}, \cos\bar{\beta}, \cos\bar{\gamma}, 1). \tag{12.22}$$

From Eqns. (21) and (22) it follows that

$$M = \frac{h\nu}{c} B. \tag{12.23}$$

But the equation

$$E = mc^2 \tag{12.24}$$

following from Eqn. (12) is always valid in every ICS and can serve for the determination of mass if the energy is known. Therefore, this equation is valid for a photon too, and in view of Eqn. (19), Eqn. (23) fully coincides with Eqn. (16).

It is evident from the above that a photon has all the corpuscular properties, that is, the properties of any concrete particle, namely, it possesses mass, four-quantity B, and four-momentum M in every ICS, the properties being related by the same equations as in the case of any ordinary concrete particle. Furthermore, since

$$c = \lambda\nu \tag{12.25}$$

for a monochromatic electromagnetic wave, where λ is its length, Eqn. (20) can be rewritten as

$$p = h/\lambda. \tag{12.26}$$

Recall that Eqn. (26) is the well-known de Broglie relation which is also valid for any ordinary particle. Hence in this respect also, the photon does not differ from any other particle.

One might think that there is an essential difference between the photons and ordinary particles with respect to the origin of their properties. Namely, the wave properties of the photons are primary, their corpuscular ones being secondary. On the contrary, for any ordinary particles, the corpuscular properties are primary and the wave ones are secondary. In reality, however, this is not the case. A separate photon possesses the pronounced corpuscular properties, as is shown by all the experiments involving separate photons. At the same time, it is quite impossible to discover any wave property of a separate photon experimentally. There are some experiments which make it possible to determine the energy and momentum of a separate photon, whereas its frequency and wavelength cannot immediately be established. They should be found from Eqns. (19), (20), and (26). This means that no wave property is inherent in a separate photon itself.

To determine the wavelength experimentally it is necessary to work with a very large number of separate photons having the same energies and momenta, by making use of a diffraction grating or some other appropriate device. The frequency can then be obtained from Eqn. (25). But this means that the notion of wavelength actually refers to the probability properties of an abstract photon, which are revealed through the statistical distribution of the large number of separate photons under test, and indeed have nothing to do with any separate photon. Wavelength refers, thus, to a state of an abstract photon, which is characterized by certain values of energy and momentum. It is necessary to emphasize, by the way, that frequency is an auxiliary quantity even for an abstract photon, since it is impossible to determine it by direct experiment. Therefore, Eqn. (19) has to be considered the definition of frequency, not the relation between two really existing physical quantities.

But for any ordinary particle, too, the experimental determination of the wavelength λ in Eqn. (26) requires the application of diffraction methods. The wavelength in this case also refers to the probability properties of the corresponding abstract particle, and is revealed through the statistical distribution of the concrete particles under test on the respective recording device. Equation (19) defines 'frequency' ν for any ordinary particle as well. Then it follows from Eqns. (13), (19), and (24) that

$$m_0c^2/(1-\beta^2)^{1/2} = h\nu, \tag{12.27}$$

while Eqns. (13), (12), and (26) yield

$$\lambda = h(1-\beta^2)^{1/2}/m_0c\beta. \tag{12.28}$$

Furthermore, since the phase velocity w of a wave is

$$w = \lambda\nu, \tag{12.29}$$

one gets, in view of Eqns. (27) and (28), by expressing β in terms of λ or ν

$$w = c/\beta = c(1 + m_0^2c^2\lambda^2/h^2)^{1/2} = c(1 - m_0^2c^4/h^2\nu^2)^{-1/2}. \tag{12.30}$$

So, in all mechanical respects photons behave like any ordinary particles, and we should therefore consider separate photons to be concrete particles, as was stated in Section 1. But since a separate photon is a concrete particle, we have to conclude, in virtue of what has been proved in the preceding section, that the rest mass of a concrete photon cannot be equal to zero; but then a concrete photon cannot move exactly at c. Thus, a concrete photon must have a very small, but nevertheless nonzero rest mass and move at a velocity very near c, but all the same, somewhat less than c. This means, however, that in fact, instead of a nondenumerable set of concrete photons of various frequencies, there exists a subatomic electromagnetic particle of a very small rest mass, which the author has called 'emon'. The letters 'e' and 'm' in its name indicate the *e*lectro*m*agnetic character of the entity. Since this subatomic particle can even be at rest, it should no longer be called 'photon' in itself. The motion of concrete emons at a variety of rates of nearly c determines the whole set of concrete photons. So, it is precisely a concrete emon moving at a velocity near to c which is usually called a photon.

Let us note at once that the existence of emons does not contradict the relativity theory. Indeed, as has been noted in Section 1, special relativity theory affirms solely that for particle motion there exists the ultimate velocity c, whereas its exact coincidence with the velocity of light in vacuum can be checked only experimentally. But a measurement of the velocity of light can be made to a certain accuracy only, so that the error may prove to be greater than the difference between the emon velocity and c. The error may also exceed the difference of the emon velocity from the corresponding phase velocity. Therefore, the existence of the emon does not contradict Eqn. (30) either. The following calculation illustrates this.

Denoting the emon rest mass by m_{em}, we get from Eqn. (27), in view of Eqn. (17),

$$\nu_{\text{em}} = \nu\eta, \tag{12.31}$$

where $\nu_{\text{em}} = m_{\text{em}} c^2/h \approx 1.36 \times 10^{47} m_{\text{em}}\ \text{sec}^{-1}$. It follows from Eqn. (31) that the greater the ν, the smaller the η and, hence, the less the relative difference between c and the emon velocity since, in view of Eqn. (17), for small enough η

$$1 - \beta = \eta^2/(1+\beta) \approx \tfrac{1}{2}\eta^2. \tag{12.32}$$

Let us note, by the way, an important result following from Eqn. (31), namely, there must exist a nonzero lower limit of radio wave frequencies which is equal to ν_{em} ($\eta = 1$ and $\beta = 0$, that is, the emon is at rest). Equstions (31) and (32) yield

$$1 - \beta = (\nu_{\text{em}}/\nu)^2/(1+\beta). \tag{12.33}$$

Let the least actually observed radio wave frequency be ν_0, then $\nu_{\text{em}} < \nu_0$. If, for instance, $\nu_0 \approx 10^0\ \text{sec}^{-1}$, then in view of Eqn. (33) and $\beta \approx 0$

$$1 - \beta < \nu^{-2}. \tag{12.34}$$

Equation (34) gives: for $\nu \approx 10^5\ \text{sec}^{-1}$ (long waves) – $1 - \beta < 10^{-10}$, for $\nu \approx 10^{15}\ \text{sec}^{-1}$ (visible light) – $1 - \beta < 10^{-30}$, and for $\nu \approx 10^{20}\ \text{sec}^{-1}$ (gamma rays) – $1 - \beta < 10^{-40}$. In order that the error of the velocity measurement for the long waves might be less than the difference between c and the corresponding emon velocity, the measurement should be made with an accuracy to a few millimeters/seconds, which is hardly possible at present. The case of higher frequencies is obviously entirely out of the question. It is evident that if the least actually observed radio wave frequency is less than 1 sec^{-1}, the situation becomes still worse.

A natural question now arises: how could one determine the emon rest mass experimentally? The problem is, of course, difficult, because of the extreme smallness of that mass. By now only two possibilities are sufficiently clear. The first one follows from the above noted fact that ν_{em} is the lower limit of radio wave frequencies. Should this limit become known, the calculation of m_{em} would be possible because of the connection between these two quantities which is given after Eqn. (31). By making use of the above example, where the least actually observed radio wave frequency ν_0 is supposed to be $10^0\ \text{sec}^{-1}$, one obtains $m_{\text{em}} < 0.74 \times 10^{-47}$ g. The upper estimate of this quantity should, however, decrease if the ν_0 does. The other possibility may be based on the velocity measurements for very long waves. If ν_1 is the frequency of such a wave and β_1 is its velocity [cf. Eqn. (6)], then, in view of Eqn. (31),

$$m_{\text{em}} \cong 0.74 \times 10^{-47} \nu_1 \eta_1 , \tag{12.35}$$

where η_1 is connected with β_1 by Eqn. (32). The success of such experiments obviously depends on the possibility of producing sufficiently precise measurements of velocities for very long waves.

It follows from the above that there is no need to probabilistically consider the electromagnetic field because that proves to be merely an outcome of the presence of an enormous number of emons. The wave properties of the electromagnetic field have been shown to be actually those of the abstract emon, since with respect to wave-corpuscle duality emons do not differ from any other particles. The explanation of the specific features of electromagnetic and other related phenomena in terms of the emon will require, however, a revision of many aspects of physics. No contradiction with the special relativity theory will be caused, since no interaction conditioned by the presence of emons can propagate with a speed greater than their velocities, that is, greater than c.

In the preceding chapter we have shown that the emission of a photon by an atomic system is like a decay process. The notion of the emon fully conforms to this fact too. This notion does not introduce any difficulties, either, in the understanding of processes involving photons, such as the Compton effect, photoelectric effect, and others, where only the mechanical properties of photons are essential.

12.5. Relativistic Aberration and Doppler Effect

The relativistic aberration and Doppler effect can also be easily explained in terms of the emon by considering the free motion of a concrete emon in two different ICS. The reasoning is as follows.

Let the four-velocity of a free particle in two ICS, K and K', be

$$u = (\beta_x/(1-\beta^2)^{1/2}, \beta_y/(1-\beta^2)^{1/2}, \beta_z/(1-\beta^2)^{1/2}, 1/(1-\beta^2)^{1/2}) \tag{12.36}$$

and

$$u' = (\beta'_x/(1-\beta'^2)^{1/2}, \beta'_y/(1-\beta'^2)^{1/2}, \beta'_z/(1-\beta'^2)^{1/2}, 1/(1-\beta'^2)^{1/2}) \tag{12.37}$$

respectively. Furthermore, let the respective axes of these ICS be parallel, the K' system moving along the x-axis with the velocity $b = V/c$. Then, in accordance with Eqn. (3), the components of u and u' are related by the equations

$$\begin{aligned} \frac{\beta_x}{(1-\beta^2)^{1/2}} &= \frac{\beta'_x + b}{(1-\beta'^2)^{1/2}(1-b^2)^{1/2}} ; \quad \frac{\beta_y}{(1-\beta^2)^{1/2}} = \frac{\beta'_y}{(1-\beta'^2)^{1/2}} ; \\ \frac{\beta_z}{(1-\beta^2)^{1/2}} &= \frac{\beta'_z}{(1-\beta'^2)^{1/2}} ; \quad \frac{1}{(1-\beta^2)^{1/2}} = \frac{1 + b\beta'_x}{(1-\beta'^2)^{1/2}(1-b^2)^{1/2}} . \end{aligned} \tag{12.38}$$

By dividing all the components of u by u_4 we obtain the four-quantity

$$B = (\beta_x, \beta_y, \beta_z, 1) = \left(\frac{\beta_x' + b}{1 + b\beta_x'}, \frac{\beta_y'(1 - b^2)^{1/2}}{1 + b\beta_x'}, \frac{\beta_z'(1 - b^2)^{1/2}}{1 + b\beta_x'}, 1\right), \quad (12.39)$$

whose first three components are the corresponding velocities of the particle in the K system. Note, by the way, that Eqn. (39) represents the well-known Einstein velocity summation.

Let now β_z be zero. Then

$$\begin{aligned} &\beta_x = \beta \cos \bar{\alpha}, \qquad \beta_y = \beta \sin \bar{\alpha}; \\ &\beta_x' = \beta' \cos \bar{\alpha}', \qquad \beta_y' = \beta' \sin \bar{\alpha}', \quad \beta_z' = 0, \end{aligned} \quad (12.40)$$

where $\bar{\alpha}$ and $\bar{\alpha}'$ are the corresponding direction angles of the particle velocity in the K- and K'-systems. For a concrete photon, that is, for a concrete emon moving with a velocity close to c, we now get, in view of Eqns. (39) and (40),

$$\begin{aligned} &\cos \bar{\alpha} = \frac{\cos \bar{\alpha}' + b}{1 + b \cos \bar{\alpha}'}, \qquad \sin \bar{\alpha} = \frac{\sin \bar{\alpha}'(1 - b^2)^{1/2}}{1 + b \cos \bar{\alpha}'}, \\ &\tan \bar{\alpha} = \frac{\sin \bar{\alpha}'(1 - b^2)^{1/2}}{\cos \bar{\alpha}' + b}. \end{aligned} \quad (12.41)$$

Equation (41) represents the well-known relations for the relativistic aberration.

To get the Doppler effect in its usual form it is convenient to rewrite the last of Eqns. (38) in the form

$$\frac{1}{(1 - \beta'^2)^{1/2}} = \frac{1 - b\beta_x}{(1 - \beta^2)^{1/2}(1 - b^2)^{1/2}}. \quad (12.42)$$

For any concrete particle with the rest mass m_0 we obtain, by multiplying Eqn. (42) by m_0c^2,

$$m_0c^2/(1 - \beta'^2)^{1/2} = m_0c^2(1 - b\beta_x)/(1 - \beta^2)^{1/2}(1 - b^2)^{1/2}. \quad (12.43)$$

In view of Eqn. (27), this gives

$$\nu' = \nu \frac{1 - b\beta_x}{(1 - b^2)^{1/2}}. \quad (12.44)$$

For a concrete emon moving with a velocity close to c we finally get from Eqn. (44)

$$\nu' = \nu \frac{1 - b}{(1 - b^2)^{1/2}}, \quad (12.45)$$

that is, the well-known relation for the Doppler effect.

Problems

12.1. A particle of rest mass m_0 is emitted from the surface of a spherical heavenly body of mass M and radius r_0 in the reverse-to-gravity direction, the initial velocity of the particle (in the center-of-mass ICS) being v_0. Figure out, by using the Newton's gravitation law, whether the particle will go to infinity or come back after reaching some finite point.

12.2. Consider the preceding problem for the particular case of the emon.

Suggested References

1. Landau, L. D., and E. M. Lifshitz: 1975, *The Classical Theory of Fields*, tr. by M. Hamermesh, 4th rev. edn., Pergamon Press, New York. Consideration of aberration and the Doppler effect can in particular be found in this book.
2. Mayants, L. S.: 1981, *Found. Phys.* **11**, 577. In this paper the problem of the existence of zero-rest-mass particles is examined and the notion of *emon* is introduced for the first time.
3. Schiff, L. I.: 1968, *Quantum Mechanics*, 3rd edn., McGraw-Hill, New York. The quantization of the electromagnetic field is considered in particular in this book.
4. Semat, H.: 1946, *Introduction to Atomic Physics*, Rinehart, New York. This book contains information about the corpuscular properties of light.
5. Synge, J. L.: 1960, *Classical Dynamics*, Springer, Berlin. The elements of special relativity can be found in this book.

Part III

Methodological Problems

In contrast to Parts I and II, which present the basic ideas and methods of probabilistics and probabilistic physics with some applications, this part contains discussion of some separate problems connected with the material of those two parts. The choice of the problems under discussion is determined mainly by the present author's intention to clear up, to the best of his power, the widespread misunderstandings caused by an inadequate construction of both conventional probability theory and conventional quantum mechanics as well as classical statistical mechanics, and thereby to give the reader a better understanding of all these matters.

Some well-known issues, which might also seem to deserve discussion, are however not considered here for certain reasons. For instance, the problem of the time irreversibility in both classical statistical mechanics and quantum mechanics is not specially treated here because its solution is too simple. The reversibility of time is valid solely for every *concrete closed* mechanical system, whereas both disciplines mentioned, being the interconnected domains of probabilistic physics, deal merely with *abstract* (*pseudoclosed*) mechanical systems and, hence, have nothing to do with the problem under consideration at all. For a similar reason, the issue of causality in quantum mechanics is not discussed here. It suffices to note that according to the formulation of the causality principle given in Section 11.1, it is the concrete processes occurring in concrete physical systems which are completely causally conditioned. But probabilistic physics does not deal with them, and, hence, the validity of the causality principle for concrete physical systems can in no respects contradict its particular domain – quantum physics.

The key methodological principle underlying the whole of this book is also of especial significance for the treatment of problems related to humanistic sciences, such as, for instance, sociology, history, ethics, etc., as well as to sciences dealing with literature, arts, etc., etc. The principal mistakes in the construction of such sciences, not infrequently even fatal ones, are caused by the fact that one completely forgets the really existing concrete people and contemplates them, their features, and their mutual relations in terms of some far-fetched abstractions which have nothing to do with the complicated concrete problems of the existence of concrete human beings in some concrete society.

To be sure, it would be possible to discuss also at least some of the most important problems of this kind here by making use of the aforementioned methodological principle. But we shall not even touch upon any of them herein because the treatment of these problems in the present book might seem alien to its basic material.

Chapter 13

Problems Related to Probability

13.1. Basic Phenomenon of Probabilistics

As has been mentioned in the Preface, probabilistics, like any other natural science, is founded on its basic phenomenon which may be described in the following words: When a very large number of random tests have been carried out on a set of concrete objects, the number of tests revealing concrete objects of each subset of the set is approximately proportional to the relative measure of this subset. It is precisely this statement which has actually been taken as Axiom 1 of probabilistics (see Sect. 1.6).

This science, like other natural sciences, has both experimental and theoretical parts; the former dealing with concrete objects, whereas the latter is concerned with abstract objects. The experimental part of probabilistics is experimental statistics, and its theoretical part is probability theory.

As is well-known, the correctness of the theoretical conclusions of any natural science must be checked by appropriate experiments. In this respect, too, probabilistics does not differ from other natural sciences. An essential distinction between probabilistics and many other sciences is, however, in the specific features of the experiments that can be performed in probabilistics, which are conditioned by the peculiarity of the basic phenomenon of this science.

In the natural sciences, a basic phenomenon reveals some properties of the corresponding concrete objects and may be discovered on any of these concrete objects as a result of a single observation. In these sciences, the theoretical conclusions can also be verified, therefore, on any concrete objects involved by means of a single measurement.

For instance, the basic phenomenon of electrostatics, which reveals the charge as a property of a concrete physical body, may be observed on any two concrete charged bodies, and the fulfillment of the inverse square interaction law can be verified (perhaps only approximately) by single measurements on any two pertinent concrete bodies.

In probabilistics, the basic phenomenon is related to a set of concrete objects and reveals the property of any such set to have a definite relative measure of every one of its subsets, the relative measure of the whole set being unity. To discover the phenomenon it is necessary to make use of a specific statistical method (see Sect. 1.6), the tests performed must necessarily meet two requirements: (1) they have to be random; (2) their number should be large enough.

The notion of randomness has already been explained in Section 1.6, so we may omit its discussion here. We will only note in addition that this notion is altogether irrelevant to probability theory, since it obviously refers to the condition of carrying out the statistical tests and, hence, concerns merely experimental statistics. For this reason, all the attempts to define 'randomness' formally, that is, purely mathematically have proved to be unsuccessful. The meaning of a 'large enough number' of tests has been explained in Section 4.2.

Due to the peculiarity of the phenomenon involved the theoretical conclusions of probabilistics, that is, the results of any probabilistic treatment, can also be verified only on pertinent sets of concrete objects by making use of the statistical method. On the other hand, a set of concrete objects by itself may be considered a concrete object, too, whose only property of interest to probabilistics is the manifold of relative measures of all its subsets. We will call a concrete object of this kind a '*probabilistics concrete object*'. Every probabilistics concrete object has a definite value of this property, that is, a manifold of definite values of the corresponding relative measures. A single measurement of this property is provided by the statistical method, which follows from the basic phenomenon of probabilistics, and it consists of a large enough number of random tests. We will call a measurement of this kind a ***statistical measurement***.

Thus, in probabilistics, too, its theoretical conclusions can be verified on any probabilistics concrete objects involved by means of a single statistical measurement.

In order to avoid any misunderstanding, it is necessary to keep in mind the distinction between the notions of probabilistics concrete object and of (ordinary) concrete object as well as that between the notions of single statistical measurement and of a random test.

The above considerations, in view of the connection of the relative measure of a subset of a set of concrete objects with the probability of the corresponding (random) event (see Sects. 1.9 and 2.1), and in view of the meaning of (statistical) frequency (see Sect. 1.7), allows us to give one more wording of the basic phenomenon of probabilistics: the frequency of an event in a large enough number of random tests is approximately equal to its probability.

Thus, the statement that the frequency of an event in a large enough number of random tests is the measured value of the corresponding probability may be considered to follow directly from the basic phenomenon of probabilistics.

13.2. Mises' Definition of Probability

We have just seen that probabilistics does not differ from other natural sciences in some respects of principle; there is one more such a respect. In any science, a single measurement of any property of a concrete object can be made, as is well known, to some accuracy only, and different single measurements generally give different results. This means that no measurement can give the exact value of the property the concrete object under test really possesses.

The same is true to probabilistics. Indeed, a single statistical measurement, which is supposed to consist of a large enough number of random tests, yields the frequencies of all the pertinent events at once, that is, the approximate values of all the corresponding probabilities. So, a single statistical measurement of probabilities can also be made to some accuracy only, different single measurements generally yielding different approximate results. This means likewise that no statistical measurement can give the exact value of the property the probabilistics concrete object under test really possesses, and hence, no such measurement can give the exact values of the corresponding probabilities. We may thus state that no matter how large the number N of random tests, the frequency of an event equals its probability generally only approximately.

Let us now consider in this connection Mises definition of probability, which reads that probability is the limit the corresponding frequency (cf. Sect. 1.7) approaches when the number of tests goes to infinity (see, for instance, Ref. [5]). Even if the definition were correct, the above statement would not change, since any statistical measurement can consist of a finite number of tests only. But it is incorrect for the following reasons.

First of all, the supposition that the frequency tends to some limit, when the number of tests increases without limit, does not follow from anything and cannot, therefore, be proven. One might think that it should follow from the law of large numbers (see Sect. 2.5). Indeed, when random tests are performed, for instance, in accordance with Bernoulli's scheme (see Sect. 2.4), Eqn. (2.46), which expresses Bernoulli's law of large numbers, might suggest that in every single statistical measurement of such kind, $\lim_{n\to\infty} (m/n) = p$ should be valid, since the probability of any deviation of the frequency, m/n, from the probability, p, which we will denote by P_d, tends to zero when $n \to \infty$: $\lim_{n\to\infty} P_d = 0$. But this supposition is incorrect.

As has already been mentioned in Section 2.5, a law of large numbers deals with probabilities of abstract events and, hence, has nothing to do with concrete events occurring in a single statistical measurement. From $\lim_{n\to\infty} P_d = 0$ it does not follow, therefore, that $\lim_{n\to\infty} (m/n) = p$ for a single statistical measurement. On the contrary, the idea of m/n having a limit with respect to n can be demonstrated to be incompatible with both the requirements of the statistical method and Eqn. (2.46).

Indeed, the frequency m/n is an experimental result of the use of the statistical method. If m/n had a limit p with respect to n, it would mean that to any small number ϵ there corresponds some large integer M, such that the inequality $|m/n - p| < \epsilon$ holds valid for any $n \geqslant M$. But this contradicts the essential requirement of the statistical method – the one of the randomness of the tests. Furthermore, let N_0 be a large enough number of single statistical measurements, and let ϵ be some definite small number. The integer M in the aforementioned definition of the limit of m/n, provided the limit exists, may prove to be different in different series of random tests (different single statistical measurements), but we may choose the maximum of them, M_{max}, as the common one for all the series. Then, if m/n had the limit p, it would mean that for all the N_0 single statistical measurements $|m/n - p| < \epsilon$ would be valid for any $n \geqslant M_{max}$, and none of these measurements could give $|m/n - p| \geqslant \epsilon$. But this means that the probability P_d equals zero precisely for any $n \geqslant M_{max}$, which contradicts Eqn. (2.46), since it follows from it that for any finite ϵ

$$P_d = 1 - 2\pi^{-1/2} \int_0^{z_0} \exp(-z^2)\, dz \neq 0,$$

where $z_0 = (M_{max})^{1/2}(2pq)^{-1/2}\epsilon$ (see Sect. 2.5).

Thus, the frequency of an event does not tend to its probability when the number of tests tends to infinity. Hence, Mises' definition of probability is fallacious in principle.

The above considerations confirm once again, by the way, that a law of large numbers does not establish a relationship between probability theory and experimental statistics (see Sect. 2.5).

Despite the fallacy of Mises' definition of probability, his point of view is actually widely accepted by many natural scientists, and his theory does practically work well. The explanation of this fact seems quite simple. The frequencies of events Mises theory is actually based on, being the measured approximate values of the corresponding probabilities, to a great accuracy obey all the rules of probability calculus, which follow directly from the basic phenomenon of probabilistics, and only this is often necessary for the practical use of probabilities.

Thus, the practical applicability of Mises' theory results in the last analysis from the fact that it actually implicitly proceeds from the basic phenomenon of probabilistics.

13.3. Kolmogorov's Probability Theory

Kolmogorov's axiomatic probability theory (ordinary probability theory) has been built as a purely mathematical discipline. It avoids any direct definition of probability and proceeds from a few axioms the probabilities should obey. To be sure, the axioms were not made up of thin air – they expressed the same rules of probability calculus as those used in van Mises theory, which were discovered in experimental statistics. So, the axioms of ordinary probability theory and, hence, this discipline itself also resulted in the last analysis from the implicitly used basic phenomenon of probabilistics. The fact that this phenomenon reveals the property of a probabilistics concrete object to have a manifold of definite relative measures of all its subsets explains why the axioms of ordinary probability theory had actually to be related to measure theory. At the same time, even though the title of this theory contains the word 'probability', it cannot be the theoretical part of a science of probability, since it is a purely mathematical discipline.

But no branch of pure mathematics may be in principle the theoretical part of any concrete science. Indeed, both parts of a science, theoretical and experimental, should be connected by its specific subject and must, hence, make use of notions particularly relevant to it, whereas pure mathematics is rather a universal language and a universal tool of science, whose every branch may generally be applied, as is well known, to altogether different concrete sciences.

In this connection, we will note the following inconsistency in Kolmogorov's probability theory, as it appears in his fundamental work (see Ref. [3]). Kolmogorov believes his system of axioms to be not complete, '... for in various problems in the theory of probability different fields of probability have to be examined' (loc. cit., § 1.1). This statement actually implies that the axioms lack one more which would make it possible to solve real probabilistic problems. But from a purely mathematical standpoint, there is no need for such an extra axiom at all, since the system of axioms is quite sufficient for forming a particular mathematical discipline. Thus, the above statement contradicts the pure mathematical character of Kolmogorov probability theory claimed by its author.

Since ordinary probability theory is a mathematical discipline, it cannot be linked directly to experimental statistics, from which it actually was derived, without some special premises. What are these premises? The answer is given in Kolmogorov's words: 'In establishing the premises necessary for the applicability of the theory of probability to the world of actual events, the author has used, in large measure, the work of R. v. Mises ...' (loc. cit., § 1.2). So, ordinary probability theory does practically work well because it is linked to experimental statistics by the pertinent results of von Mises.

From the results of the preceding section and the above considerations we may conclude the following. In application to practical problems, von Mises' theory does not need, as a rule, the considerations of Kolmogorov's one, whereas the latter cannot manage without some of the former's reasoning. In some respects, therefore, Mises' theory is closer to probabilistics than ordinary probability theory. At the same time, the latter, being in the main a branch of measure theory which is relevant for the treatment of probabilities, should exhibit certain correspondence with the theoretical part of probabilistics. Let us consider some points of this conformity.

A concrete result of a single random test carried out over a set A of concrete objects and revealing some concrete object of a certain subset of set A may be called a concrete event. A large enough number of random tests leads to the same large number of concrete events, and concrete objects of subsets of equal measure are revealed approximately an equal number of times. Hence the structure of a large enough set of concrete events practically does not differ from the structure of set A: Approximately one and the same number of classes of corresponding concrete events refers to every class of concrete objects. A random event is actually an abstract event – the image of a union of some classes of possible concrete events. Therefore, any random event and a certain set of definite abstract objects stand in a one-to-one correspondence. In particular, each elementary event of a Kolmogorov set U (in our notation) of elementary events and a definite abstract object $f_{12 \ldots n} \in \theta_{12 \ldots n}$ are in one-to-one correspondence; every random event, a subset of U, is in one-to-one correspondence with a definite subset of $\theta_{12 \ldots n}$ and, hence, also with a definite subset of A. The set U itself evidently stands in one-to-one correspondence with $\theta_{12 \ldots n}$ and, hence, also with A.

It is precisely these considerations which were used at the end of Section 1.9 and in Chapter 2 in order to express the results of probabilistics in terms of conventional probability theory. It is also easy to see, then, that all the Kolmogorov axioms (Ref. [3], § 1.1) are either the corollaries of the very definition of probability (see Sect. 1.9) or easily provable theorems (see Sect. 2.2) in probabilistics. Hence all the other mathematical results of Kolmogorov probability theory may also be obtained in probabilistics by proceeding from the definition of probability.

An essential difference between probabilistics and ordinary probability theory may be found in the definition of a random variable. In the latter, the notion of random variable combines two absolutely different mathematical quantities having quite different mathematical meaning, namely, the possible values and their probabilities (or probability densities), in other words, the spectrum of a random variable and the probability distribution corresponding to it. This makes the definition unnecessarily complex and inconsistent with the common usage of words; besides, it complicates both the consideration of functions of random variables and the introduction of the quantum approach to the treatment of random variables, since the spectrum of a Hermitian operator corresponding to a random variable and the state vector describing a particular state of the abstract object under consideration actually merge in one notion.

The definition of a random variable in probabilistics does not have the aforementioned defects (see Sect. 3.1) and works well, as may be seen from the results of its further use in the present book.

One more significant advantage of probabilistics over ordinary probability theory is that in probabilistics, as is seen from Definition 2a of probability, one need not distinguish 'absolute' from conditional probabilities. The necessity of such a modification of probability axiomatics which takes conditional rather than absolute probability as a fundamental quantity was realized long ago and was dictated, in particular, by the mathematical requirements of physics (see, for instance, Ref. [4]). But the main advantage of probabilistics over ordinary probability theory which involves all the others is that the former is a real science of probability founded on its basic phenomenon, whereas the latter is a branch of pure mathematics which should be linked to probability and experimental statistics with the aid of some special premises, which are not pertinent to ordinary probability theory in itself.

13.4. Bertrand Paradox

Bertrand's problem has already been solved by different methods in the present book. The theoretical considerations based both on the choice of different, independent sets adequate to the set of chords in a circle (Sects. 1.8–1.9) and on the application of the probability density $\rho(l)$ of chord length, l (Sect. 4.3), led to one and the same result, $P(l > 3^{1/2}) = 1/4$, which was confirmed by the numerical statistical experiment that yielded $P(l > 3^{1/2}) = 0.26$ (Sect. 4.3). So, one can hardly doubt the correctness of the solution obtained. The apparent paradox discovered by Bertrand was explained to be due to a choice of different sets used for solving the problem, of which only the set of points in the circle was adequate to the set of chords (Sect. 1.9).

At the same time, many mathematicians engaged in probability theory believe all the solutions obtained by Bertrand to be correct (see, for instance, Refs. [1] and [2]); they do not see any paradox in Bertrand's results at all. Since this statement contradicts the aforementioned theoretical and experimental results obtained here, we should dwell on this matter in more detail.

The Bertrand problem has been formulated in Section 1.9. Bertrand himself gave three different solutions of it, which we reproduce here following Ref. [1].

(1) By considerations of symmetry, we may fix the direction of the chord in advance. Let us draw the diameter perpendicular to this direction. It is obvious that only the chords which intersect the diameter in the interval from one quarter to three quarters of its length will exceed the side of the equilateral triangle. Thus the probability needed is 1/2.

(2) By considerations of symmetry, one end-point of the chord may be fixed in advance. The tangent to the circle at this point and the two sides of the inscribed equilateral triangle with vertex at this point form three 60° angles. Only the chords which fall within the middle angle satisfy the condition of the problem. Thus, this method of computation gives for the probability sought the value of 1/3.

(3) In order to fix the position of the chord, it suffices to give its midpoint. For the chord to satisfy the condition of the problem, it is necessary that its midpoint lie within the concentric circle with radius half that of the given circle. The area of this circle is one-forth the area of the given circle. Hence, the probability sought for is 1/4.

The third solution has already been shown to be correct. The question is whether the first two might also be correct. To answer the question let us consider a simpler problem: What is the probability that the distance of a point in a given circle of radius R from the center is less than $R/2$?

This problem is obviously completely tantamount to the Bertrand one. Indeed, every point of the given circle is in one-to-one correspondence with the chord whose midpoint it is, while every point whose distance from the center is less than $R/2$ is in one-to-one correspondence with the pertinent chord whose length exceeds the side of an inscribed equilateral triangle. At the same time, a point is the simplest geometrical object and there cannot be any doubt as to how to determine its position – the problem that is considered a confusing one in case of chords (see Ref. [2], § 1.3).

The solution of the new problem is well-known: the probability sought equals the relative area of the concentric circle of radius $R/2$, that is, 1/4. But we may try to solve it, for instance, by a method similar to the first one employed above to solve the Bertrand problem. The reasoning is as follows.

By considerations of symmetry, we may fix the direction of the diameter a point lies on in advance. It is obvious that only the points which are in the interval from one quarter to three quarters of the length of the diameter will satisfy the condition of the problem. Thus the probability turns out to be 1/2.

However, this result obviously contradicts the correct solution of the problem, which is 1/4, and hence the method used must be considered fallacious. We now try to explore why it is wrong.

To that end, let us introduce the polar coordinates r and φ in order to determine the position of a point. By the mention of symmetry it is implied that not only the points lying on the diameter chosen, but also all the other points in the circle are taken into consideration. It is also meant that the probability for a point to be within an interval $(r, r + dr)$ is independent of its coordinates r and φ, which is true since the length of this interval does not depend of them. But the method used actually involves one more assumption, namely, that the probability for a point to lie within an interval $(\varphi, \varphi + d\varphi)$ does not depend on r and φ. However, this supposition is wrong, since that probability, being independent of φ, does depend on r – it is proportional to the length of that interval which is $r\, d\varphi$, as is well known.

That is the gist of the affair. Instead of solving the problem formulated above, we have solved another problem: What is the probability that the distance of a point lying on the given diameter from the center is less than half a radius of the circle? This problem obviously is concerned with conditional probability, since it contains an additional condition requiring a point to lie on the given diameter, and its solution – the probability equals 1/2 – is correct. Certainly, we may use the method under consideration for solving the problem we wanted to as well. But then we have to make use of the correct mode of calculation of the total probability of an event (see Sect. 2.3).

For the case under consideration, we get from the reasoning of that section the following equation for the required probability:

$$P(r > R/2) = \int_0^{2\pi} \int_0^{R/2} r\, dr\, d\varphi \Big/ \int_0^{2\pi} \int_0^{R} r\, dr\, d\varphi = \tfrac{1}{4}.$$

Thus, the proper use of the method we have employed leads to the correct result.

We may now conclude from what we have examined that the first two solutions of the Bertrand problem are really wrong. Although they seemingly involve all the chords satisfying the condition of the problem, they actually take into consideration merely those obeying one additional condition: in the first method, the condition that the chords should be perpendicular to a given direction; in the second method, the condition that they should have one common end-point at a given point of the circumference.

The conventional way of settling the Bertrand paradox by some experts in probability theory is, thus, erroneous. It has nothing to do with the supposed specificity of the so-called geometrical probability. As a matter of fact, the apparent paradox is due to a misunderstanding in the logic of its consideration, and it is caused by the misuse of the probabilistic treatment.

Bibliography

1. Gnedenko, B. V.: 1962, *The Theory of Probability*, tr. by B. D. Seckler, Chelsea, New York.
2. Kendall, M. G., and P. A. P. Moran: 1963, *Geometrical Probability*, C. Griffin and Co., London.
3. Kolmogorov, A. N.: 1950, *Foundations of the Theory of Probability*, tr. by N. Morrison, Chelsea, New York.
4. Renyi, A.: 1955, *Acta Math. Acad. Sci. Hung.* 6, 285.
5. von Mises, R.: 1931, *Wahrscheinlichkeitsrechnung*, Fr. Deuticke, Leipzig and Vienna.

Chapter 14

Problems Related to Physics

14.1. Gibbs' Paradox and Indistinguishability of Particles

Gibbs' paradox concerns the theoretical entropy of mixing of two ideal gases at constant temperature and pressure. This entropy proved to depend only on the numbers of molecules in each of the gases. So, it seemed to exist also in the case when both gases were identical, which result was clearly wrong and, hence, paradoxical. To eliminate the paradox, it proved enough to divide the original partition functions of each of the gases by a pertinent factor which was the factorial of the number of molecules in it. But why is it so?

The classical method of the probabilistic treatment of pseudoclosed mechanical systems has led, in particular, to Eqn. (9.91) for the entropy of a big macroscopic system consisting of N identical mechanical systems, the equation which has already contained the factor $(N!)^{-1}$, while the related reasoning has answered the above question (cf. Sect. 9.4). However, some physicists do believe that Gibbs' paradox cannot be avoided in classical statistical mechanics without a special premise unjustifiable there, one which can be conceived from the quantum mechanical standpoint only (cf., for instance, Refs. [22] and [23]). This discrepancy is worth exploring, hence we shall deal with this matter in more detail.

Gibbs' paradox leads, in view of Eqn. (9.56), to a similar paradox concerning Helmholtz free energy F. Hence, we may consider the Helmholtz free energy of mixing two ideal gases, instead of the corresponding entropy. Furthermore, since the entropy of mixing depends only on the numbers of the molecules in the gases, so does the Helmholtz free energy. This fact demonstrates, by the way, that the paradox under consideration is related to the translational parts of the correponding thermodynamic quantities (cf. Sect. 9.5), regardless of the structure of the molecules involved. First, this means that the paradox has nothing to do with the quantum peculiarities of molecules, and secondly, that it can be considered on the simplest example of monatomic ideal gases.

In order to calculate F, by Eqn. (9.68), for a pure monatomic gas containing N atoms of mass m, we should first find the partition function Z for it. If we make use of Eqn. (9.35), then we obtain, in view of Eqns. (9.40) and (9.103),

$$Z = [(2\pi mkT)^{3/2} V_0 N]^N, \tag{14.1}$$

where V_0 is the gas volume per atom at temperature T and pressure P. Equation (9.68) now yields

$$F = -NkT \ln C(m, T, P) - NkT \ln N, \tag{14.2}$$

where

$$C(m, T, P) = (2\pi mkT)^{3/2} V_0$$

is obviously a constant for constant T and P.

Suppose now that two portions of the gas under consideration containing N_1 and N_2 atoms, respectively, are combined at constant T and P. The 'mixture' has the same T and P. Hence, it follows from Eqn. (2) that ΔF of 'mixing' is

$$\Delta F = -kT[N_1 \ln (1 + N_2/N_1) + N_2 \ln (1 + N_1/N_2)]. \tag{14.3}$$

This result is paradoxical (it is obviously connected with Gibbs' paradox), since in reality no mixing occurred in the case under consideration. This means that Eqn. (2) cannot be correct. Incidentally, this is easily seen from Eqn. (2) itself. Indeed, the additivity of Helmholtz free energy requires F to be proportional to the number of atoms, N, in the case under consideration, whereas the second term in Eqn. (2) prevents it from satisfying this condition.

By tracing back Eqn. (2), we come to the conclusion that something was wrong with the application of Eqn. (9.35) to the case of pure ideal gas. But this equation resulted from Eqn. (9.33), hence we have to reconsider the method of application of the latter to this particular case. We may reason as follows.

Equation (9.33) determines the probability density of the coordinates and momenta of all N separate atoms of the ideal gas. But the thermodynamic properties of a concrete macroscopic system are determined by the probability distribution of its microstructures (cf. Sect. 9.4), and a microstructure described by definite values of (q, p) will not change upon performing any permutation among the identical atoms of the pure gas, that is, among N different indices assigned to the atoms. Since the total number of permutations among N objects is $N!$, the probability density, $\rho_0(q, p)$, of a microstructure is, in view of Eqn. (9.33),

$$\rho_0(q, p) = N!\rho(q, p) = Z_0^{-1} \exp(-\epsilon/kT), \tag{14.4}$$

where $Z_0 = Z/N!$. This gives, in view of Eqn. (9.34),

$$Z_0 = \int_{\Gamma_0} \exp(-\epsilon/kT)\, d\Gamma_0 \tag{14.5}$$

instead of Eqn. (9.35). Here the integration is over the pertinent $(N!)^{-1}$th part of the whole phase space of the ideal gas. Instead of Eqn. (9.36), we now get

$$\rho(\epsilon) = Z_0^{-1} \exp(-\epsilon/kT)\omega_0(\epsilon), \tag{14.6}$$

where $\omega_0(\epsilon) = \omega(\epsilon)/N!$. This results in

$$Z_0 = \int_\epsilon \exp(-\epsilon/kT)\omega_0(\epsilon)\, d\epsilon, \tag{14.7}$$

instead of Eqn. (9.37).

Equations (9.66) and (9.67) should now be replaced by

$$U = \bar{\epsilon} = \frac{\int_0^\infty \epsilon \exp(-\epsilon/kT)\omega_0(\epsilon)\, d\epsilon}{\int_0^\infty \exp(-\epsilon/kT)\omega_0(\epsilon)\, d\epsilon} \tag{14.8}$$

and

$$U = kT^2 \left(\frac{\partial \ln Z_0}{\partial T}\right)_V \tag{14.9}$$

respectively, which finally leads to

$$F = -kT \ln Z_0, \tag{14.10}$$

instead of Eqn. (9.68).

When N is large enough, as it should be for macroscopic portions of gas, we get, in view of Eqns. (1) and (2.37),

$$Z_0 = [(2\pi mkT)^{3/2} V_0 e]^N, \tag{14.11}$$

whence

$$F = -NkT \ln C(m, T, P)e. \tag{14.12}$$

Here F is proportional to N, as it should be, and no Gibbs' paradox appears.

The true expressions for Z_0 and F differ from those obtained here (cf. Sect. 9.5), but for the consideration of Gibbs' paradox that is inessential. Though traced on the example of a monatomic ideal gas, the above reasoning is obviously valid for the consideration of the translational part of thermodynamic functions of any pure ideal gas, regardless of the complexity of its molecules (cf. also Sect. 9.5). This means that the appearance of the factor $(N!)^{-1}$ in the calculation of the partition function of a pure ideal gas should not be ascribed to the so-called 'indistinguishability' of particles.

Indeed, different concrete molecules of a pure gas can even be in different rotation–vibration states, and hence, they may by no means be regarded as being indistinguishable. At the same time, this fact changes nothing in the above reasoning concerning the calculation of the translational part of Helmholtz free energy for a pure ideal gas.

Thus, the introduction of the factor $(N!)^{-1}$, which obviates Gibbs' paradox, is required in the last analysis by the additivity of the thermodynamic state functions

(F and S, in particular), and only a proper probabilistic treatment of the problem may explain the necessity of that factor.

We may conclude, from what we discussed both above and in Section 9.4, that in equilibrium classical statistical mechanics, one should deal, in order to get correct results, with the probability distributions of the microstructures of a macroscopic system only, that is, with ones which do not change by any permutation of the identical systems. Now, since the probability distributions of quantum mechanics go over into the corresponding ones of classical statistical mechanics in the classical limit (cf. Sect. 9.6 and Chap. 10), a similar proposition is also valid in quantum mechanics, which may be worded as follows.

A quantum mechanical probability distribution related to a system of identical particles should not change by any permutation of the particles.

If a state is described by a ψ-function, then this proposition leads, in view of Eqn. (5.106), to the following statement.

A ψ-function describing a state of a system of identical particles can be either symmetric or antisymmetric with respect to a permutation of any two particles, that is, the result of such a permutation is either $\psi \to \psi$, or $\psi \to -\psi$.

This fact is well known and is believed to be due to the indistinguishability of identical particles in quantum mechanics (cf., for instance, Ref. [38]). We have just seen, however, that it has a merely probabilistic explanation and, therefore, bears no relation to the possibility to discern different identical particles experimentally. To clarify this statement, it is necessary to recall that the probabilistic treatment of physical systems concerns abstract physical systems, whereas any experiment is carried out on concrete physical systems.

So, the term 'indistinguishability' is actually altogether irrelevant for the matter under discussion. However, if one wants to retain it, one has to keep in mind its real probabilistic meaning.

14.2. Classical Limit of Quantum Mechanics

Quantum mechanics and classical statistical mechanics are two interconnected domains of probabilistic physics, and at large enough quantum numbers the probability distributions for microsystems go over into the corresponding probability distributions for classical systems (macrosystems). This means that the classical limit of quantum mechanics is classical statistical mechanics, which fact has been repeatedly noted and illustrated by examples in the present book (cf. Chap. 10).

However, because of some historical reasons, the assertion that the transition from quantum to classical mechanics involves the transition from quantum-mechanical equations to the equations of classical mechanics is widely spread among physicists. It therefore makes sense to identify the error in the reasoning that leads to this common conclusion (cf. Ref. [29]).

One traditionally maintains that classical mechanics is obtained from quantum mechanics as $\hbar$ tends to zero and that the Schrödinger equation simultaneously goes over into the Hamilton–Jacobi equation. The statement '$\hbar$ tends to zero' should, of course, not be understood literally, since a constant cannot vanish; and one can hardly doubt that this reservation is well understood by anyone supporting the aforementioned hypothesis. Nevertheless, the statement conceals a latent danger which accompanies its uncritical application. Indeed, various powers of $\hbar$ appear as factors in some terms of all quantum-mechanical equations; and thus a comparative estimation of the absolute value of any particular term cannot be made without a more detailed analysis of all the terms, especially as $\hbar$ is a denominate quantity whose numerical value depends on the choice of units. It is therefore impossible to regard beforehand the terms containing higher powers of $\hbar$ as small in comparison with those containing lower powers of $\hbar$ (this circumstance is noted in Ref. [18] as well).

The Hamilton–Jacobi equation

$$\partial S/\partial t + H(q_j, \partial S/\partial q_j, t) = 0 \tag{14.13}$$

arises, as is known (cf., for instance, Ref. [20]), in the course of canonical transformation from coordinates q_j and momenta p_j at time t to their values at the time zero. This canonical transformation gives an explicit dependence of coordinates and momenta on time and, hence, settles completely the problem of motion of a mechanical system. In Eqn. (13) H is the original Hamiltonian where the original momenta p_j are replaced by $\partial S/\partial q_j$; S is a function of the old coordinates q_j, time t, and the new momenta P_j, the latter being regarded as constants. If the function S is found – it turns out to be a physical quantity, the action – its partial derivatives with respect to the original coordinates are the original momenta:

$$\partial S/\partial q_j = p_j \quad (j = 1, 2, \ldots, n). \tag{14.14}$$

Substitution of Eqn. (14) into Eqn. (13) yields

$$\partial S/\partial t + H(q_j, p_j, t) = 0. \tag{14.15}$$

By virtue of the mathematical equivalence of coordinates and time, one may expect, by analogy with Eqn. (14), that

$$\partial S/\partial t = p_0 = -E. \tag{14.16}$$

And this guess is in fact correct, since Eqn. (15), in combination with Eqn. (16), gives Eqn. (8.4).

Thus, a function S satisfying Eqn. (13) corresponds to real motion of a system, that is, to motion for which equality of energy and the Hamilton function is fulfilled at every instant. In Eqn. (15), $-\partial S/\partial t$ and H are different physical quantities – merely their numerical values coincide for real motion of a system – and in no way can these quantities be considered identical, as is usually done. From Eqn. (16) it is evident that $\partial S/\partial t$ should be identical with $-E$. This conclusion becomes especially obvious in the particular case of conservative systems. In this instance a solution of Eqn. (13) is of the form

$$S = W(q_1, q_2, \ldots, q_n) - Et, \tag{14.17}$$

where E is a constant value of the energy of a system, and the validity of Eqn. (16) seems to be quite evident.

The conventional statement that the Schrödinger wave equation goes over into the Hamilton–Jacobi equation is exactly the corollary of the view that terms containing higher powers of $\hbar$ can be neglected in comparison with the terms containing lower powers. In the case that a particle moves in an external field $U(x, y, z)$, this assertion is founded on the following reasoning (cf., for instance, Ref. [25]).

Let a wave function be written as

$$\psi = a \exp (iS/\hbar), \tag{14.18}$$

where a and S are real. By substituting this function into the Schrödinger equation, one obtains from the real part of the result the equation

$$\partial S/\partial t + (1/2m)\,[(\partial S/\partial x)^2 + (\partial S/\partial y)^2 + (\partial S/\partial z)^2] + U - (\hbar^2/2ma)\nabla^2 a = 0. \tag{14.19}$$

The neglect of the term containing $\hbar^2$ leads to the equation

$$\partial S/\partial t + (1/2m)\,[(\partial S/\partial x)^2 + (\partial S/\partial y)^2 + (\partial S/\partial z)^2] + U = 0, \tag{14.20}$$

that is, to the Hamilton–Jacobi equation; the function S is thereby identified with the action.

If this sort of rreasoning correctly demonstrates that the Schrödinger wave equation goes over into the Hamilton–Jacobi equation, then it must be valid in the most general case, that is, for any wave function ψ of the form (18), which describes a real state of any physical system conforming to the model for one-particle motion in an external field. We may therefore consider, for example, a wave function with a real spatial part for a stationary state with an energy value E. In this case, $S = -Et$ and $a = a(x, y, z)$. Since $\partial S/\partial x = \partial S/\partial y = \partial S/\partial z = 0$, and $\partial S/\partial t = -E$, Eqn. (20) then yields

$$E = U \tag{14.21}$$

– a result that is obviously incorrect. At the same time, the lack of dependence of S on space coordinates indicates that S is not the action.

Thus, on the whole, S in Eqn. (20) is not the action, and the equation itself is wrong. But since Eqn. (20) follows on deletion of the term containing $\hbar^2$ in the correct Eqn. (19), one should not neglect the $\hbar^2$ term in the general case. But when this term is retained in the example under consideration, multiplication of both sides of Eqn. (19) by a leads to the equation

$$\frac{\partial S}{\partial t} a + \left(U - \frac{\hbar^2}{2m} \nabla^2 \right) a = 0. \tag{14.22}$$

Since

$$\left(U - \frac{\hbar^2}{2m} \nabla^2 \right) a = \hat{H} a \tag{14.23}$$

and $\partial S/\partial t = -E$, Eqn. (22) represents again the Schrödinger equation, but for the spatial part $a(x, y, z)$ of the wave function alone.

Hence, the Schrödinger wave equation does not go over into the Hamilton–Jacobi equation, and the procedure, based on neglect of the term proportional to $\hbar^2$, which purports to prove the correctness of this transition, turns out to be fallacious.

14.3. Energy and Time

The choice of the Hermitian operators for energy and time (cf. Sect. 10.2.2) has been founded on the considerations of Chapter 6. It follows from the results of that chapter that energy E and the Hamiltonian function H are different physical quantities, the former (with the sign reversed) being the momentum canonically conjugate to time t. This conclusion is also confirmed by the reasoning of the preceding section concerning the Hamilton–Jacobi equation. So, one may hardly doubt the correctness of the results obtained in Section 10.2.2, the more so, because they lead directly to the experimentally confirmed Schrödinger equation (cf. Sect. 10.3).

Meanwhile, physicists offer different operators for energy; and, accompanying this, there is a fair amount of jumble in the literature, leading to a number of unsolvable contradictions in the whole range of problems related to the notions of energy and time.

In particular, many physicists, thinking that energy and the Hamiltonian function H are identical, have regarded $\hat{H}$ as the energy operator. Some authors (cf., for instance, Ref. [38]), choosing at first the correct operator as given in Eqn. (10.16) for the energy, replace it later on by $\hat{H}$, in the mistaken assumption that $\hat{E}$ and $\hat{H}$ are equivalent. Confusion in the literature about the appropriate choice of the energy operator (cf. also Ref. [27]) has its origin in classical mechanics, where there is a widespread conviction that energy and the Hamiltonian function are the same. It is routinely asserted that for conservative systems, for instance, energy is the constant value of the corresponding Hamiltonian function (cf., for instance, Ref. [20]). Whereas in classical mechanics such a misconception is apparently not too dangerous – as it does not lead to any serious misunderstandings – it is absolutely inadmissible in quantum physics.

We shall dwell briefly on one of the typical mistakes in commonly accepted quantum physics, which is connected with the confusion of energy and the Hamiltonian function. It concerns attempts to derive the Schrödinger wave equation by requiring the linearity of the operator that determines wave function variations with the time (cf. Refs. [25] and [12]). In the deduction presented in Ref. [12] by Dirac, the reasoning is quite logical until the point is reached where an energy operator has to be chosen. Here the energy is incorrectly replaced by the Hamiltonian function, that is, $\hat{H}$ is taken as the operator for E. In Ref. [25], Landau and Lifshitz assume, when going over to the classical limit, that $\partial S/\partial t = -H$, and therefore $\hat{H}$ is again taken as an operator for energy. Both references reach the invalid conclusion that the linear operator determining wave function variations with the time is $\hat{H}$, and in this manner they obtain the Schrödinger wave equation. As a matter of fact, however, their deductions lead only to identities, and the operator sought after turns out to be, as it should, $\hat{E}$, that is, an operator proportional to $\partial/\partial t$.

The fact that, in a number of cases, energy spectra of conservative systems are discrete, whereas the spectrum of $\hat{E}$ is continuous, is usually advanced as an argument for the choice of $\hat{H}$ rather than of $\hat{E}$ as the operator of energy. However, this argument can easily be invalidated. For conservative systems, the possible energy values are in fact selected from the continuous spectrum of $\hat{E}$ by the condition $E = H$ [Eqn. (8.4)]. Therefore, when the spectrum of $\hat{H}$ is discrete, the possible energy values should also be discrete. Hence, it is the eigenvalues of $\hat{H}$ that are immediately quantized, whereas the eigenvalues of $\hat{E}$ become quantized only by virtue of Eqn. (8.4) (cf. also Sects. 6.4 and 10.3).

One more obstacle standing in the way of the recognition of $\hat{E}$, as given in Eqn. (10.16), as the operator for energy is the refusal of physicists to introduce an operator corresponding to time. This attitude is connected with the widespread belief that time is merely an ordinary number. Argument for and against the introduction of a time operator and debates as to the form of the energy operator are associated with the names of prominent scientists, such as Mandelstam (Ref. [27]), Pauli (Ref. [34]), Weyl (Ref. [43]), etc. In fact there are, however, no solid grounds for controversies on this subject. The question whether or not an operator should be introduced for time is seen to reduce to the question whether time is a random variable, since a Hermitian operator can be introduced for any random variable. But in accordance with the definition of a random variable in probabilistics (cf. Sect. 3.1), time, like any other generalized coordinate, is indubitably a random variable. Consequently, there can be no doubt that an operator $\hat{t}$ should be employed for time t if it is indeed required by the statement of a problem – this operator being of the form given in Eqn. (10.15) in the coordinate-time representation. It is therefore relevant to discover the mistaken notions of those physicists who argue that it is impossible to introduce a time operator. Leaving aside some of the objections (for their discussion cf. Ref. [29]), let us briefly consider merely the objection of Pauli (cf. Ref. [34]), whose reasoning is as follows:

If the operator $\hat{t}$ existed, then the relation

$$\hat{H}\hat{t} - \hat{t}\hat{H} = -ih\hat{I} \tag{14.24}$$

($\hat{I}$ is the unity operator), derived from the general propositions of quantum physics, should be satisfied. Owing to the fact that it is generally impossible to form a Hermitian operator $\hat{H}$ satisfying Eqn. (24), one must conclude that introduction of $\hat{t}$ is forbidden by the rules and postulates of quantum theory; hence, time in quantum mechanics is to be regarded as an ordinary number (c-number).

However, the objection of Pauli and thus also his final conclusion is invalid because it is based on Eqn. (24), which is fallacious. The correct equation, obtained by substitution of $\hat{t}$ for $\hat{F}$ in Eqn. (10.145), is

$$\dot{\hat{t}} \equiv \mathrm{d}\hat{t}/\mathrm{d}t = i\hbar^{-1}\{[-\hat{E}, \hat{t}] + [\hat{H}, \hat{t}]\} = \hat{I}, \tag{14.24a}$$

just as one should expect, and by no means Eqn. (24).

Some other issues related to the notions of energy and time will be considered later on.

14.4. One-Particle Relativistic Equation

In Section 10.4 we considered the Schrödinger equation for a free relativistic particle. It proved to work well and gave quite reasonable results for all the quantities involved, in particular for the particle velocity [cf. Eqn. (10.199)]. At the same time, the Dirac equation is usually believed to be the only correct one, at least for the electron and for the positron. This equation gives, however, altogether different results for a free relativistic particle. We will therefore explore this matter more thoroughly.

If for a physical system H does not depend on time explicitly, the operators $\hat{E}$ and $\hat{H}$ for it commute. Then, starting from Eqn. (10.161), one may obtain any number of other equations of the form

$$\hat{E}^n \psi = \hat{H}^n \psi, \tag{14.25}$$

where n is any arbitrary positive integer. Solution of all these equations results in the same eigenfunctions of $\hat{E}$ and $\hat{H}$ as does the solution of Eqn. (10.161). However, now a number of energy values correspond to every eigenfunction. In particular, for Eqn. (25) the values of E corresponding to one function ψ are

$$E_j = H \exp(i2\pi j/n) \quad (j = 0, 1, 2, \ldots, n-1). \tag{14.26}$$

Only one of all these values, namely, E_0, satisfies condition (10.167) for real motion and hence Eqn. (10.161). Therefore, only E_0 has physical meaning. Thus in cases when for the sake of making the solution of the problem easier one replaces the initial equation (10.161) by some other auxiliary equation, one must discard all values of E except that which satisfies condition (10.167).

Equation (25) with $n = 2$, in view of Eqn. (10.195), gives for a relativistic free particle the equation

$$\hat{E}^2 \psi = (m^2 c^4 \hat{I} + c^2 \hat{p}^2)\psi \tag{14.27}$$

known as the relativistic Schrödinger equation (cf. Ref. [38]) or the Klein–Gordon equation (cf. Refs. [39] and [40]). This equation historically was proposed instead of Eqn. (10.196), because of the unattractiveness of the latter from a mathematical point of view. Equation (27) follows immediately from the familiar relation

$$E^2 = m^2 c^4 + c^2 p^2 \tag{14.28}$$

valid for a relativistic free particle. In considering the negative energy values which are obtained from Eqn. (27), one usually reasons as follows (cf., for instance, Ref. [38]). Equation (28) immediately leads to

$$E = \pm(m^2 c^4 + c^2 p^2)^{1/2}. \tag{14.29}$$

Therefore, the negative energy values in principle appear also in classical relativistic mechanics. However, since in classical mechanics the energy changes continuously, the transition from the positive ($E \geqslant mc^2$) to the negative ($E \leqslant -mc^2$) energy values is impossible there. But in quantum mechanics, where energy may change discontinuously, such a transition is quite possible.

This reasoning is, however, fundamentally wrong. Equation (29) indeed follows immediately from Eqn. (28). But condition (10.167) requires that only one sign in Eqn. (29) should be chosen, since the Hamiltonian function H is a single-valued function of its variables, namely the plus sign. Condition (10.167), as has been noted above, is the condition of reality of motion – it connects the generalized momentum ($-E$) corresponding to time with the remaining generalized momenta and coordinates for real motion. The operator equation (10.161) arises just from this condition, which is valid for every concrete physical system, the solution of this equation describing real states of the corresponding physical system. Therefore, it is impossible to disregard condition (10.167), and the minus sign in Eqn. (29) must be rejected both for classical particles and for microparticles. Thus, Eqn. (27) is also an auxiliary equation, and the negative energy values obtained from it have to be discarded.

The Dirac equation for a free particle with spin $\frac{1}{2}$ is obtained (cf. Ref. [12]) when H in Eqn. (10.161) is replaced by the corresponding Dirac operator

$$\hat{H}_{D(0)} = \alpha_1 \otimes cp_1 + \alpha_2 \otimes cp_2 + \alpha_3 \otimes cp_3 + \beta \otimes mc^2 \hat{I}. \tag{14.30}$$

Here α_j ($j = 1, 2, 3$) and β are four by four matrices which can conveniently be written as

$$\alpha_j = \sigma_1 \otimes \sigma_j \quad (j = 1, 2, 3); \qquad \beta = \sigma_3 \otimes I_2, \tag{14.31}$$

where σ_j ($j = 1, 2, 3$) are the Pauli matrices (cf. Eqn. (10.158); I_2 is the two by two unit matrix; p_j ($j = 1, 2, 3$) are the respective particle momenta; indices 1, 2, 3 are used instead of x, y, z, respectively. (The fact that it is necessary to use the explicit notation of the direct (Kronecker, or tensor) product (cf. Ref. [5]) will become clear subsequently.)

Eigenvalues of $\hat{H}_{D(0)}$ are

$$H_{D(0)} = \pm(m^2c^4 + c^2p^2)^{1/2} \tag{14.32}$$

as a consequence of which the energy of a free particle, in view of condition (10.167), takes on not only positive but also negative values. This result immediately follows by using $\hat{H}_{D(0)}$ as the operator for H and cannot formally contradict the aforementioned results, since they are corollaries of the use of the operator $\hat{H}$ as given in Eqn. (10.195). However, this result cannot be correct, for a number of reasons, it arises from the fact that the eigenvalues of $H_{D(0)}$ can be negative. But this in turn contradicts Eqn. (10.194), according to which H cannot take negative values for a free particle.

Even this circumstance alone makes one doubt the correctness of the choice of $\hat{H}_{D(0)}$ as an operator for the Hamiltonian function of a free particle. There is, however, just one more essential circumstance which shows unambiguously that this choice is indeed wrong. Using $\hat{H}_{D(0)}$ as $\hat{H}$ in Eqn. (10.145), one easily obtains the operators $\hat{v}_j$ (j = 1, 2, 3) for the Cartesian components of velocity of a free particle in the form

$$\hat{v}_j = \alpha_j \otimes c\hat{I}. \tag{14.33}$$

Eigenvalues of the matrices σ_j (j = 1, 2, 3) are ± 1. Therefore, eigenvalues of the matrices α_j (j = 1, 2, 3) are also ± 1, and eigenvalues of $\hat{v}_j$ (j = 1, 2, 3) are $\pm c$. Eigenvalues of a direct product of matrices (or operators) are products of eigenvalues of the factors (cf. Ref. [5]). Hence, the choice of $\hat{H}_{D(0)}$ as the operator for the Hamiltonian function of a free particle necessarily leads to the conclusion that such a particle must move with the velocity of light in vacuum. But this conclusion contradicts experimental data and is physically meaningless, since a particle cannot in principle move with the velocity of light in vacuum (cf. Section 12.3).

Dirac tried to avoid this difficulty, drawing on the imaginary measurement procedure for the velocity of a relativistic particle (cf. Ref. [12]). But this reasoning cannot be considered correct. Indeed, the relativistic theory includes the nonrelativistic theory as a particular case. General conclusions which follow from the relativistic theory have to be valid also in the nonrelativistic case. The transition from the relativistic to the nonrelativistic particle can be realized by a proper choice of the inertial coordinate system. In particular, there exists a system in which the particle is at rest. But should a free particle move at c in some inertial coordinate system, it should move at c in any other such system. This contradicts the basic propositions of relativity theory for a particle of nonzero rest mass [cf. Eqn. (12.39)]. Thus, the choice of $\hat{H}_{D(0)}$ as the operator for the Hamiltonian function of a free particle leads to conclusions which contradict both the basic principles of relativity theory and the experimental data, and so it is incorrect.

Note, further, that $\hat{H}_{D(0)}$ has the same form in all inertial coordinate systems and does not give, therefore, the Schrödinger $\hat{H}$ for a free particle in those inertial coordinate systems where the particle is nonrelativistic. Finally, there is another surprising circumstance related to the choice of $\hat{H}_{D(0)}$ as the operator for H. $\hat{H}_{D(0)}$ was originally taken instead of $\hat{H}$ as given in Eqn. (10.195) in order that Eqn. (10.161) might become linear in all the $\hat{p}_j$ and at the same time lead to Eqn. (27). Although no other restrictions were imposed on it, the equation, which was obtained by Dirac on the basis of these requirements and which conventionally has the form

$$\hat{E}\psi = \hat{H}_{D(0)}\psi \tag{14.34}$$

is considered applicable only for particles with spin $\frac{1}{2}$ (in fact, only for the electron and positron). But why is this the case, and how should one regard Eqn. (34)?

In order to answer these questions let us consider a somewhat more general problem. Given operators $\hat{B}_j$ ($j = 1, 2, \ldots, n$) which have a common complete set of eigenvectors, there exists the operator

$$\hat{D} = \left[\sum_{j=1}^{n} \hat{B}_j^2\right]^{1/2}. \tag{14.35}$$

We now ask: Is it possible to find eigenvalues and eigenvectors of $\hat{D}$ by making use of some other operator $\hat{D}'$ which is linear in $\hat{B}_j$ ($j = 1, 2, \ldots, n$)?

To solve the problem let us try to form the operator

$$\hat{D}' = \sum_{j=1}^{n} a_j^{(m)} \otimes \hat{B}_j \tag{14.36}$$

in such a way that its eigenvalues coincide with those of the operator

$$\hat{D}'' = a^{(m)} \otimes \hat{D}, \tag{14.37}$$

where $a_j^{(m)}$ and $a^{(m)}$ are m by m matrices. (This form of the operator $\hat{D}''$ is determined by the requirement that $\hat{D}'$ and $\hat{D}''$ must be defined in the same space.) Generally this cannot be done for $m = 1$, as one may easily see by considering the new problem in terms of the common complete set of eigenvectors of $\hat{B}_j$. To solve this problem for $m \neq 1$ it is sufficient to require that $\hat{D}'^2 = \hat{D}''^2$, that is,

$$a^{(m)2} \otimes \hat{D}^2 = \sum_{j=1}^{n} a_j^{(m)2} \otimes \hat{B}_j^2 + \sum_{\substack{j,k=1 \\ (j \neq k)}}^{n} (a_j^{(m)}a_k^{(m)} + a_k^{(m)}a_j^{(m)}) \otimes \hat{B}_j\hat{B}_k. \tag{14.38}$$

Further, in order that Eqn. (38) should be satisfied it is sufficient to require that the matrices $a_j^{(m)}$ and $a^{(m)}$ satisfy the conditions

$$a_j^{(m)2} = a^{(m)2} = I_m \quad (j = 1, 2, \ldots, n), \tag{14.39}$$

where I_m is the m by m unit matrix, and

$$a_j^{(m)}a_k^{(m)} = -a_k^{(m)}a_j^{(m)}. \tag{14.40}$$

It follows from Eqn. (39) that eigenvalues of $a_j^{(m)}$ and $a^{(m)}$ are ± 1. We see from Eqn. (40) that m is an even number, and from Eqns. (40) and (39) that the trace of $a_j^{(m)}$ $(j = 1, 2, \ldots, n)$ is equal to zero. Therefore, $m/2$ eigenvalues of $a^{(m)}$ are equal to $+1$ and $m/2$ eigenvalues are equal to -1. We now conclude from Eqn. (36) that the trace of $\hat{D}'$ vanishes. Since eigenvalues of $\hat{D}'$ and $\hat{D}''$ must coincide, the trace of $a^{(m)}$ is also equal to zero, in consequence of which any of the matrices $a_j^{(m)}$ $(j = 1, 2, \ldots, n)$ can be taken as $a^{(m)}$.

Thus, the problem of forming the operator $\hat{D}'$ [Eqn. (36)] whose eigenvalues coincide with those of operator $\hat{D}''$ [Eqn. (37)] reduces to constructing n matrices $a_j^{(m)}$ satisfying conditions (39) and (40). This problem has an infinite number of solutions. Indeed, let f_m matrices $a_j^{(m)}$ $(j = 1, 2, \ldots, f_m)$ and $f_{m'}$ matrices $a_k^{(m')}$ $(k = 1, 2, \ldots, f_{m'})$ satisfy conditions (39) and (40). Then $f_{m'} + f_m - 1$ matrices $a_k^{(mm')}$ $(k = 1, 2, \ldots, f_{m'})$, $a_{f_{m'}+s}^{(mm')} = b_j^{(mm')}$ $(s = 1, 2, \ldots, f_m - 1)$, which are defined by

$$\begin{aligned} a_k^{(mm')} &= a_i^{(m)} \otimes a_k^{(m')} \qquad (i \text{ fixed}; k = 1, 2, \ldots, f_{m'}) \\ b_j^{(mm')} &= a_j^{(m)} \otimes I_{m'} \qquad (j \neq i) \end{aligned} \tag{14.41}$$

also satisfy those conditions. Since at least three Pauli matrices, which are Hermitian, satisfy the necessary conditions for $m = 2$, it follows that at least $2l + 1$ Hermitian matrices obtained from the Pauli matrices by iterative use of relationship (41) also satisfy them at $m = 2^l$ (l is a positive integer). Hence, for any n it is possible to solve the original problem using just the Hermitian matrices $a_j^{(2^l)}$ formed from the Pauli matrices, if $n \leqslant 2l + 1$. But since one may take an arbitrary large l, the number of such solutions is infinite.

It follows from Eqn. (37) that eigenvalues of $\hat{D}''$ repeat each eigenvalue of $\hat{D}$ a total of $m/2$ times with the plus sign and $m/2$ times with the minus sign (recall that $a^{(m)}$ has $m/2$-fold degenerate eigenvalues $+1$ and -1). Hence, these are also the eigenvalues of $\hat{D}'$. If ψ_λ is an eigenvector of $\hat{D}$ corresponding to its eigenvalue λ, then eigenvectors of $\hat{D}'$ corresponding to the $m/2$-fold degenerate eigenvalue λ of the latter are

$$u_{\lambda_j} = \tau_{\lambda_j} \otimes \psi_\lambda \qquad (j = 1, 2, \ldots, m/2), \tag{14.42}$$

where τ_{λ_j} is an unnormalized m by 1 matrix defined for every λ_j. Thus, it is sufficient to form operator $\hat{D}'$ [Eqn. (36)] using any of the infinite number of ways available and to find its eigenvalues and eigenvectors instead of finding directly the eigenvalues and eigenvectors of $\hat{D}$ [Eqn. (35)]. Then it is necessary to take just one of the positive $m/2$-fold set of the degenerate eigenvalues of $\hat{D}'$ as the eigenvalue of $\hat{D}$ and as its eigenvector the right-hand factor in the direct product which determines the corresponding eigenvector of $\hat{D}'$.

Now the way one has to regard Eqn. (34) becomes clear. First, from the mathematical standpoint it is written incorrectly; it should have been written as

$$(I_4 \otimes \hat{E})\psi = \hat{H}_{D(0)}\psi. \tag{14.43}$$

It is true that some authors note that in Eqn. (34) multiplication of $\hat{E}$ by the four by four unit matrix is implied, but actually this circumstance is not always taken into account. Thus, for instance, one usually supposes that Eqn. (27) immediately follows from Eqn. (34). But, as a matter of fact, Eqn. (43) yields

$$(I_4 \otimes \hat{E}^2)\psi = [I_4 \otimes (m^2c^4\hat{I} + c^2\hat{p}^2)]\,\psi, \tag{14.44}$$

and obviously this equation is different from Eqn. (27).

Further, $\hat{H}_{D(0)}$ is just one of the possible auxiliary operators $\hat{D}'$ as given in Eqn. (36) for $n = 4$; $\hat{B}_j = c\hat{p}_j$ $(j = 1, 2, 3)$ and $\hat{B}_4 = mc^2\hat{I}$; $a_j^{(4)} = \alpha_j$ $(j = 1, 2, 3)$ and $a_4^{(4)} = \beta$; where the α_j's and β are the Dirac matrices. It is not the operator for the Hamiltonian function of a relativistic free particle and as such cannot be used to find the particle velocity operator. Hence, the Dirac equation for a free particle is also an auxiliary equation, from the solution of which, using the above procedure, one can obtain the solutions of the correct equation (10.196). For this purpose, from the solutions of Eqn. (43) one has to take just one of the two-fold degenerate positive values of E and the right-hand factor in the corresponding direct product of vectors.

It now becomes clear why the $\hat{H}_{D(0)}$ operator so surprisingly turns out to be applicable only to the electron and positron, although in its construction no specific physical properties of a free particle are taken into consideration. The explanation of this mysterious circumstance is the following. It is the fact that matrices $a_j^{(m)}$ (including the Pauli matrices), by means of which the operators $\hat{D}'$ (including $\hat{H}_{D(0)}$) are formed, have only two different eigenvalues (equal to ± 1) in accordance with the eigenvalues of the electron and positron spin components, which leads to the above-mentioned false conclusion. But, as a matter of fact, the operator $\hat{H}_{D(0)}$ is not an operator for the Hamiltonian function of any free particle (including, of course, the electron and positron). The Pauli matrices, by means of which this operator is formed, have a purely mathematical meaning here which has no relation either to spin or to any other physical quantity. It is also clear that, being an auxiliary operator, $\hat{H}_{D(0)}$ does not have to go over into the Schrödinger $\hat{H}$ for a free particle when the particle becomes nonrelativistic.

We shall now consider a charged particle in the electromagnetic field. (For the reasons explained in Sect. 12.4, this cannot quite be a consistent consideration, but we have no other choice for the present.) Let a particle of rest mass m and charge e be in the electromagnetic field characterized by potentials (the vector potential **A** and the scalar potential φ) corresponding to every inertial coordinate system. Then, as is generally known, the Hamiltonian function for this particle is

$$H = [m^2c^4 + (c\mathbf{p} - e\mathbf{A})^2]^{1/2} + e\varphi, \tag{14.45}$$

where $(c\mathbf{p} - e\mathbf{A})^2 = \Sigma_{j=1}^3 (cp_j - eA_j)^2$; p_j (j = 1, 2, 3) are the Cartesian components of the generalized momentum of the particle. Hence, we have in this case

$$\hat{H} = [m^2c^4\hat{I} + (c\hat{p} - e\hat{A})^2]^{1/2} + e\hat{\varphi}, \tag{14.46}$$

and Eqn. (10.161) yields

$$\hat{E}\psi = \{[m^2c^4\hat{I} + (c\hat{p} - e\hat{A})^2]^{1/2} + e\hat{\varphi}\}\psi. \tag{14.47}$$

If the electromagnetic field vanishes, Eqns. (46) and (47) go over, respectively, into Eqns. (10.195) and (10.196) at any value of e, as they should. In the nonrelativistic approximation Eqns. (46) and (47) yield the corresponding correct expressions.

If the electromagnetic field does not depend on time, neither does H. In this case $\hat{E}$ and $\hat{H}$ commute and the solution of Eqn. (47) reduces to finding the eigenvalues and eigenfunctions of $\hat{H}$. The energy values are then determined by condition (10.167). The probability density of coordinates in a stationary state is independent of time (as for a classical particle) and is determined by the modulus squared of the particular eigenfunction of $\hat{H}$.

And again Eqn. (47), primarily due to its unattractiveness from the mathematical standpoint, was replaced, as is generally known, by the equation

$$(\hat{E} - e\hat{\varphi})^2\psi = [m^2c^4\hat{I} + (c\hat{p} - e\hat{A})^2]\psi. \tag{14.48}$$

This equation, readily obtained from the familiar relation

$$(E - e\varphi)^2 = m^2c^4 + (c\mathbf{p} - e\mathbf{A})^2, \tag{14.49}$$

which is valid for a charged relativistic particle in the electromagnetic field, is known by the same names as Eqn. (27). Equation (49) yields

$$E = \pm[m^2c^4 + (c\mathbf{p} - e\mathbf{A})]^{1/2} + e\varphi, \tag{14.50}$$

and one usually concludes from this fact that negative energy values would appear this time for a charged, zero-spin particle described by Eqn. (48).

However, the minus sign in Eqn. (50) must be rejected, since condition (10.167) has to hold for real motion, while H is a single-valued function of its variables. Thus, one could treat Eqn. (48) as an auxiliary equation that makes it possible to 'find' real states and the energy values for the case under consideration. This conclusion would seem to be confirmed by the fact that Eqn. (48) goes over into Eqn. (27) if the electromagnetic field vanishes. But, as is readily seen, Eqn. (48) is not sufficiently accurate.

Indeed, it determines the states for which mathematical expectations of the random variables $(E - e\varphi)^2$ and $(H - e\varphi)^2$ are equal. But these states are not real, since for real states the mathematical expectations of E and H must be equal, and what is more, in stationary states the values of E and H themselves have to be equal. One can easily make sure that Eqn. (48) cannot be obtained from the correct Eqn. (47). For this purpose write Eqn. (47) as

$$(\hat{E} - e\hat{\varphi})\psi = [m^2c^4\hat{I} + (c\hat{p} - e\hat{A})^2]^{1/2}\psi. \tag{14.51}$$

Multiplying it by the operators contained on its left- and right-hand sides and adding the results, one gets

$$(\hat{E} - e\hat{\varphi})^2\psi = \{m^2c^4\hat{I} + (c\hat{p} - e\hat{A})^2 + e[(m^2c^4\hat{I} + (c\hat{p} - e\hat{A})^2)^{1/2}, \hat{\varphi}]\}\psi. \tag{14.52}$$

Equation (52) is the correct auxiliary equation and differs from Eqn. (48) by the existence of the third term, which does not vanish in the presence of an electromagnetic field, since $\hat{p}_j$ $(j = 1, 2, 3)$ and $\hat{\varphi}$ do not commute. Condon and Shortley (cf. Ref. [11]) incorrectly assert that the operators $\hat{p}^2$ and $r^{-1}\hat{I}$ commute. The difference between Eqns. (52) and (48) can be seen in the case of the Coulomb field. Here one has $\mathbf{A} = 0$ and $e\varphi = -Ze^2/r$, if the proper inertial coordinate system is chosen.

Take $E_0 = mc^2$ as the unit of energy, $a_0 = \hbar^2/mZe^2$ as the unit of length, and $p_0 = mc$ as the unit of momentum. Then, dividing Eqn. (46) for the Coulomb field by E_0 and using the dimensionless quantity $\alpha = Ze^2/\hbar c$ ($\alpha = Z/137$ if e is the electronic charge), we obtain, in view of $mc = \hbar/a_0\alpha$, the equality

$$\hat{H} = (\hat{I} + \alpha^2\hat{p}^2)^{1/2} - (\alpha^2/\tau)\hat{I}. \tag{14.53}$$

Here all the quantities are dimensionless, $\hat{p}_j = -i\,\partial/\partial x_j$ $(j = 1, 2, 3)$, and $\tau = (x_1^2 + x_2^2 + x_3^2)^{1/2}$ (cf. Sect. 10.7).

It follows from Eqn. (53) that the eigenfunctions and eigenvalues of $\hat{H}$, and hence also the values of E, are functions of α^2 only. Therefore, E can be expressed as a power series in α^2:

$$E = 1 + b_0\alpha^2 + b_1\alpha^4 + b_2\alpha^6 + \cdots \tag{14.54}$$

The terms $b_0\alpha^2$, $b_1\alpha^4$, $b_2\alpha^6$, etc., correspond, respectively, to the nonrelativistic, first relativistic, second relativistic approximation, etc. In the units we have chosen Eqns. (52) and (48) now go over, respectively, into

$$\left(\hat{E} + \frac{\alpha^2}{\tau}\hat{I}\right)^2\psi = \left\{\hat{I} + \alpha^2\hat{p}^2 + \alpha^2\left[\frac{1}{\tau}\hat{I}, (\hat{I} + \alpha^2\hat{p}^2)^{1/2}\right]\right\}\psi, \tag{14.55}$$

$$\left(\hat{E} + \frac{\alpha^2}{\tau}\hat{I}\right)^2\psi = (\hat{I} + \alpha^2\hat{p}^2)\psi. \tag{14.56}$$

The power series expansion in α^2 of the third term on the right-hand side of Eqn. (55) starts with a term proportional to α^4. But since the real parts of diagonal matrix elements of a commutator of any two Hermitian operators are equal to zero, the values of E obtained from Eqns. (55) and (56) coincide up to and including the first relativistic approximation; they differ beginning with the second relativistic approximation.

The case of a relativistic particle with an intrinsic magnetic moment is more complicated. In the case that the Coulomb field arises from the charge $-Ze$, an approximate expression for the Hamiltonian function, based on the ideas proposed in Ref. [41], is

$$H = (m^2c^4 + c^2\mathbf{p}^2)^{1/2} - \frac{Ze^2}{r} + \frac{Ze^2}{4m^2c^2r^3}\, g(\mathbf{s}, \mathcal{M}), \tag{14.57}$$

where g is a factor specified for every particle, s is the spin of the particle, and $\mathcal{M}$ is its orbital angular momentum (cf. Sect. 10.7).

To the product $s_j\mathcal{M}_j$ of the quantities s_j and $\mathcal{M}_j$ ($j = 1, 2, 3$) there corresponds the direct products $\hat{s}_j \otimes \hat{\mathcal{M}}_j$ of the operators for them. Since $\hat{s}_j$ is represented by an appropriate n by n matrix S_j (cf. Sect. 10.2.6), there corresponds to the approximate Hamiltonian function (57) (using the above dimensionless quantities) the operator

$$\hat{H} = I_n \otimes \left[(\hat{I} + \alpha^2\hat{p}^2)^{1/2} - \frac{\alpha^2}{\tau}\hat{I}\right] + \frac{\alpha^4}{4\tau^3}\, gS \otimes \hat{\mathcal{M}}, \tag{14.58}$$

where

$$S \otimes \hat{\mathcal{M}} = \sum_{j=1}^{3} S_j \otimes \hat{\mathcal{M}}_j. \tag{14.59}$$

Note, first of all, that for E expression (54) is again valid, while the existence of the last term on right-hand side of (58) changes the coefficients b_i beginning only with $i = 1$. This means that the nonrelativistic correction in $\hat{H}$ caused by the particle's intrinsic magnetic moment, which is expressed by this term, changes the energy values in (54) beginning only with the term corresponding to the first relativistic approximation. Therefore, one may expect that the accurate treatment of the effect of the particle's intrinsic magnetic moment on the energy values, if such a treatment is possible at all, would change them some more (in the usual cases), beginning only with the second relativistic approximation.

Further, one may easily verify, using the conventional commutation relations for S_j ($j = 1, 2, 3$) and $\mathcal{M}_j$ ($j = 1, 2, 3$) (cf. Sects. 10.2.4 and 10.2.6), that neither s_j nor $\mathcal{M}_j$ can have definite values in stationary states, since $\hat{H}$ [Eqn. (58)] does not commute either with $S_j \otimes \hat{I}$ or with $I_n \otimes \hat{\mathcal{M}}_j$ ($j = 1, 2, 3$). But the sum $M_j = \mathcal{M}_j + s_j$ ($j = 1, 2, 3$), which corresponds to the total angular momentum of a particle, and hence $M^2 = \Sigma_{j=1}^{3} M_j^2$, can have definite values in such states, because $\hat{H}$ [Eqn. (58)] commutes with $\hat{M}_j = I_n \otimes \hat{\mathcal{M}}_j + S_j \otimes \hat{I}$ ($j = 1, 2, 3$).

There have been no reservations above concerning the spin of a particle. Therefore, Eqns. (57) and (58), in spite of their approximate character, are universal, and the results obtained by using them are also universal. For the electron (spin $\frac{1}{2}$, $g = 2$) the use of $\hat{H}$ [Eqn. (58)] for $n = 2$ leads to the same energy values (up to and including the term proportional to α^4) as those given by the Pauli theory for hydrogen-like atoms. This follows from the fact that the energy values obtained from Eqns. (55) and (56) coincide within this accuracy. But the results of the Pauli theory agree with the experimental findings with respect to the energy also within this accuracy, as is generally known. Therefore, one may believe that the correctness of the choice of the approximate operator $\hat{H}$ [Eqn. (58)] is also confirmed experimentally with the same accuracy (at least for the electron).

Let us now proceed to the Dirac equation for a charged particle in the electromagnetic field, which is considered applicable only to the electron and positron. This equation, correctly written, has the form

$$(I_4 \otimes \hat{E})\psi = \hat{H}_D \psi, \tag{14.60}$$

where

$$\hat{H}_D = \sum_{j=1}^{3} \alpha_j \otimes (c\hat{p}_j - e\hat{A}_j) + \beta \otimes mc^2 \hat{I} + I_4 \otimes e\hat{\varphi}. \tag{14.61}$$

If the electromagnetic field vanishes, $\hat{H}_D$ goes over into $\hat{H}_{D(0)}$. It has been shown above that $\hat{H}_{D(0)}$ is not the operator for the Hamiltonian function of a free particle, and Eqn. (43) is only the auxiliary equation through which, making use of a certain procedure, one can find solutions of the correct equation (10.196). Also for these same reasons $\hat{H}_D$ is not the operator for the Hamiltonian function of the electron in the electromagnetic field, and thus Eqn. (60) is only an auxiliary equation. But this time one cannot assert that this equation makes it possible to obtain the solutions of a certain correct equation, since the correct equation so far is not known. It can only be claimed that Eqn. (60) corresponds but approximately to the Pauli theory (cf., for instance, Ref. [6]) and, therefore, the positive energy values obtained from it, for example in the case of a hydrogen-like atom, coincide with those obtained from the correct (though approximate) equation which is based on $\hat{H}$ [Eqn. (58)], right up to and including the terms proportional to α^4 in Eqn. (54).

14.5. Possible States of a Conservative System

In Section 8.4 it has been demonstrated that in probabilistic physics the conservation laws, which refer to the physical quantities conserved due to the general properties of space and time, are the same for both abstract classical systems and microsystems. They may be worded as follows: An abstract system can be only in states determined by the totalities of definite values of all the quantities conserved in common.

This fact led, in particular, to the conclusion that a conservative system can be only in stationary states, that is, in states with definite energy values, described by wave functions of the type as given in Eqn. (10.171) (cf. Sect. 10.3). At the same time, the opinion that a conservative system can be in nonstationary states described by wave functions of the type as given in Eqn. (10.169) is generally accepted. Such functions are widely used in theoretical physics for describing states of conservative systems, in particular, when decay processes of these systems are considered. Since the subject under consideration is of rather great importance, let us offer one more reason in support of the above statement.

For large enough quantum numbers, the probability distribution of any physical quantity belonging to a microsystem goes over into the corresponding probability distribution for a classical system. If a conservative system were in a nonstationary state described by a function of the type (10.169), the probability density for x would depend on time, and this dependence had to persist also when the transition to corresponding classical system occurs. Since for conservative systems H is independent of time, however, the conclusion obtained contradicts the general statement that the probability distribution of any physical quantity inherent in a classical system cannot in principle depend on a quantity whose change does not affect the Hamiltonian function H (cf. Sect. 6.2).

The above considerations lead to one more conclusion. The whole range of questions associated with the notion of 'wave packet' turns out to be far-fetched. For a freely moving particle, wave packets in any case simply do not exist at all. This assertion immediately derives from the fact that an abstract freely moving particle can only be in stationary states with definite values of the momentum. Therefore, it hardly makes any sense to discuss any questions concerning wave packets.

14.6. Conventional Decay Theories

The well-grounded thesis of probabilistic physics asserting that an abstract conservative system can only be in stationary states differs, as has been mentioned in the preceding section, from the conventional opinion of quantum physicists. One usually believes that no process can occur in a stationary state of a conservative system at all, and hence it is necessary to use nonstationary states in order to treat a decay process, for instance.

In Chapter 11, a universal theory of physical transformations – theory of transitional configuration – has been outlined, which is based on the aforementioned thesis. The applicability of the theory in principle to the consideration of decay processes and of intramolecular rearrangements has also been demonstrated there. We shall now show that the premises of conventional decay theories are wrong in principle.

The Gamov theory (cf. Ref. [19]) is based on a quite reasonable physical idea that decay is due to a particle leaving a potential well through a potential barrier. But this theory employes an assertion devoid of physical meaning, namely, that eigenvalues of the energy operator and, hence, also of the Hamiltonian, belonging to a decaying conservative system are complex (cf. Ref. [19]). In this theory the Schrödinger equation is regarded as purely mathematical. In fact, however, this equation has a profound physical meaning – it determines real states where the energy and the Hamiltonian function of a conservative system have definite, equal values. Therefore, the Gamov theory is only an artificial mathematical construction which allows one in some cases to estimate the decay rate constants. We shall not discuss it in more detail, but we shall see further why such estimation by the theory is possible, in principle.

Another decay theory is founded on the explicit assumption that a decaying conservative system is in a state which is not stationary (cf. Refs. [24, 26, 36]). According to this theory, a decay law is fully determined by the energy distribution $\omega(E)$ in the initial state of a system which is not yet decayed. The theory leads to a decay law that is not exponential in principle, the necessary condition for an approximately exponential decay law being the existence of one pole of the function $\omega(E)$ in the lower half of the complex E plane (if there are several such poles, the exponential character of the decay becomes apparent only after a sufficiently long period of time). Further, in this theory it is asserted that, for the case when a particle leaves a potential well, the poles of $\omega(E)$ are Gamow's 'complex eigenvalues' corresponding to the 'quasi-steady states' in which the decay occurs; the imaginary part of these eigenvalues $(-\Gamma)$ is connected with Gamow's decay constant κ_G by the relation

$$\kappa_G = 2\Gamma/\hbar, \tag{14.62}$$

and coincides in absolute magnitude with the 'half-width' of the function $\omega(E)$, which is representable by means of the dispersion formula

$$\omega(E) = \pi^{-1}\Gamma\,[(E - E_j)^2 + \Gamma^2]^{-1}. \tag{14.63}$$

The decay theory under consideration leads, as has been shown in Ref. [20], to physically absurd conclusions. One of them is the vanishing of the decay rate constant $\kappa(t)$ when $t \to \infty$, since it has the form of

$$\kappa(t) = 2pt^{-1}, \tag{14.64}$$

where $p > 0$. This means that after a long enough time the decaying system is becoming gradually more and more stable, and the more time passed since its origin, the more stable it becomes. Another conclusion is that small time intervals should exist wherein this constant is negative, that is, wherein the decay of a system is reversed, the rate of restoration of the undecayed state being capable of exceeding the rate of decay by many orders.

Now, it will be shown that for the example considered in Section 11.4, for which Gamov's complex eigenvalues undoubtedly exist, the theory leads to a function $\omega(E)$ that does not have the necessary poles and is not of the form given in Eqn. (63); and this proves once again the complete inadequacy of the theory. Indeed, for the example mentioned, Γ is calculated from the conditions governing the joining of the wave functions at $r = r_1$ and at $r = r_2$, provided that only outgoing waves exist when $r > r_2$. Then, in our notation, one obtains for κ_G the expression

$$\kappa_G = 8k_j'^2(k_j^2 + k_j'^2)^{-1}\kappa_0(E_j), \tag{14.65}$$

where $\kappa_0(E_j)$ is as given in Eqn. (11.34). Therefore, should the theory under consideration be true, the function $\omega(E)$ for the initial wave function $\psi^0_{E_j}(r)$, which describes the presence of a particle inside the well with the energy value E_j, would have to have poles near E_j whose imaginary parts are of the order of magnitude of $\exp(-2k'L)$. But $\psi^0_{E_j}(r)$ coincides with $\psi_{E_j}(r)$ inside the interval $(0, r_2)$ and vanishes outside it. Hence, the function $\omega(E)$ is given by

$$\omega(E) = |J(E, E_j)|^2 (\mathrm{d}k/\mathrm{d}E)|D'|^{-2}, \tag{14.66}$$

where

$$J(E, E_j) = \int_0^{r_1} \sin kr \sin k_j r \, \mathrm{d}r + C_j \int_0^L \{B \exp[(k' - k_j')r] + \\ + C \exp[-(k' + k_j')r]\} \, \mathrm{d}r, \tag{14.67}$$

$$|D'|^2 = (\pi/2k^2)\{(k^2 + k'^2)[B^2 \exp(2k'L) + \\ + C^2 \exp(-2k'L)] + 2(k^2 - k'^2)BC\}. \tag{14.68}$$

The factor $|J(E, E)|^2 (\mathrm{d}k/\mathrm{d}E)$ is obviously an analytic function of the complex E, and only the factor $|D'|^{-2}$ might exhibit the necessary pole. This means that the right-hand side of Eqn. (68) would have to vanish at a certain value $E = E_j' = E_j - i\alpha \exp(-2k'L)$, where α is the unknown. Since B and C are analytic functions of E, one may with sufficient accuracy write

$$B(E_j') = -i\beta'\alpha \exp(-2k_j'L), \qquad C(E_j') = C_j - ic'\alpha \exp(-2k_j'L), \tag{14.69}$$

where $\beta' = (\mathrm{d}B/\mathrm{d}E)_{E=E_j}$ and $c' = (\mathrm{d}C/\mathrm{d}E)_{E=E_j}$. Substituting Eqn. (69) into Eqn. (68), one obtains

$$\beta'^2\alpha^2 + C_j^2 = 0, \tag{14.70}$$

whereas from the imaginary part it follows that

$$(k_j^2 - k_j'^2)\beta'\alpha C_j = 0. \tag{14.71}$$

In the general case $k_j \neq k_j'$, and the equations obtained are incompatible. Therefore, the function $\omega(E)$ involved indeed does not possess the necessary pole. Hence it follows, further, that this function is not representable in the form given in Eqn. (63) either.

Nevertheless, it is of some interest to calculate the 'width' Δ_j' of the function $\omega(E)$, defined as follows: $\Delta_j' = \delta_j'^+ - \delta_j'^-$, where $\delta_j'^+$ and $\delta_j'^-$ are, respectively, the positive and the negative values of δ_j' that satisfy the equation

$$\omega(E_j + \delta_j') = 2^{-1}\omega(E_j). \tag{14.72}$$

As is easily seen, all the factors in Eqn. (66), except $|D'|^{-2}$, can be considered constant when E changes only slightly. Therefore, to the necessary accuracy, Eqn. (72) is equivalent to the equation

$$|D'(E_j + \delta_j')|^2 = 2|D'(E_j)|^2 \tag{14.73}$$

or, in view of Eqn. (68), to the equation

$$(k_j^2 + k_j'^2)[(\delta' B_j)^2 \exp(2k_j'L) + C_j^2 \exp(-2k_j'L)] + 2C_j(k_j^2 - k_j'^2)\delta' B_j$$
$$= 2(k_j^2 + k_j'^2)C_j^2 \exp(-2k_j'L). \tag{14.74}$$

Writing $\delta' B_j = b_j' C_j \exp(-2k_j' L)$ and making use of Eqns. (11.36) and (11.34), we obtain finally

$$\Delta_j' = 4[1 + (1 - 2E_j^2 U_1^{-2})]^{1/2}\hbar\kappa_0(E_j). \tag{14.75}$$

Thus, Δ_j' is obviously different from $2\Gamma = \hbar\kappa_G$ but can turn out to be near both this quantity and the 'width' of the 'relative stability interval' [Eqn. (11.38)]. Since in addition κ_G can be of the same order of magnitude as κ_0, it becomes clear why the conventional quantum theories of decay can give correct estimates of some decay rate constants despite their incorrectness in principle, the theory based on the assertion that a decay law is fully determined by the function $\omega(E)$ not only contradicting the Gamow theory but also falling short in both physical and mathematical respects.

This is not surprising at all because the only premise of this theory is the assertion that a conservative system exists in a nonstationary state. But this premise is altogether wrong, as has already been proven before, and the failure of the theory of decay based on it may be regarded as one more indirect corroboration of the statement of probabilistic physics that a conservative system can exist merely in stationary states.

14.7. Time–Energy Uncertainty Relation

The time–energy uncertainty relation gave rise to never-ending debates (cf. for instance, Refs. [1, 2, 16, 17]). The difference in the opinions of scientists on the treatment of energy and time in quantum mechanics has been one of the causes which prevented the debates from terminating. But the main reason for such development has been a lack of awareness of what quantum physics is in reality. From the standpoint of probabilistic physics, whose domain is quantum physics, the issue is settled simply enough.

First of all, it makes sense to discuss this relation only when in a theoretical problem, or in an experiment under consideration, instants of time are regarded as random or are chosen at random, that is, when the 'randomness' of time really reveals itself. Since time and the negative of energy are canonically conjugate quantities, the theoretical development of the uncertainty relation for energy and time will not differ from that for any other pair of canonically conjugate variables; and the form of the relation obtained will also be the same (cf. Sect. 6.5). Therefore, the need to introduce the special Bohr uncertainty relation (principle) for energy and time disappears. In a stationary state of a conservative system the product of the standard deviations of energy (ΔE) and of time (Δt) vanishes. Since conservative systems can exist only in stationary states, the relation $\Delta E\,\Delta t = 0$ is always fulfilled for these systems. Hence, the product $\Delta E\,\Delta t$ can be nonzero merely for physical systems which are not conservative. A probabilistic problem concerning probability distributions of energy and time should, of course, be raised quite distinctly, just as in any other case, that is, one should have in mind some definite abstract physical object that is in a definite state, which is characterized, for instance, by specific values of all coordinates q_j ($j = 1, 2, \ldots, n$). It hardly makes sense to speak in general about an energy–time uncertainty relation, if it is not understood that there exist quite definite probability distributions for energy and time.

So, the solution of the problem under consideration seems quite clear. Nevertheless, it might be of interest to consider a special opinion concerning this subject, which states that the time–energy uncertainty relation can be obtained only for the relativistic case (cf. Ref. [37]), whereas in a nonrelativistic case merely the coordinate–momentum uncertainty relation of the type

$$\Delta p\,\Delta q \geqslant \hbar/2 \tag{14.76}$$

can be proven strictly for the general case, that is, for any state of a system. Here Δq and Δp are the standard deviations of the generalized coordinate q and of the canonically conjugate to it momentum p, respectively.

The results of Section 6.5, in view of the considerations of Section 8.2, demonstrate, however, that the general uncertainty relation for any pair of canonically conjugate quantities, in particular for coordinate–momentum is

$$\Delta p\,\Delta q \geqslant \frac{\hbar}{2}\left|\left\{1 - [|\psi(q)|^2 q f(q)]\Big|_a^b\right\}\right|. \tag{14.77}$$

Here $\psi(q)$ is the wave function describing the state involved, and $f(q)$ is the weight function, which appears in the definition of the scalar product, similar to that given in Eqn. (6.87); (a, b) is the range of q-values.

Relation (76) is thus a particular case of relation (77) and is valid only when

$$|\psi(a)|^2 af(a) = |\psi(b)|^2 bf(b) \tag{14.78}$$

[cf. also Eqn. (6.133)]. The common belief that it is relation (76) which is the correct one for the general case is based on a purely mathematical mistake.

In conventional derivations of the coordinate–momentum uncertainty relation, it is the supposed hermiticity of the operator $\hat{p}$ which leads to the final result as given in Eqn. (76). But when performing the derivation one forgets that the definition of Hermitian operator in the functional space indispensably contains a reference to the set of functions the operator is Hermitian on (hermiticity range). Besides, the notion of a Hermitian (or self-adjoint) operator results, in the last analysis, from the notion of a scalar product of vectors. But the scalar product in a functional space generally contains, by definition, a real weight function. Hence, the form of a Hermitian operator should necessarily depend on the weight function too.

In Section 10.2.1, it has been shown that the Hermitian operator $\hat{p}$ *must* be of the form given in Eqn. (10.2), and it is Hermitian *only* on a set of functions, each pair ψ_1 and ψ_2 of which obey Eqn. (10.7) (cf. also Refs. [31, 32]). So, not every function of the corresponding functional space belongs to the hermiticity range of the operator $\hat{p}$ of the form given in Eqn. (10.2). Out of all eigenfunctions of $\hat{p}$ which are

$$\psi_\beta = \text{const} \exp{(i\beta q)} f(q)^{-1/2}, \tag{14.79}$$

where β is any real number, only those satisfying Eqn. (10.7) belong to the hermiticity range of $\hat{p}$. Therefore, the normalized to unity eigenfunctions of the *Hermitian* operator p *must* be of the form

$$\psi_m = (b-a)^{-1/2} \exp{[im(b-a)^{-1} 2\pi q]} f(q)^{-1/2} \tag{14.80}$$

$(m = 0, \pm 1, \pm 2, \ldots)$ [cf. also Eqn. (10.13)].

We can now see that relation (77), in view of Eqn. (80), is in agreement with the fact that in any state with a definite value of momentum p, where Δp is exactly zero, the product $\Delta q\ \Delta p$ must also vanish; this is in full accord with the probabilistic meaning of the quantities involved. This is true, in particular, for the case of the canonically conjugate pair φ, $\mathcal{M}_z$, where φ is the angular spherical coordinate corresponding to the rotation about z-axis, in any state of the hydrogen atom with a definite value of the angular momentum $\mathcal{M}_z$.

If both a function ψ and the function $q\psi$ belong to the hermiticity range of $\hat{p}$, Eqn. (78) is satisfied because it then coincides with Eqn. (10.7), and relation (76) is valid. But functions of the form $q\psi_m$ *must not*, in view of Eqn. (80), satisfy Eqn. (10.7) for any $m = 0, \pm 1, \pm 2, \ldots$ and, hence, *do not* belong to the hermiticity range of $\hat{p}$. It is necessary to note that this result is a purely mathematical one which has no relation to 'observables', measurements, etc. It is a particular case of Proposition (1) in Section 5.5. That proposition elucidates some misunderstandings resulting from an illicit use of the hermiticity property of operators. It eliminates, in particular, at least one of the 'surprises in theoretical physics' by Peierls (cf. Ref. [35]). The 'surprise' is formulated as follows.

Let θ denote both the angular plane coordinate and the operator for it. Then

$$\mathscr{M} = -i\,\partial/\partial\theta \tag{14.81}$$

is the operator for the angular momentum canonically conjugate to θ, in terms of $\hbar$. These two operators obviously obey

$$[\mathscr{M}, \theta] = -i. \tag{14.82}$$

In a basis formed by the eigenfunctions of $\mathscr{M}$, Eqn. (82) is represented, in Dirac notation, by

$$(\mathscr{M} - \mathscr{M}')\langle\mathscr{M}|\theta|\mathscr{M}'\rangle = -i\,\delta_{\mathscr{M}\mathscr{M}'}, \tag{14.83}$$

which is an impossible relation, because for $\mathscr{M} = \mathscr{M}'$ the left-hand side is zero, whereas the right-hand side is $-i$.

To resolve this paradox Peierls resorts to some artificial physical considerations, which are far from being correct, whereas the paradox arises merely from a mathematical mistake. Equation (83) is wrong, since to get it from Eqn. (82) it is necessary to make use of the hermiticity of $\mathscr{M}$. But the aforementioned proposition forbids this.

It is necessary to notice, by the way, that the generally useful and beautiful Dirac notation is misleading at this particular point – consideration of the hermiticity property of operators. This is caused by the fact that Dirac's mathematics seems similar, in many respects, to that of the matrix calculus, and perhaps because of that, it does not contain the notion of hermiticity range. In addition, definition of a Hermitian operator proceeding from the matrix representation of operators is incorrect in the general case, as has been demonstrated in Section 5.5.

Another error in Ref. [37] is the assertion that there should be a strict distinction, in mathematical respect, between spatial coordinates and time. This has been disproved by the reasoning in Sections 6.1 and 8.2. It follows from that reasoning, in view of Eqn. (6.134), that the commutativity relation

$$[-\hat{E}, \hat{t}] = -i\hbar\hat{I} \tag{14.84}$$

should be satisfied, which is a particular case of the general relations

$$[\hat{p}_j, \hat{q}_l] = -i\hbar\hat{I}\,\delta_{jl}; \qquad [\hat{p}_j, \hat{p}_l] = [\hat{q}_j, \hat{q}_l] = 0 \tag{14.85}$$

($j, l = 0, 1, \ldots, n$), for $q_0 = t$ and $p_0 = -E$.

Recall now that when deriving uncertainty relation (76) in the conventional way (cf., for instance, Ref. [42]), the particular form of the operators for q and p is of no significance at all. The only relation really utilized is one of the form given in Eqn. (85). Therefore, in view of the aforesaid, the conventional derivation of the time–energy uncertainty relation should not differ from that of any coordinate–momentum uncertainty relation. Besides, Eqns. (85) are not of a specific physical nature – they result from the corresponding mathematical relations valid for any canonically conjugate pair of Hamiltonian random variables (cf. Sects. 6.2–6.4). Hence they are equally related to both nonrelativistic and relativistic cases. But this means that the conventional derivation of an uncertainty relation for any canonically conjugate pair of physical quantities, including time–energy, would have to be the same, regardless of whether it concerns relativistic or nonrelativistic cases.

A general mistake in Ref. [37], is common for most physicists, experts in quantum physics; it is taken for granted that quantum physics deals with the peculiarities of measuring 'observables' (that is, physical quantities) on microsystems. But there are no reasonable arguments for such a statement. Henry Margenau actually still disapproved of it in 1935) (cf. Ref. [28]). The most general, experimentally revealed feature of these measuremnts is the statistical character of the data obtained. The only conclusion that can be drawn from this fact is that quantum physics should deal with probabilistic treatment of microsystems, which means that it is to be based on probabilistic consideration of these systems. Besides, no true physical theory can in principle deal with peculiarities of measuring, since the measurements are always carried out on concrete objects, whereas any theory concerns abstract objects – images of the corresponding sets of concrete objects (cf. Chap. 1 and Ref. [33]).

Thus, the way in which the energy–time uncertainty relation is considered in Ref. [37] is altogether wrong and has nothing to do with the correct treatment of the problem.

14.8. Measurement and Related Problems

The issue of the role of measurements in quantum physics is worth discussing in more detail. We should be reminded again and again of the fact that measurements are always carried out on concrete physical objects, regardless of what the objects are. Such objects may be particles or sets of particles, they may be gaseous, fluid, or solid bodies, and so on; but they are always necessarily concrete. This rather trivial fact is essential for our correct understanding of the possibility of measuring and the role of measurement of physical quantities in probabilistic physics. If unavoidable errors in concrete measurements are disregarded and it is assumed, as usual, that every such measurement is absolutely exact, then the result of the measurement is just that value of the measured quantity which is possessed by the concrete physical object under investigation. Thus, the common term 'observable' is unnecessary, and may only give rise to confusion, since it actually coincides with the terms 'physical property' or 'physical quantity'. Further, as follows from the very essence of probabilistic physics, the role of measurements in probabilistic physics consists in the accumulation of a large enough number of statistical data, which permits us to make certain probabilistic statements or to verify the correctness of probabilistic predictions founded on theoretical considerations (cf. also Sect. 13.1).

In probabilistic physics, a single measurement is obviously of no significance. Therefore, the totality of issues surrounding the procedure of measurement and the role of a device in quantum physics falls away entirely. The part which a device plays is one and the same in all physics: It consists in the experimental determination of the value of the measured physical quantity that belongs to the concrete physical system under investigation.

Finally, it is immaterial, of course, in what manner the required statistical data are accumulated. It is only necessary that the statistical data be appropriate to just that problem for whose solution they are intended. Consequently, any problem in particular concerning the possibility of simultaneous measurement of two physical quantities likewise falls away in probabilistic physics.

The issue raised here has no relation to regularities of probabilistic physics as such. Perhaps the question made sense when stated as follows: Is it possible to construct a device which could allow one to measure simultaneously two physical quantities on one concrete physical object (supposing, of course, that the physical quantities are not functions of one another)? However, such a question, first, is the same both for classical physical objects and for microobjects; and, secondly, the answer to it can in no way be obtained from probabilistic physics, which deals with abstract physical objects, but by no means with separate concrete physical objects.

Let us note, by the way, that the common interpretation of the results of thought experiments dealing with the simultaneous measurement of two canonically conjugate quantities, turns out, on closer scrutiny, to be founded on a misunderstanding. As is known, one concludes from these Gedankenexperiments that it is impossible to measure simultaneously the values of any generalized coordinate q and its canonically conjugate momentum p with absolute precision – the inaccuracies Δq and Δp in these measured values statisfying Heisenberg's uncertainty principle

$$\Delta q \, \Delta p \gtrsim h. \tag{14.86}$$

However, in all the imaginary experiments under discussion one device alone is utilized, and it measures, as a matter of fact, only one quantity, say q, on a single concrete particle with an inaccuracy Δq. Thus, in these thought experiments an attempt is not even made to measure both quantities at the same time, and the uncertainty Δp arises already solely as a result of the q measurement. Therefore, the uncertainties Δq and Δp have clearly different physical meanings: One of them is the inaccuracy in the measured value of one quantity, whereas the other is the uncertainty introduced by this measurement in the value of the remaining quantity. Hence, the usual interpretation of relation (86) is definitely incorrect.

Besides, since the measurement is carried out on a concrete particle, relation (86) is likewise restricted to concrete particles and is in no way connected with probability distributions for q and p. Therefore, the meaning of the deltas in Eqn. (86) is different from that of the deltas in the theoretically obtained Heisenberg uncertainty relation given in Eqn. (76) (cf. also the discussion of the relation in the preceding section). Indeed, the Δ's in Eqn. (76) denote standard deviations characterizing the probability distributions for q and p, respectively. The difference in meaning between the deltas in Eqns. (86) and (76) becomes apparent also from the fact that the quantities on the left-hand side of Eqn. (86) are not viewed as depending on the quantum state of the particle under investigation (a state is not taken into account at all in these thought experiments), whereas the left-hand side of Eqn. (76) is fully dependent on the quantum state; it equals, for example, $(n + \frac{1}{2})\hbar$ for the nth stationary state of a one-dimensional harmonic oscillator [cf. Eqn. (6.123)].

Discussions concerning possible simultaneous measurement of two physical quantities usually involve the questions of the commutativity of the corresponding operators as well as of the existence of a joint probability distribution for these quantities. It is believed that if operators for two physical quantities do not commute, then a joint probability distribution does not exist for these quantities, since they cannot have simultaneously quite definite values and, hence, cannot be measured simultaneously. These statements are, however, erroneous, as can be demonstrated on the examples of a free particle and a particle in a central field.

An abstract free particle is the image of the set of all concrete free particles which have concrete values of various physical quantities, specifically the momentum components and coordinates, in every inertial coordinate system. A state of a free particle can be described with respect to the jth momentum component and the jth coordinate, for instance, by the joint probability density of these quantities. For a free particle the probability density of any of its Cartesian coordinates is constant and, in particular, does not depend on the values of its Cartesian momentum components. Therefore, in a stationary state of an abstract free particle having the definite value y' of the jth Cartesian momentum component the joint probability density of this momentum and the jth Cartesian coordinate [the probability density of the two-dimensional random variable (p_j, q_j)] is given by

$$\rho(y_j, x_j) = C\,\delta(y_j - y_j'). \tag{14.87}$$

Here C is a constant which depends on the length of the space interval in which the particle moves freely. If the length of this interval increases without limit, C vanishes, but this does not prevent one from computing all mathematical expectations needed, by Eqn. (87) (cf. Sect. 10.4).

Similarly, the probability density of the spherical coordinate φ for an abstract particle in the central field is constant and does not depend on the value of the component $\mathcal{M}_z$ of the angular momentum, in particular. Hence, in a stationary state of such a particle having the definite value m' of the random variable $\mathcal{M}_z$ the joint probability density of φ and $\mathcal{M}_z$ is

$$\rho(\varphi, m) = (1/2\pi)\,\delta(m - m'). \tag{14.88}$$

The examples given enable one to write the explicit expressions for the joint probability density of each pair of canonically conjugate coordinates and momenta. These examples are therefore counterexamples disproving the common belief that the joint probability of the canonically conjugate coordinate and momentum does not exist at all. The reasons which prevent one from writing explicit expressions for the joint probability density of the canonically conjugate coordinate and momentum in other cases were cleared up earlier (cf. Sect. 6.4). In the examples given here these reasons are nonessential, due to the fact that the probability distributions of the coordinates and momenta in this situation are mutually independent.

As was mentioned above, it is usually believed that if operators for two physical quantities fail to commute, the joint probability for these quantities does not exist. This statement is motivated, in particular, by the belief that such physical quantities cannot be measured simultaneously, since they cannot have definite values simultaneously. The counterexamples that were just given show that this is not so and makes one look into the problem more carefully. First, it is necessary to re-emphasize that any measurements are made on concrete physical systems, whereas probabilities relate to the corresponding abstract physical systems. Therefore, the question whether it is possible to measure simultaneously two physical quantities has no direct bearing on the existence of a joint probability for these physical quantities. This question concerns only the possibility of direct experimental verification of the correctness of the results of the probabilistic treatment. However, even if it were found that in no case is it possible to measure simultaneously (quite accurately, of course) two physical quantities for a concrete physical system, this would not mean at all that the correctness of the results of the probabilistic approach cannot be tested in a different way.

For instance, in the case of a free particle there is no need at all to measure simultaneously and accurately any coordinate and momentum component of each concrete particle. One could make sure that the probability density of a coordinate is approximately constant and does not depend on the energy (and, hence, on the momentum) by carrying out at different times a sufficiently large number of measurements on different concrete particles (the accuracy of the energy measurement is not essential). This alone should prove that the probability distributions of the coordinate and momontum are mutually independent, and Eqn. (87) is valid in the respective states.

If operators for two physical quantities commute, these physical quantities have definite values in the states described by common eigenfunctions of the operators. It is usually stated that such physical quantities are measurable simultaneously. But what does it actually mean? Does it mean, for example, that one can design an instrument which would make it possible to measure simultaneously the energy and the total angular momentum of every concrete hydrogen atom? Of course not! The question whether it is possible to construct an instrument of this type has no bearing on the actual existence of the states of the abstract hydrogen atom in which the energy and the total angular momentum have definite values. To test the correctness of this result of a probabilistic treatment of the hydrogen atom it would be sufficient to have two instruments: one would measure only the energy and the other would measure only the total angular momentum. The results of these separate measurements carried out at different times (simultaneity is again of no significance here) have to be identical for the whole set of the concrete hydrogen atoms involved, which conforms to the abstract hydrogen atom in the corresponding state.

Let us ask one further question. If the operators for two physical quantities commute, does this mean that the probability distributions of the quantities are mutually independent and that the joint probability density of these quantities is equal to the product of their probability densities? The answer to this question is also negative. One may see this, say, by the example of a particle in a central field. In a stationary state having definite values of the total angular momentum and of its component along a certain axis, the joint probability density of the particle's Cartesian coordinates cannot be represented as the product of their respective probability densities, although the operators for the Cartesian coordinates involved commute.

Thus, commutativity of operators for physical quantities is connected neither with the existence of the joint probability of these physical quantities nor with their simultaneous measurability for every concrete physical system. It has no relation either to whether or not the probability distributions of these physical quantities are mutually independent. This should be clear, since physical quantities in probabilistic physics are quite ordinary random variables, and the commutativity concept is alien to them. This concept concerns, of course, only the random variable operators. But in probabilistics these operators appear if a state of an abstract object is described with respect to some random variables, say, by a state vector or a density operator. Hence, not only the state vector and density operator, but also the operator for a physical quantity are only auxiliary mathematical quantities which have no independent physical meaning. The form of the operator for a physical quantity depends, in particular, also on the way the state is described. If the state is described with respect to a certain number of physical quantities, at least as many operators may appear. All these operators commute, since each operates on its own variable, and to any function of the physical quantities with respect to which the state is described there corresponds the same function of their operators. Noncommuting operators can appear only when the number of operators involved exceeds the number of physical quantities with respect to which the state is described, as can easily be seen from many examples (cf. also Sect. 6.4).

The common assertion that the joint probability of canonically conjugate coordinate and momentum does not exist because the operators for these quantities are noncommuting turns out to be wrong, as shown above. As a matter of fact, the situation is quite the opposite: These operators can be (and, indeed, are) noncommuting because the states for which they are introduced are described with respect to only one of these physical quantities, that is, the space where both operators are defined consists of functions of only one physical quantity. The explanation of this fact has been given in Section 6.4.

One usually believes that it is the operators that are the self-contained quantities which have quite definite physical meaning; they are often identified with 'observables'. It is also held that the properties of operators determine the properties of physical systems. And, what is more, one even asserts that an 'observable' does not exist at all, if among its eigenvectors there are some corresponding to states which are not realizable. Therefore, it is especially important to stress once again that it is the physical quantities that are self-contained in probabilistic physics, where they play the part of ordinary random variables (of probabilistics). We also recall the fact that all the peculiarities of quantum mechanics result directly from the mathematical properties of the Hamiltonian random variables. This is one more additional argument for the assertion that quantum mechanics has nothing to do with the specificity of measurements of physical quantities on microsystems.

14.9. Wave-Corpuscle Duality and Reality of Motion

The fact that, depending on circumstances, particles behave in mechanical respects as macroscopic bodies (that is, classical particles) or as waves, is known as the wave-corpuscle duality of particle properties. In the first case the particles move in sufficiently definite paths, revealing thereby their corpuscular properties, whereas in the second case they undergo, for instance, diffraction, which shows that they have definite wave properties. Scrupulously carried out experiments have shown that a diffraction pattern which appears when particles pass through a diffraction grating does not depend upon the intensity of the beam – the pattern remains practically unchanged even when the particles pass one by one through the apparatus with large enough intervals between them (cf. Ref. [7]). It is generally concluded from this fact that wave properties and, therefore, the wave-corpuscle duality of properties are inherent in each separate particle. It is just this conclusion that seems paradoxical from the point of view of classical physics. Indeed, it is quite clear how a particle as a whole can move. It is also obvious how a wave propagates. But how something can simultaneously move as a particle and propagate as a wave, is indeed impossible to understand. It is this circumstance that has apparently led most physicists to an actual refusal to recognize the reality of particle motion along space–time paths; and this is also the basic reason why quantum mechanics came to be interpreted as a theory of measurement of 'observables'.

However, it turns out that the rationale of the wave-corpuscle duality is actually rather simple. Every concrete particle moves in a continuous (as the particle velocity is finite) concrete path in space-time – hence the corpuscular properties are those of concrete particles. By contrast the 'wave properties' concern an abstract particle – the image of the whole class of corresponding concrete particles – and represent its probabilistic properties. It is necessary to stress once again that the motion of every concrete particle in a path by no means contradicts the probabilistic (that is, the so-called 'wave') properties of the corresponding abstract particle (cf. Sect. 4.2).

Let us now demonstrate that this statement is in complete accord with experiments, while the usual interpretation of the diffraction pattern of particles passing through two slits is wrong. We note that if particles are allowed to pass through a diffraction grating, one after another, then every concrete particle makes its own mark on a screen located behind the grating. As long as the number of particles is small, a diffraction pattern does not appear – nothing but marks, which are made by each particle, are seen. The more or less distinct diffraction pattern appears only when a sufficient number of particles has passed, that is, when the statistical distribution of the particle imprints can be approximated, with some degree of accuracy, by means of a continuous probability distribution for the coordinates of the corresponding abstract particle. Thus, the experimental facts agree fully both with the statement that every concrete particle moves along its own path (as verified by the mark that every particle leaves on the screen) as well as with the assertion that the diffraction pattern resulting from a large enough number of particles is determined by the probability distribution for the coordinates of the corresponding abstract particle.

We now consider a thought experiment in which particles pass through two slits A and B. The distribution of particles over the screen located behind the slits in the case that both slits are open differs from the simple superposition of the particle distributions which one obtains by opening the slits one at a time. Instead of a simple superposition of the latter distributions, a typical interference pattern appears. It can even happen that, when both slits are open, particles avoid some screen locations which they used to reach in numbers when one slit alone was open. This result is discussed in the following manner.

Suppose a particle were moving along a definite path. Then it would reach a particular spot on the screen after passing through one definite slit; and it must naturally hit the same spot of the screen with the same certainty if the second slit, too, is open. The fact, that in reality particles stop impinging on the initial spot when both slits are open, proves that the supposition of a particle moving along any path is wrong in the case under consideration and must therefore be discarded.

Another conclusion that is customarily drawn from the two-slit experiment is that the addition rules for probabilities in this experiment and, hence, also in quntum physics generally – differ from those adopted in conventional probability theory (cf., for instance, Refs. [14, 15, 21]). This assertion is motivated roughly as follows. Let the probability for a particle passing through slit A and impinging on a particular point of the screen be $P(\mathrm{A})$ when the slit A is open and the slit B closed, and let the corresponding probability be $P(\mathrm{B})$ when the slit B is open and slit A closed. Since the simultaneous passage of a particle through both slits is an inonceivable event, the probability $P(\mathrm{A+B})$ of a particle passing through either slit and impinging on the same point of the screen as before, when both slits are open, should equal $P(\mathrm{A}) + P(\mathrm{B})$, according to the rules of ordinary probability theory. From the experiment under consideration it follows, however, that

$$P(\mathrm{A + B}) \neq P(\mathrm{A}) + P(\mathrm{B}). \tag{14.89}$$

But it is easy to detect logical flaws in the foregoing arguments (cf. Ref. [30]). Indeed, the thought experiment under consideration is actually a triple experiment, which is composed of three experiments performed under the following different conditions. Experiment (A): slit A is open, slit B is closed; Experiment (B): slit B is open, slit A is closed; Experiment (AB): both slits are open.

The addition rules of customary probability theory are valid when there is *one invariable complex of conditions* (under which random tests are carried out) determining the probabilities of various events (cf. the end of Sect. 2.2). What are the events we are interested in? They are obviously the impingements of particles on various parts of the screen. An impingement of a particle on some area of the screen is a random event, and the statistical distribution of the concrete impingements over the screen, when an enormous number of concrete particles are employed in the experiment, reproduces approximately the probability distribution of the abstract particle coordinates on the screen. For each of the three experiments under consideration, (A), (B), and (AB), there exists a definite probability distribution for the abstract particle coordinates on the screen, and the addition rules of conventional probability theory are undoubtedly fulfilled for them.

Indeed, for any two incompatible events, that is, impingements of particles on two different non-intersecting areas, say 1 and 2, of the screen, the probability P_{12} that either occurs equals the sum of probabilities (P_1 and P_2) of each of them: $P_{12} = P_1 + P_2$. But the complexes of conditions under which random tests are carried out, being invariable during the performance of each series of tests in each experiment, are different for all the three experiments (A), (B), and (AB), and therefore the probability distributions on the screen we are interested in are different for them too. Thus, the fact that the sum of probabilities in the experiments (A) and (B) is not equal to the corresponding probability in the experiment (AB) has nothing to do with the addition rules of the conventional probability theory and, hence, cannot contradict it.

It is necessary to emphasize, by the way, that there is also a confusion of events in the common treatment of the results of the triple experiment under consideration. When one says that the simultaneous passage of a particle through both slits is an inconceivable event, one actually considers just the passage through the slits A and B as the incompatible random events. Then it would be more logical to discuss the probabilities of *these* events, that is, the probability for a particle to pass through slit A or slit B. It is evident that in experiment (A) the probabilities are: $P_A = 1$, $P_B = 0$; in experiment (B) they are: $P_A = 0, P_B = 1$; in experiment

(AB), if the slits A and B are symmetrically disposed with respect to the particle source, these probabilities are $P_A = 1/2$ and $P_B = 1/2$. In any case, however, also for these events the addition rules of the customary probability theory are valid, and $P_A + P_B = 1$ always. But the incompatibility of the events Q_A and Q_B, which are the passages of a particle through slits A and B, respectively, has nothing to do with the incompatibility of the events Q_1 and Q_2 that are the impingements of particles on different non-intersecting areas, 1 and 2, of the screen.

We can conclude from the thought experiment under consideration, first, that it is a diffraction experiment and, at the same time, a statistical experiment, since the diffraction pattern on the screen is made up of the enormous number of imprints left on it by the concrete particles impinging on it at random, and secondly, that the diffraction pattern or, what is equivalent to it, the probability distribution of particles on the screen depends on the diffraction grating used.

We now return to the aforementioned common assertion that the supposition of a particle moving along a path is wrong in the case under consideration. This was drawn from the fact that the supposition of a particle moving along a *definite* path led to a contradiction. But the negation of this supposition is merely the statement that a particle does *not* move along a *definite* path rather than the assertion that it does not move in a path at all. And this means that the results of the above, essentially statistical experiment, are fully compatible with the idea that a concrete particle moves in a concrete random path, since the only conclusion one may draw from that experiment is, in fact, the following: The totality of real random paths along which a concrete particle can move is determined by the conditions under which the statistical experiment is carried out, that is, by whether the slits A and B are open together or one at a time.

The above considerations confirm that recognition of the reality of motion of concrete physical systems in space-time does not contradict the experimentally proven existence of 'wave' properties of the corresponding abstract systems. After these considerations, the question whether or not quantum mechanics is compatible with the existence of 'hidden variables' of physical systems can now easily be settled. As a matter of fact, the debates on this issue are due to a misunderstanding, since the problem itself turns out to be far-fetched. Indeed, it is sufficient to remember that quantum physics, being a division of probabilistic physics, deals exclusively with abstract physical systems, which, generally speaking, do not have definite values of any physical quantities, except those that determine the states of these physical systems. Therefore, quantum physics cannot in principle include any parameters characterizing concrete motion of the corresponding concrete physical systems.

At the same time, the motion of concrete physical systems occurs along concrete paths (in CPS; cf. Sect. 8.2). It is therefore also true that at any instant a concrete system has concrete values of coordinates and momenta. However, it hardly makes sense to call these properties 'hidden variables'; such a term might only lead to misunderstandings, since the coordinate and momentum values do not exist at all for the corresponding abstract physical system considered in quantum physics.

14.10. Reality of Motion and Potential Energy

The statement of the reality of motion of a concrete microparticle seems to contradict the well-known result of quantum mechanics, such as, for instance, the possibility of penetration through a potential barrier ('tunnel effect'). Indeed, inside the potential barrier, the potential energy of a concrete particle would have to be greater than its total energy, and therefore, should its energy be equal to the sum of potential and kinetic energies, the latter would have to be negative, which is obviously impossible.

This paradoxical result is a particular case of a more general curious one arising when Hamiltonian random variables are treated by means of the quantum approach, which has been discussed in Section 6.4. We shall dwell here on two significant physical examples: (1) a one-dimensional harmonic oscillator; (2) a particle in a central field.

(1) The condition for the potential energy of a one-dimensional harmonic oscillator to exceed its energy is, in view of Eqns. (8.9), (8.10), and (6.109), $|\eta| > 2n + 1$. Hence, all the concrete one-dimensional harmonic oscillators corresponding to the abstract one occupying a stationary state with the energy value $E_d = 2n + 1$, which satisfy this condition, would have to exhibit the aforementioned paradox. Their relative measure is, in view of Eqn. (6.110),

$$P_{\text{par}} = 1 - 2 \int_0^{2n+1} |\psi_n(x)|^2 \, dx. \tag{14.90}$$

This gives for the ground state: $P_{\text{par}} = 0.157$.

(2) For a particle in a central field with the potential energy as given by Eqn. (10.245), the above condition is, in view of Eqns. (10.266) and (10.245), $\eta > 2n^2$. By using Eqn. (10.268), we get

$$P_{\text{par}} = 1 - \int_0^{2n^2} \rho(\eta) \, d\eta \tag{14.91}$$

for the relative measure of the supposedly paradoxical concrete systems in this case. This yields for the ground state: $P_{\text{par}} = 0.238$.

Incidentally, it is easy to see that in both cases the relative measures of the paradoxical concrete systems vanish when the corresponding quantum numbers increase without limit, which is also in accord with the fact that the classical limit of quantum mechanics is classical statistical mechanics.

Let us now try to explore the particular cause of the paradoxical results obtained for physical microsystems. One usually believes that these results exclude any possibility for microparticles to move in space along trajectories. But such a conclusion is inadmissible, since that is the only way in which a concrete particle can move in space. Moreover, a concrete particle trajectory, though belonging to a pertinent set of possible random trajectories, should be continuous because the velocity of a particle cannot exceed that of light in vacuum. Hence, we have to find another source of the above paradoxical results, and we may suspect *that* to be the representation of energy as a sum of two terms, of which potential energy can give rise to a particular doubt.

Indeed, what is the origin of the potential energy? In the first example considered above, the simplest physical system, which corresponds to the mathematical model of a one-dimensional oscillator, is the ion H_2^+. If we could solve the Schrödinger equation for this ion precisely, we would get the values of energy for its stationary states and the respective wave functions describing the states with respect to the coordinates of both the two protons and the only electron. No potential energy, a function of the distance between the protons, would appear in this case at all. But one cannot get the exact solution of the Schrödinger equation even for this simplest case and uses the Born–Oppenheimer approximation to that end. In this approximation, the potential energy of two protons arises, which is the approximately separated electronic energy that depends on the distance between the protons. Hence follows that an exact solution of the Schrödinger equation for any molecule would avoid the above paradoxical results connected with the introduction of the pottential energy related to the atomic cores, since no potential energy of this kind could appear at all. Nevertheless, the paradoxical situation does not entirely disappear because the electrostatic potential energy of the cores and electrons a molecule consists of still remains, which should lead to the paradoxical results similar to those considered in example (2) of a particle in a central field.

This kind of potential energy, which seems to be due to the so-called action-at-a-distance, contradicts, however, special relativity theory and, hence, cannot actually exist. Thus, we have encountered the following situation. On the one hand, the Schrödinger equation which utilizes this kind of potential energy works well enough (within the corresponding accuracy). On the other hand, this kind of potential energy does not exist in reality and, hence, should not be used. How should we understand this situation?

In classical physics, a pertinent field is usually introduced, and action-at-a-distance is replaced by close-range interaction which is consistent with special relativity theory. In Chapter 12 it has been shown, however, that the electromagnetic field does not exist as a physical entity and proves to be an outcome of the presence of an enormous number of emons. Therefore, the electrostatic field should also be connected with emons and does not exist as a physical entity either.

The construction of the theory of electromagnetic phenomena, proceeding from the idea that the emons should be employed to that end, is a task for the future. For the present, one may suppose that the origin of the electrostatic potential energy of charged particles is in a sense similar to that of the potential energy of cores in a molecule. The latter is, we recall, the distance dependent electronic energy of a molecule, which appears as a result of the Born–Oppenheimer approximation. So, the former may prove to be the distance dependent emonic energy, which would also appear as a result of some approximate solution of the pertinent (still unknown) quantum equation that does not contain any potential energy and whose exact solution would avoid the potential energy related to charged particles.

This might be the answer to the above question. Similar ideas might also be valid, in particular, for the case of the gravitational field, where the corresponding subatomic particle which could be called the *gravon* would have to be introduced.

The considerations of this section eliminate the aforementioned objection against the recognition of the reality of motion, which has been indicated at the beginning.

14.11. Einstein–Podolsky–Rosen Paradox

As has been repeatedly noted, quantum mechanics concerns merely abstract physical systems and has nothing to do with measurements which can be made on concrete physical systems only. It has also been shown that no 'hidden variables' which are supposed to be related to the corresponding concrete physical systems may appear in quantum mechanics in principle (cf. Sect. 9). In the meantime, the Einstein–Podolsky–Rosen's (EPR's) arguments, which claim the quantum theory to be incomplete, might seem to suggest the necessity of introducing into it some hidden variables in order to make it complete (cf. Ref. [13]). The discussion of the EPR's paradox and of related topics has not terminated as yet, hence, it might be of interest to explore the issue from the standpoint of probabilistic physics. A correct solution of this paradox can also be found in Ref. [28].

The paradox arises in the example of an imaginary system consisting of two separated particles, which is in the state described by the wave function $\psi = \delta(x_1 - x_2 - l_0)$ (our notation). Here x_1 and x_2 are the coordinates of the first and second particles, respectively, and l_0 is a constant. This ψ-function is obviously an eigenfunction of the operator $\hat{x}_1 - \hat{x}_2$ corresponding to its eigenvalue l_0. It is also an eigenfunction of the operator $\hat{p}_1 + \hat{p}_2$ corresponding to its eigenvalue 0, where p_1 and p_2 are the momenta of the respective particles. So, the system is in a state, for which the difference $x_1 - x_2$ of the particle positions is the constant l_0, and the sum of their momenta is the constant 0.

That ψ-function does not allow one to predict the positions and momenta of either particle. At the same time, if one knows, for instance, the coordinate x_1 and the momentum p_1 of the first particle as a result of two different measurements on it, one can predict the coordinate x_2 and the momentum p_2 of the second particle, which have quite definite values: $x_2 = x_1 - l_0$ and $p_2 = -p_1$. Thus, two pertinent measurements made on either particle allow one to obtain the simultaneous values of the coordinate and the momentum of the other particle, which quantities, in accordance with conventional quantum mechanics, can have no definite values simultaneously, neither can those values be predicted by the ψ-function involved. This is a paradoxical situation, and it suggests, in EPR's opinion, that conventional quantum mechanics does not give a complete description of physical reality.

We now examine a more general case from the standpoint of probabilistic physics. Let a state of the two-particle system under consideration be described with respect to two random variables, the coordinate x of its center-of-mass and the distance l between the particles, by the ψ-function

$$\psi = \delta(l - l_0) \exp(ixP_0/\hbar). \tag{14.92}$$

Here $l = x_1 - x_2$, $x = (m_1x_1 + m_2x_2)/m$, while m_1 and m_2 are the corresponding particle masses, m is their sum, and P_0 is the constant momentum of the system. The ψ-function used by EPR is obviously a particular case of this one for $P_0 = 0$.

The energy of the system in the state under consideration, $E_0 = P_0^2/2m$, is also constant. Hence, we deal with a stationary state of a freely moving system consisting of two free particles, which state is determined by the constant value E_0 of the energy.

To the abstract system being in this state there corresponds an infinite set of concrete systems, each of which has the same value E_0 of energy and the same value P_0 of momentum. Every concrete system also has one and the same value l_0 of the distance between the particles. Hence, every concrete system moves freely along a definite axis of coordinates at the rate $v_0 = P_0/m$, keeping the distance between the particles in it constant. Different concrete systems differ in values of x only.

The momenta and energies (E_1 and E_2, respectively) of the particles in every concrete system are also constant, namely: $p_1 = (m_1/m)P_0$, $p_2 = (m_2/m)P_0$; $E_1 = (m_1/m)E_0$, $E_2 = (m_2/m)E_0$. For the EPR case, $P_0 = 0$ and, hence, $p_1 = p_2 = 0$, and $E_1 = E_2 = 0$. This means that every concrete system and, therefore, the particles forming it are at rest in this case.

The momenta p_1, p_2, and P_0 relating to every concrete system are interconnected and can be expressed in terms of each other. It suffices, therefore, to know one of them in order to obtain the remaining two; the same is true about x_1, x_2, and x.

Let us now consider the possible measurements on these systems. We cannot obviously measure the coordinate x of the center-of-mass of a concrete system directly. In order to get it we have to measure, first, either x_1 or x_2, and then to calculate x. To be sure, after we have measured, say, x_1 for one concrete system, we can 'predict' x_2 for it. But it is impossible to verify the prediction by a measurement, for the following reason: The measurements of a quantity in probabilistic physics must be of random character in principle, since the statistical method is to be used in order to find the probability distribution for the quantity experimentally (cf. Sects. 1.6, 4.2, and 13.1). Hence, the value of x_1, which we have obtained as a result of a separate measurement, is related to a chance concrete system that turned up at random. The next separate measurement either of x_1 or of x_2 would be related, therefore, to a different chance concrete system, and this fact does not allow one to verify the prediction made for the first chance system. Any attempt to perform two consecutive measurements, of x_1 and then of x_2, on one concrete system would disturb the randomness of the tests, which is a necessary condition for obtaining a correct probability distribution by experiment. Thus, the only theoretical way to verify such a 'prediction' is to measure both x_1 and x_2 in one random test.

The above reasoning refers not only to the system under consideration, but also to any physical system consisting of two separate subsystems. To verify theoretical conclusions concerning the possible measurement results for interconnected quantities related to both subsystems, it is necessary to ensure that the measurements indeed refer to one and the same concrete system. To that end, both measurements should be performed simultaneously in one random test.

Matters stand differently with the measurements of p_1 and p_2 for the system under consideration. First of all, they need not be measured at all, since they can easily be calculated from the value of P_0 given in Eqn. (92). However, if one wishes to measure p_1 and p_2, one can do so without being afraid of confusing different concrete systems, since they all have the same values of p_1 and p_2. In the EPR case, in particular, they all are zero.

The above reasoning thus shows that no paradox is contained in the results of the EPR's thought experiment if they are treated from the standpoint of probabilistic physics. Incidentally, being actually related to a freely moving system of two free particles, the example under consideration can be treated, with the same result, by the classical method of probabilistics.

14.12. Bell's Theorem

It is widely believed that a series of specific inequalities contradicting the predictions of quantum mechanics, of which the first one was obtained by Bell (cf. Ref. [3]), have been strictly proved for the so-called 'local realistic theories'. Since quantum mechanics generally works well, whereas none of those inequalities has been corroborated experimentally, one might infer that a realistic view of the physical world must be abandoned, or some kind of action-at-a-distance should be accepted (cf. Ref. [10]).

This issue has attracted attention of many theorists and experimenters, but has not yet found any resolution. It is therefore interesting to consider it from the standpoint of probabilistic physics.

Bell's theorem involves a system consisting of two spatially separated spin-$\frac{1}{2}$ particles (1 and 2), which is in a singlet state ($S = 0$). This instance was originally considered by Bohm as a new version of the EPR's thought experiment (cf. Ref. [8]). Hence, we start with the discussion of Bohm's arguments.

His reasoning may be stated as follows. The function ψ describing the spin state $S = 0$ of the system is of the form

$$\psi = 2^{-1/2} [v_n^+(1) \otimes v_n^-(2) - v_n^-(1) \otimes v_n^+(2)], \tag{14.93}$$

where $v_n^{\pm}(i)$ (i = 1, 2) are the wave spin functions describing states, in which the ith particle has spin $+\frac{1}{2}$ or $-\frac{1}{2}$, respectively, along the direction **n** that may be chosen arbitrarily. One cannot predict the result of a measurement of the spin of particle 1 along, say, the z-axis. But if this result has turned out to be $+\frac{1}{2}$, then one can predict that the result of measuring the spin of particle 2 along the same axis will be $-\frac{1}{2}$. Similarly, the result of measuring a spin component of either particle along any direction can be predicted by the experimenter, after the result of measuring the spin component of the other particle along the same direction has become known. Thus, the values of any spin components of each particle are definite, which conclusion obviously contradicts quantum mechanics and suggests its incompleteness.

Let us now consider Bohm's Gedankenexperiment from the standpoint of probabilistic physics. The wave spin function given in Eqn. (93) describes a state of the pertinent abstract system. All the concrete systems corresponding to it have one and the same spin: $S = s_1 + s_2 = 0$. Hence $s_2 = -s_1$ is valid for any concrete system involved. For the reasons mentioned in the preceding section, a measurement of s_1 is made on a chance concrete system which is turned up at random. Therefore, even though the value of s_2 for the concrete system for which the value s_1 has been measured can be predicted to be $s_2 = -s_1$, this 'prediction' cannot be verified by a measurement, since any following measurement would refer to a different chance concrete system. Such a kind of prediction can be, thus, verified by measurements, only if they are performed for s_1 and s_2 simultaneously, in one random test. For the same reason, measurement could by no means yield different 'spin components' of one concrete particle, even if they existed. But the point is that they do not.

Indeed, it follows from the considerations of Sections 8.4 and 10.2.4 that the angular momentum of even a concrete macrosystem can be referred to only one axis at a time. This means that even in classical mechanics the angular momentum of a concrete system is not a vector and, hence, does not have components along different axes simultaneously. This peculiarity of the angular momentum of a concrete classical system leads to the following significant inference. Suppose it is possible to measure angular momentum along any axis. Then simultaneous measurements of angular momenta along two different axes are unrealizable in principle. This is due to the fact, we recall, that if one axis has already been chosen, it is impossible to simultaneously choose another one. We thus see that this peculiarity of the angular momentum, which is usually believed to occur only in quantum mechanics, is the same in classical mechanics, and it refers to every concrete system, regardless of whether it is macroscopic or microscopic. So, we stress once again, simultaneous different components of spin of a *concrete* particle do not exist in reality – a concrete particle has only one spin component at a time, which is related to the particular axis chosen that time.

Furthermore, a concrete spin-$\frac{1}{2}$ particle, which has the spin value $+\frac{1}{2}$ along a definite axis, may have the spin value either $+\frac{1}{2}$ or $-\frac{1}{2}$ along any other direction. Although it is impossible to predict that value for every concrete particle, it is quite possible to find the probabilities of those values for the corresponding abstract particle. Simple calculations, based on quantum-mechanical considerations, that is, on probabilistic physics (cf. Ref. [12]), yield the following results:

$$P(b^+/a^+) = \cos^2(\gamma/2); \qquad P(b^-/a^+) = \sin^2(\gamma/2);$$
$$P(b^+/a^-) = \sin^2(\gamma/2); \qquad P(b^-/a^-) = \cos^2(\gamma/2). \tag{14.94}$$

Here a and b denote two different axes, γ being the angle between them; $P(b^+/a^+)$ is the probability for an abstract particle to have spin value $+\frac{1}{2}$ along the b-axis, provided it is $+\frac{1}{2}$ along the a-axis, etc.

Hence, if all the concrete particles have a definite spin value along, say, the a-axis, they have different spin values along any other axis, the relative measures of those having values $+\frac{1}{2}$ and $-\frac{1}{2}$ being determined by Eqn. (94). For an abstract pair of particles in a state $S = 0$, Eqn. (94) uniquely determines the probabilities of the different spin values of the abstract particle 2 along any b-axis, provided the abstract particle 1 is in a state determined by its definite spin value along a chosen a-axis, since for every concrete pair of particles, $s_2 = -s_1$ and, hence, the relative measure of concrete particles 2 having spin value $\pm\frac{1}{2}$ equals that of concrete particles 1 which have spin value $\mp\frac{1}{2}$, respectively.

The above considerations show that no difficulty arises at all when interpreting the results of Bohm's Gedankenexperiment from the standpoint of probabilistic physics.

We now turn to the discussion of Bell's theorem, which states that a realistic local theory is incompatible with quantum mechanics. Its proof is based on the calculation of the mathematical expectation Ma_1b_2 (cf. Sect. 3.4) of the product of two random variables a_1 and b_2 in the aforementioned instance. Here a_1 is the spin of particle 1 along the a-axis and b_2 is the spin of particle 2 along the b-axis (in terms of $\hbar/2$). This mathematical expectation is, in view of Eqns. (5.109) and (93), and the considerations of Section 10.2.6,

$$Ma_1b_2 = (\sigma_{1a} \otimes \sigma_{2b}\psi, \psi) = -\cos\gamma_{ab}, \tag{14.95}$$

where $\sigma_{ln} = \sigma_x^{(l)} \cos(\mathbf{i}, \mathbf{n}) + \sigma_y^{(l)} \cos(\mathbf{j}, \mathbf{n}) + \sigma_z^{(l)} \cos(\mathbf{k}, \mathbf{n})$ $(l = 1, 2)$; $\mathbf{n}$ is the unit vector in the n-direction ($\mathbf{n} = \mathbf{a}, \mathbf{b}$); γ_{ab} is the angle between $\mathbf{a}$ and $\mathbf{b}$. Conventional quantum mechanics yields the same result: $Ma_1b_2 = -(\mathbf{a}, \mathbf{b})$.

Considering Bohm's Gedankenexperiment, one may infer that the result of measuring any chosen spin component of particle 2 must be predetermined by previously measuring the same spin component of particle 1. Since the ψ-function given in Eqn. (93) does not determine the result of an individual measurement, this predetermination implies that there should exist a more complete specification of the state.

Let λ denote the specific states, in which the result of an individual measurement is determinate. Let the result of measuring a_1 be determined by $\mathbf{a}$ and λ, and the result of measuring b_2 be determined by $\mathbf{b}$ and λ. These states are then realistic (even deterministic) and satisfy the requirement of locality, since no action-at-a-distance is supposed. Bell's theorem states that a supposition of the existence of such states leads to an inequality which contradicts quantum mechanics. The demonstration of the theorem is given by the following reasoning.

Let $A_x^{(\lambda)}$ and $B_x^{(\lambda)}$ denote the result of measuring x_1 and x_2, respectively, in state λ, where x_i is the x-component of spin of the ith particle. Suppose now that in addition to a-direction and b-direction of Bohm's instance some c-direction is also chosen. Then the mathematical expectations of $a_1 b_2$ and $a_1 c_2$ can be represented, respectively, as

$$Ma_1 b_2 = \int_\lambda A_a^{(\lambda)} B_b^{(\lambda)} \rho(\lambda)\, \mathrm{d}\lambda \tag{14.96}$$

and

$$Ma_1 c_2 = \int_\lambda A_a^{(\lambda)} B_c^{(\lambda)} \rho(\lambda)\, \mathrm{d}\lambda, \tag{14.97}$$

where $\rho(\lambda)$ is the probability density of λ (that is, of the λ-state), and the integration is over all the λ-values. One now gets from these equations

$$Ma_1 b_2 - Ma_1 c_2 = \int_\lambda A_a^{(\lambda)} [B_b^{(\lambda)} - B_c^{(\lambda)}] \rho(\lambda)\, \mathrm{d}\lambda. \tag{14.98}$$

Since $A_x^{(\lambda)} B_x^{(\lambda)} = -1$, for any x, Eqn. (98) can be rewritten as

$$Ma_1 b_2 - Ma_1 c_2 = \int_\lambda A_a^{(\lambda)} B_b^{(\lambda)} [1 + A_b^{(\lambda)} B_c^{(\lambda)}] \rho(\lambda)\, \mathrm{d}\lambda. \tag{14.99}$$

But $A, B = \pm 1$ at any λ and x. Therefore, Eqn. (99) yields, in view of

$$\int_\lambda A_b^{(\lambda)} B_c^{(\lambda)} \rho(\lambda)\, \mathrm{d}\lambda = Mb_1 c_2, \tag{14.100}$$

and $\int_\lambda \rho(\lambda)\, \mathrm{d}\lambda = 1$,

$$|Ma_1 b_2 - Ma_1 c_2| \leqslant 1 + Mb_1 c_2. \tag{14.101}$$

But this inequality contradicts Eqn. (95) of quantum mechanics. Indeed, if the mathematical expectations involved are calculated by Eqn. (95), then the left-hand side of Eqn. (101) can exceed its right-hand side. This is the case, for instance, when $\gamma_{ac} = 2\pi/3$, $\gamma_{ab} = \pi/3$, and $\gamma_{bc} = \pi/3$, and thereby Bell's theorem is finally proven.

If this theorem were correct, it would indeed cause a terrible ordeal to a realistic point of view in physics. But we shall see that it is incorrect.

In order to explore what is wrong with it, let us consider, for instance, a detailed calculation of $Ma_1 b_2$ by means of simple probabilistic considerations. We denote the state λ, in which the random variables a_1 and b_2 have the values $+1$ and -1, respectively, by $(+-)^{ab}$. In the state $\lambda = (-+)^{ab}$ a_1 and b_2 have, respectively, values

-1 and $+1$; etc. So we have only four different states λ: $(++)^{ab}, (+-)^{ab}, (-+)^{ab}$, and $(--)^{ab}$. The values of the random variable $a_1 b_2$ in these states are obviously: $+1, -1, -1$, and $+1$, respectively.

The probability of a state determined by the value $+1$ of a_1 equals $1/2$. The same is the probability of the state, in which a_1 has the value -1. Hence, the probabilities of the above states λ are, in view of Eqn. (94) and the fact that $b_2 = -b_1 : P(++)^{ab} = 2^{-1} \sin^2 (\gamma_{ab}/2)$, $P(+-)^{ab} = 2^{-1} \cos^2 (\gamma_{ab}/2)$, $P(-+)^{ab} = 2^{-1} \cos^2 (\gamma_{ab}/2)$, $P(--)^{ab} = 2^{-1} \sin^2 (\gamma_{ab}/2)$.

We now get from Eqn. (96), by replacing integration by summation,

$$Ma_1 b_2 = 2^{-1} \left(1 \times \sin^2 \frac{\gamma_{ab}}{2} - 1 \times \cos^2 \frac{\gamma_{ab}}{2} - 1 \times \cos^2 \frac{\gamma_{ab}}{2} + 1 \times \sin^2 \frac{\gamma_{ab}}{2}\right)$$
$$= -\cos \gamma_{ab}. \tag{14.102}$$

This result coincides with that of quantum mechanics given in Eqn. (95), as it should. Incidentally, we note that quantum approach was needed only in order to get Eqn. (94). All the other calculations were based on the usual considerations of probability theory only.

Similarly, from Eqn. (97), we obtain for the random variable $a_1 c_2$:

$$Ma_1 c_2 = 2^{-1} \left(1 \times \sin^2 \frac{\gamma_{ac}}{2} - 1 \times \cos^2 \frac{\gamma_{ac}}{2} - 1 \times \cos^2 \frac{\gamma_{ac}}{2} + 1 \times \sin^2 \frac{\gamma_{ac}}{2}\right)$$
$$= -\cos \gamma_{ac}, \tag{14.103}$$

that is also the quantum-mechanical result. It is obvious that Eqns. (102) and (103) obtained from Eqns. (96) and (97), respectively, cannot lead to an inequality of the form (101), since the set of states λ as well as their probabilities are different in Eqns. (102) and (103).

It now becomes clear what a fallacy is in the proof of Bell's theorem. That is the implicit supposition that there exist realistic (or deterministic) states λ common for different directions chosen for particle 2, that is, states λ, in which spin of particle 2 has definite components along different axes simultaneously. But that is a wrong supposition, as has been shown above in our discussion of Bohm's instance. No common states λ may exist in reality for different directions, along which the spin components of a particle are considered. In the state $\lambda = (++)^{ab}$, for instance, all the concrete particles 2 have the definite spin value $+1$ along the b-direction. At the same time, each of these concrete particles also has a definite spin value along the c-direction. But which one of the values ± 1? This is not predeterminated for any choice of c-axis by the state $\lambda = (++)^{ab}$, as was discussed above. Thus, the premise of Bell's theorem, which actually supposes that the result of measuring the random variable x_2 in a state λ is determined only by λ and the choice of the x-direction is not realistic – it contradicts the real peculiarities of the spin of a concrete particle.

In a realistic theory a result of measuring any quantity should reveal the actual value of the quantity. Hence, if a concrete closed system consists of two spatially separated subsystems, and there is a conservation law for an additive quantity inherent in them, then the value of the quantity for either subsystem and, therefore, the result of its measurement is determined by the values of the quantity for the whole system and for the other subsystem. This is true for any closed physical system, regardless of whether it is macroscopic or microscopic, and for any conserved quantity inherent in it. This kind of interconnection between the values of a conserved quantity, which are related to two spatially separated subsystems, is thus conditioned by the pertinent conservation law, and it has nothing to do with action-at-a-distance.

For this reason, the fact that the spin value of a concrete particle 2 along any direction, in Bohm's instance, is opposite to that of particle 1 along the same direction has nothing to do with action-at-a-distance either. We have thus demonstrated that a true 'local realistic theory' provided by probabilistic physics is in accord with quantum physics, as it should be, whereas Bell's theorem, being based on a non-realistic premise contradicting the real features of a concrete spin particle, has been shown to be mistaken.

Other versions of Bell's theorem (cf. Refs. [4, 9, 44], for instance) are actually grounded on the same fallacious premise and are, therefore, wrong as well. It is hardly necessary to disprove them here. However, one can verify that a correct account of realistic states λ and their probabilities does not lead to the corresponding claimed versions of Bell's inequalities – it leads, as it should, to the pertinent quantum-mechanical results.

14.13. Second Quantization

The quantum approach to the consideration of Hamiltonian random fields is outlined in Section 7.4. In quantum physics the same method (often called second quantization) is employed for the treatment of wave functions ψ which are the solutions to the pertinent one-particle wave equations, for instance, of the form

$$i\,\frac{\partial\psi}{\partial t} = \hbar^{-1}\hat{H}\psi \tag{14.104}$$

(cf. Ref. [39]). The function ψ in Eqn. (104) generally is a complex function: $\psi = \psi' + i\psi''$. It is considered a field, while Eqn. (104) itself is regarded as one that this field obeys. If $\hat{H}$ does not depend on time explicitly, this 'random field' is the Hamiltonian one (cf. Sect. 7.4).

Since Eqn. (104) looks like Eqn. (7.60), it yields equations similar to Eqns. (7.64), namely

$$\dot{a}_j' = \lambda_j a_j''; \qquad \dot{a}_j'' = -\lambda_j a_j', \tag{14.105}$$

where a_j' and a_j'' are the coefficients of the expansion of ψ' and ψ'', respectively, in the eigenfunctions φ_j of the operator $\hbar^{-1}\hat{H}$:

$$\psi' = \sum_j a_j'(t)\varphi_j(x, y, z); \qquad \psi'' = \sum_j a_j''(t)\varphi_j(x, y, z), \tag{14.106}$$

while $\hbar\lambda_j = E_j$ are eigenvalues of $\hat{H}$ having the meaning of energy values in states described by the corresponding function φ_j. The variables a_j' and a_j'' are canonically conjugate, and the corresponding Hamiltonian function is of the form

$$H_j = 2^{-1}\lambda_j(a_j'^2 + a_j''^2). \tag{14.107}$$

When using the quantum approach one gets from Eqn. (105) once again an equation looking like the Schrödinger equation for a linear harmonic oscillator with the Hamiltonian

$$\hat{H}_j = 2^{-1}\lambda_j(\hat{a}_j'^2 + \hat{a}_j''^2). \tag{14.108}$$

Therefore, there is a denumerable set of energy values

$$E_n^{(j)} = (n + \tfrac{1}{2})E_j \quad (n = 0, 1, 2, \ldots) \tag{14.109}$$

for every j.

However, ψ actually is not a random field at all because it describes a state of the corresponding abstract particle – it does not refer to a concrete particle and cannot be the value of a random function on any concrete objects. Hence, the second quantization actually bears no relation to the consideration of any random field – it should be regarded as a purely artificial mode allowing one to solve certain probabilistic problems for sets of many identical particles. Its artificiality becomes apparent in many ways. First, for any time-independent operator $\hat{H}$ in Eqn. (104) the second quantization always results in equations of one and the same form with the Hamiltonian as given in Eqn. (108). Further, when making use of the second quantization, one does not later utilize the solutions of the equations obtained (the corresponding wave functions). Besides, depending on whether a solution of the Schrödinger equation (both nonrelativistic and relativistic) or that of the Dirac equation is subjected to quantization, one finds that the operators for canonically conjugate variables a_j' and a_j'' obey the equations

$$[\hat{a}_j'\hat{a}_k''] = \delta_{jk}\hat{I}; \qquad [\hat{a}_j'\hat{a}_k'] = [\hat{a}_j''\hat{a}_k''] = 0, \tag{14.110}$$

or

$$[\hat{a}_j'\hat{a}_k'']_+ = \hat{a}_j'\hat{a}_k'' + \hat{a}_k''\hat{a}_j' = \delta_{jk}\hat{I}; \qquad [\hat{a}_j'\hat{a}_k']_+ = [\hat{a}_j''\hat{a}_k'']_+ = 0 \tag{14.111}$$

respectively. Meanwhile, for any canonically conjugate Hamiltonian random variables, regardless of their origin and only in virtue of the fact that they obey Eqns. (105), the corresponding Hermitian operators must also be canonically conjugate and, hence, satisfy Eqns. (110). Since a_j' and a_j'' would have to be random variables of exactly this kind, the operators for them necessarily should obey Eqns. (110), too, but by no means any other ones. This circumstance also confirms that second quantization is a purely artificial mode only.

Various reasons substantiating the necessity of using different conditions – either (110) or (111) – for the Schrödinger and Dirac equations, respectively, when quantizing their solutions also turn out to be subject to criticism. For instance, it is believed that only Eqns. (111) are permitted to be used for the Dirac equation, since should Eqns. (110) be valid for it, the existence of the negative energy levels would force all the particles to fall down on them, and this contradicts the experiment; at the same time, Eqns. (111), being equivalent to the Pauli exclusion principle, allow one to understand, by using the assumption that all the negative energy levels are occupied, why there exist electrons with positive energy values. But, as has been shown in Section 4, the Dirac equation actually is no more than an auxiliary equation, and negative energy levels for the electron do not exist at all, because of which the reason adduced is groundless.

Since second quantization is an artificial mode, one can hardly give full credence to all the results obtained by its help, without verification of them by means of another sufficiently substantiated probabilistic method.

Bibliography

1. Aharonov, Y., and D. Bohm: 1961, *Phys. Rev.* **122**, 1649.
2. Aharonov, Y., and D. Bohm: 1964, *Phys. Rev.* **B134**, 1417.
3. Bell, J. S.: 1965, *Physics* **1**, 195.
4. Bell, J. S.: 1971, in B. D'Espagnat (ed.), *Foundations of Quantum Mechanics*, Academic Press, New York.
5. Bellman, R.: 1960 (1st edn.) and 1970 (2nd edn.), *Introduction to Matrix Analysis*, McGraw-Hill, New York.
6. Bethe, H. A., and E. E. Salpeter: 1957, *Quantum Mechanics of One- and Two-Electron Atoms*, Springer, Berlin.
7. Biberman, L., N. Sushkin, and V. Fabricant: 1949, *DAN SSSR* **66**, 185.
8. Bohm, D.: 1951, *Quantum Theory*, Prentice Hall, Englewood Cliffs.
9. Clauser, J. F., and M. A. Horne: 1974, *Phys. Rev.* **D10**, 526.
10. Clauser, J. F., and A. Shimony: *Rep. Prog. Phys.* **41**, 1981.
11. Condon, E. U., and G. H. Shortley: 1953, *The Theory of Atomic Spectra*, Cambridge U.P., Cambridge.
12. Dirac, P. A. M.: 1981, *The Principles of Quantum Mechanics*, 4th edn. rev., Clarendon Press, Oxford.
13. Einstein, A., B. Podolsky, and N. Rosen: 1935, *Phys. Rev.* **47**, 777.
14. Feynman, R. P.: 1948, *Rev. Mod. Phys.* **20**, 367.
15. Feynman, R. P., R. B. Leighton, and M. Sands: 1963, *The Feynman Lectures on Physics*, Vol. 1, Addison-Wesley, Reading.
16. Fock, V. A.: 1962, *Zh. Eksp. Teor. Fiz.* **42**, 1135.
17. Fock, V. A.: 1965, *Usp. Fiz. Nauk* **86**, 363.
18. Frenkel, Y. I.: 1950, *Wave Mechanics; Advanced General Theory*, Dover, New York.
19. Gamov, G. A.: 1928, *Z. Phys.* **51**, 204.
20. Goldstein, H.: 1980, *Classical Mechanics*, 2nd edn., Addison-Wesley, Reading.

21. Hibbs, A. R.: 1959, 'Quantum Mechanics' in M. Kac (ed.), *Probability and Related Topics in Physical Sciences*, Interscience, New York.
22. Huang, K.: 1963, *Statistical Mechanics*, Wiley, New York.
23. Kittel, C.: 1961, *Elementary Statistical Physics*, Wiley, New York.
24. Krylov, N. S., and V. A. Fock: 1947, *Zh. Eksp. Teor. Fiz.* 17, 93.
25. Landau, L. D., and E. M. Lifshitz: 1977, *Quantum Mechanics, Non-Relativistic Theory*, tr. by J. B. Sykes and J. S. Bell, 3rd edn., Pergamon Press, New York.
26. Levy, M.: 1959, *Nuovo Cimento* 14, 612.
27. Mandelstam, L. I.: 1950, *Poln. Sobran. Trudov* 3 (in Russian), Izd. Akad. Nauk SSSR, Moscow.
28. Margenau, H.: 1935, *Phys. Rev.* 49, 240.
29. Mayants, L. S.: 1967, *Quantum Mechanics and Transformations of Matter* (in Russian), VINITI, Moscow.
30. Mayants, L. S., 1970, *Zap. Erev. Gos. Univ.* 3, 156.
31. Mayants, L. S.: 1974, *Found. Phys.* 4, 335.
32. Mayants, L. S.: *Found. Phys.* 7, 3.
33. Mayants, L. S.: 1982, *Found. Phys.* 12, 797.
34. Pauli, W.: 1958, 'Die Allgemeinen Prinzipien der Quanten Theorie', in S. Flugge (ed.), *Handbuch der Physik*, Vol. 5, Springer, Berlin.
35. Peierls, R.: 1979, *Surprises in Theoretical Physics*, Princeton U.P., Princeton.
36. Petzold, J.: 1959, *Z. Phys.* 157, 122.
37. Prugovecki, E.: 1982, *Found. Phys.* 12, 555.
38. Schiff, L. I.: 1968, *Quantum Mechanics*, 3rd edn., McGraw-Hill, New York.
39. Schweber, S. S., H. A. Bethe, and F. Hoffmann: 1955, *Mesons and Fields*, Row, Peterson and Co., Evanston Ill., Elmsford.
40. Schweber, S. S.: 1961, *An Introduction to Relativistic Quantum Field Theory*, Row, Peterson and Co., Evanston Ill., Elmsford.
41. Thomas, L. H.: *Nature* 107, 514.
42. von Neumann, J.: 1932 and 1969, *Mathematische Grundlagen der Quantenmechanik*, Springer, Berlin.
43. Weyl, H.: 1931, *Gruppentheorie und Quantenmechanik*, 2nd edn., Hirzel, Leipzig.
44. Wigner, E. P.: 1970, *Amer. J. Phys.* 38, 1005.

Appendix 1

Proof of Equations (6.16″) and (6.16‴)

We start with the proof of Eqn. (6.16″). Since

$$\frac{\mathrm{d}F}{\mathrm{d}x_k} = \left(\frac{\partial F}{\partial y^{(1)}}\right) \frac{\mathrm{d}y^{(1)}}{\mathrm{d}x_k} + \left(\frac{\partial F}{\partial y^{(2)}}\right) \frac{\mathrm{d}y^{(2)}}{\mathrm{d}x_k},$$

Eqn. (6.16″) follows, in view of Eqn. (6.14), directly from Eqn. (6.6″). Hence, it is sufficient to prove Eqn. (6.6″) in any independent way [in Sect. 6.1, it was derived from Eqn. (6.16″)]. Writing

$$H(x_1, \ldots, x_n, x_{n+1}, \ldots, x_{2n-1}) = G(x_1, \ldots, x_n, x_{n+1}, \ldots, x_{n+k-1},$$
$$\varphi(x_{n+k}), x_{n+k+1}, \ldots, x_{2n-1}), \qquad \text{(A.1.1)}$$

and replacing in $H' = 0$ [Eqn. (6.10)] $\varphi(x_{n+k})$ by $-G_{n+k}$ [see Eqn. (6.10′)], we obtain

$$H' = x_{2n} + G(x_1, \ldots, x_n, x_{n+1}, \ldots, x_{n+k-1},$$
$$-G_{n+k}, x_{n+k+1}, \ldots, x_{2n-1}) = 0. \qquad \text{(A1.2)}$$

Since G_{n+k} does not contain x_{n+k}, H' now is a function of independent variables only.

It follows from Eqn. (1), in view of Eqn. (6.6), that

$$\frac{\partial G}{\partial \varphi} = \left(\frac{\mathrm{d}\varphi}{\mathrm{d}x_{n+k}}\right)^{-1} \frac{\mathrm{d}x_k}{\mathrm{d}x_n}. \qquad \text{(A1.3)}$$

From Eqn. (2) we obtain, in view of Eqns. (6.6) and (3),

$$\frac{\mathrm{d}H'}{\mathrm{d}\tilde{y}^{(1)}} = -\frac{\mathrm{d}y^{(2)}}{\mathrm{d}x_n} - \left(\frac{\mathrm{d}\varphi}{\mathrm{d}x_{n+k}}\right)^{-1} \frac{\mathrm{d}x_k}{\mathrm{d}x_n} \frac{\partial G_{n+k}}{\partial \tilde{y}^{(1)}} = 0 \qquad \text{(A1.4)}$$

and

$$\frac{\mathrm{d}H'}{\mathrm{d}\tilde{y}^{(2)}} = \frac{\mathrm{d}y^{(1)}}{\mathrm{d}x_n} - \left(\frac{\mathrm{d}\varphi}{\mathrm{d}x_{n+k}}\right)^{-1} \frac{\mathrm{d}x_k}{\mathrm{d}x_n} \frac{\partial G_{n+k}}{\partial \tilde{y}^{(2)}} = 0 \tag{A1.5}$$

from which Eqn. (6.6″) immediately results. Equations (6.6‴) which lead to Eqn. (6.16‴) are proven in a similar way.

Appendix 2

Derivation of Equations of Section 6.3

PROPOSITION 1. *For operators $\hat{a}$ and $\hat{b}$ obeying Eqn. (6.54) the equality*

$$\hat{a}^n \hat{b} = \hat{b}\hat{a}^n + cn\hat{a}^{n-1} \tag{A2.1}$$

is valid for any natural number n, including 0.

Proof. For $n = 1$ Eqn. (1) coincides with Eqn. (6.54), while its validity for $n = 0$ is evident. Let Eqn. (1) be true for some natural number n. By multiplying it by $\hat{a}$ from the left, we obtain

$$\hat{a}^{n+1}\hat{b} = \hat{a}\hat{b}\hat{a}^n + cn\hat{a}^n, \tag{A2.2}$$

which, in view of Eqn. (6.54), yields

$$\hat{a}^{n+1}\hat{b} = (\hat{b}\hat{a} + c\hat{I})\hat{a}^n + cn\hat{a}^n = \hat{b}\hat{a}^{n+1} + c(n+1)\hat{a}^n, \tag{A2.3}$$

that is, Eqn. (1) for $n + 1$. Thereby the proposition is finally proven.

CONSEQUENCE 1. *If $\hat{a}^{-1}$ exists, then Eqn. (1) is also valid for negative integers n.*

Proof. By multiplying

$$\hat{a}\hat{b} = \hat{b}\hat{a} + c\hat{I} \tag{A2.4}$$

by $\hat{a}^{-1}$ from the left and the right, we obtain

$$\hat{b}\hat{a}^{-1} = \hat{a}^{-1}\hat{b} + c\hat{a}^{-2}, \tag{A2.5}$$

whence

$$\hat{a}^{-1}\hat{b} = \hat{b}\hat{a}^{-1} - c\hat{a}^{-2} \tag{A2.6}$$

which coincides with Eqn. (1) for $n = -1$.

We now note that the above proof of Eqn. (1) for $n > 1$ does not change if we replace the number c in Eqn. (4) by any function of $\hat{a}$, that is, from

$$\hat{a}\hat{b} = \hat{b}\hat{a} + \hat{\varphi}(\hat{a}) \tag{A2.7}$$

follows

$$\hat{a}^n \hat{b} = \hat{b}\hat{a}^n + \hat{\varphi}(\hat{a})n\hat{a}^{n-1} \tag{A2.8}$$

for $n > 1$.

Denoting $\hat{a}^{-1} = \hat{d}$, we rewrite Eqn. (6) as

$$\hat{d}\hat{b} = \hat{b}\hat{d} - c\hat{d}^2. \tag{A2.9}$$

It now follows from Eqns. (8) and (9), for $n > 1$, that

$$\hat{d}^n\hat{b} = \hat{b}\hat{d}^n - cn\hat{d}^{n+1}. \tag{A2.10}$$

Replacing $\hat{d}$ by $\hat{a}^{-1}$, we obtain from this

$$\hat{a}^{-n}\hat{b} = \hat{b}\hat{a}^{-n} + c(-n)\hat{a}^{-(n+1)} \tag{A2.11}$$

which finally proves the consequence.

CONSEQUENCE 2. *For $\hat{a}$ and $\hat{b}$ obeying Eqn. (6.54) the equality*

$$\hat{a}\hat{b}^n = \hat{b}^n\hat{a} + cn\hat{b}^{n-1} \tag{A2.12}$$

is also valid for any natural number n, including 0, and for any negative integer n (if $\hat{b}^{-1}$ exists).

Proof. Equations (6.54′) and (1) result in

$$\hat{b}^n\hat{a} = \hat{a}\hat{b}^n + (-c)n\hat{b}^{n-1}, \tag{A2.13}$$

whence follows Eqn. (12).

Multiplying Eqn. (1) by an arbitrary number α_n and summing over n, we obtain

$$\left(\sum_{n=-k}^{m} \alpha_n\hat{a}^n\right)\hat{b} = \hat{b}\sum_{n=-k}^{m} \alpha_n\hat{a}^n + c\sum_{n=-k}^{m} n\alpha_n\hat{a}^{n-1}, \tag{A2.14}$$

where k and m are arbitrary natural numbers. Denoting

$$\sum_{n=-k}^{m} \alpha_n\hat{a}^n = \hat{P}_{k,m}(\hat{a}), \tag{A2.15}$$

we obtain from Eqn. (14), in view of Eqn. (6.64),

CONSEQUENCE 3.

$$\hat{P}_{k,m}(\hat{a})\hat{b} = \hat{b}\hat{P}_{k,m}(\hat{a}) + c\hat{P}^{(1)}_{k,m}(\hat{a}). \tag{A2.16}$$

Similarly, we get from Eqn. (12)

$$\hat{a}\hat{P}_{q,r}(\hat{b}) = \hat{P}_{q,r}(\hat{b})\hat{a} + c\hat{P}^{(1)}_{q,r}(\hat{b}) \tag{A2.17}$$

where

$$\hat{P}_{q,r}(\hat{b}) = \sum_{n=-q}^{r} \beta_n \hat{b}^n \tag{A2.18}$$

(q and r are arbitrary natural numbers; β_n are arbitrary numbers).

Let $q = 0$ and r be arbitrary large integer in Eqn. (18). If the right-hand side of Eqn. (18) for $r \to \infty$ determines a function $\hat{f}$ given in Eqn. (6.65) and

$$\hat{f}^{(1)}(\hat{b}) = \sum_{n=1}^{\infty} n\beta_n \hat{b}^{n-1}, \tag{A2.19}$$

then Eqn. (19) yields the first of Eqns. (6.66). Similarly, if for $k = 0$ and $m \to \infty$ the right-hand side of Eqn. (15) determines a function $\hat{\varphi}$ given in Eqn. (6.65) and

$$\hat{\varphi}^{(1)}(\hat{a}) = \sum_{n=1}^{\infty} n\alpha_n \hat{a}^{n-1}, \tag{A2.20}$$

then Eqn. (16) yields the second of Eqns. (6.66).

PROPOSITION 2. *For operators $\hat{a}$ and $\hat{b}$ obeying Eqn. (6.54) the equality*

$$\hat{a}^n \hat{P}_{q,r}(\hat{b}) = \sum_{l=0}^{n} c^l \binom{n}{l} \hat{P}_{q,r}^{(l)}(\hat{b})\hat{a}^{n-l} \tag{A2.21}$$

is valid for any natural number n, including 0, where $\hat{P}_{q,r}(\hat{b})$ is determined by Eqn. (18) and $\hat{P}_{q,r}^{(l)}(\hat{b})$ denotes its lth derivative.

Proof. For $n = 0$ Eqn. (21) is obviously true, and for $n = 1$ it coincides with Eqn. (17). Let it be true for some natural number n, then multiplying it by $\hat{a}$ from the left we obtain

$$\hat{a}^{n+1} \hat{P}_{q,r}(\hat{b}) = \sum_{l=0}^{n} c^l \binom{n}{l} \hat{a}\hat{P}_{q,r}^{(l)}(\hat{b})\hat{a}^{n-l}. \tag{A2.22}$$

Since $\hat{P}_{q,r}^{(l)}(\hat{b})$ has the form as given in Eqn. (18), we can write, in view of Eqn. (17),

$$\hat{a}\hat{P}_{q,r}^{(l)}(\hat{b}) = \hat{P}_{q,r}^{(l)}(\hat{b})\hat{a} + c\hat{P}_{q,r}^{(l+1)}(\hat{b}). \tag{A2.23}$$

Introducing this on the right-hand side of Eqn. (22), we find

$$\hat{a}^{n+1} \hat{P}_{q,r}(\hat{b}) = \sum_{l=0}^{n} c^l \binom{n}{l} \hat{P}_{q,r}^{(l)}(\hat{b})\hat{a}^{n+1-l} +$$

$$+ \sum_{k=0} c^{k+1} \binom{n}{k} \hat{P}_{q,r}^{(k+1)}(\hat{b})\hat{a}^{n-k}. \tag{A2.24}$$

Singling out the term with $l = 0$, in the first sum, and that with $k = n$, in the second sum, and replacing $k + 1$ by l, we obtain

$$\hat{a}^{n+1}\hat{P}_{q,r}(\hat{b}) = \hat{P}_{q,r}(\hat{b})\hat{a}^{n+1} + \sum_{l=1}^{n} c^l \left[\binom{n}{l} + \binom{n}{l-1}\right] \hat{P}_{q,r}^{(l)}(\hat{b})\hat{a}^{n+1-l} +$$

$$+ c^{n+1}\hat{P}_{q,r}^{(n+1)}(\hat{b}). \tag{A2.25}$$

Since $\binom{n}{l} + \binom{n}{l-1} = \binom{n+1}{l}$, Eqn. (22) goes over into Eqn. (21) for $n + 1$, which completes the proof.

CONSEQUENCE 1. *If in Eqn. (18) $q = 0$, Eqn. (21) yields*

$$\hat{a}^n \hat{Q}_r(\hat{b}) = \sum_{l=0}^{\min(r,n)} c^l \binom{n}{l} \hat{P}_r^{(l)}(\hat{b})\hat{a}^{n-l}, \tag{A2.26}$$

where

$$\hat{P}_r(\hat{b}) = \sum_{j=0}^{r} \beta_j \hat{b}^j. \tag{A2.27}$$

CONSEQUENCE 2. *Let*

$$\hat{P}_m(\hat{a}) = \sum_{j=0}^{m} \alpha_j \hat{a}^j, \tag{A2.28}$$

where α_j are arbitrary numbers. Then the following equality is valid:

$$\hat{P}_m(\hat{a})\hat{P}_{q,r}(\hat{b}) = \sum_{l=0}^{m} \frac{c^l}{l!} \hat{P}_{q,r}^{(l)}(\hat{b})\hat{P}_m^{(l)}(\hat{a}), \tag{A2.29}$$

where $\hat{P}_m^{(l)}(a)$ is the lth derivative of $\hat{P}_m(\hat{a})$ with respect to $\hat{a}$.

Indeed, multiplying Eqn. (21) by α_n and summing over n, we obtain, in view of Eqn. (28),

$$\hat{P}_m(\hat{a})\hat{P}_{q,r}(\hat{b}) = \sum_{n=0}^{m} \alpha_n \sum_{l=0}^{n} c^l \binom{n}{l} \hat{P}_{q,r}^{(l)}(\hat{b})\hat{a}^{n-l}. \tag{A2.30}$$

The double sum can be rewritten as

$$\sum_{l=0}^{m} \frac{c^l}{l!} \hat{P}_{q,r}^{(l)}(\hat{b}) \sum_{n=l}^{m} \frac{n!}{(n-l)!} \alpha_n \hat{a}^{n-l} = \sum_{l=0}^{m} \frac{c^l}{l!} \hat{P}_{q,r}(\hat{b})\hat{P}_m^{(l)}(\hat{a}) \tag{A2.31}$$

which leads to Eqn. (29).

If $q = 0$ in Eqn. (18), then Eqn. (29) goes over, in view of Eqn. (27), into

$$\hat{P}_m(\hat{a})\hat{P}_r(\hat{b}) = \sum_{l=0}^{\min(r,m)} \frac{c^l}{l!} \hat{P}_r^{(l)}(\hat{b})\hat{P}_m^{(l)}(\hat{a}). \tag{A2.32}$$

CONSEQUENCE 3.

$$\hat{P}_m(\hat{b})\hat{P}_{q,r}(\hat{a}) = \sum_{l=0}^{m} (-1)^l \frac{c^l}{l!} \hat{P}_{q,r}^{(l)}(\hat{a})\hat{P}_m^{(l)}(\hat{b}). \tag{A2.33}$$

This is obtained from Eqn. (29) by permutation of $\hat{a}$ and $\hat{b}$, with due account for Eqn. (6.54′).

Since both canonically conjugate operators $\hat{a}$ and $\hat{b}$ are operators of differentiation in the corresponding representations [cf. the discussion following Eqn. (6.66)], Eqns. (6.66) may be extended to any functions $\hat{f}(\hat{b})$ and $\hat{\varphi}(a)$ defined in any way, provided they have the first derivatives $\hat{f}^{(1)}(\hat{b})$ and $\hat{\varphi}^{(1)}(\hat{a})$, respectively. If these functions have all the derivatives up to and including the kth, then Eqns. (6.66) yield, respectively

$$\hat{a}\hat{f}^{(l)}(\hat{b}) = \hat{f}^{(l)}(\hat{b})\hat{a} + c\hat{f}^{(l+1)}(\hat{b}), \tag{A2.34}$$

$$\hat{\varphi}^{(l)}(\hat{a})\hat{b} = \hat{b}\hat{\varphi}^{(l)}(\hat{a}) + c\hat{\varphi}^{(l+1)}(\hat{a}) \tag{A2.35}$$

($l = 0, 1, \ldots, k-1$). We now can easily prove the following proposition generalizing Proposition 2:

PROPOSITION 3. *For operators $\hat{a}$ and $\hat{b}$ obeying Eqn. (6.54) the equality*

$$\hat{a}^n\hat{f}(\hat{b}) = \sum_{l=0}^{n} c^l \binom{n}{l} \hat{f}^{(l)}(\hat{b})\hat{a}^{n-l} \tag{A2.36}$$

is valid for any natural n, including 0, where $\hat{f}(\hat{b})$ is an arbitrary function having all the necessary derivatives with respect to $\hat{b}$.

The proof of this proposition is similar to that of Proposition 2. It is necessary only to use Eqn. (34) instead of Eqn. (23).

CONSEQUENCE 1. *Let $\hat{P}_m(\hat{a})$ be defined by Eqn. (28), and let $\hat{f}(\hat{b})$ have all derivatives with respect to $\hat{b}$, including the nth. Then the equality*

$$\hat{P}_m(\hat{a})\hat{f}(\hat{b}) = \sum_{l=0}^{m} \frac{c^l}{l!} \hat{f}^{(l)}(\hat{b})\hat{P}_m^{(l)}(\hat{a}) \tag{A2.37}$$

is valid.

The proof is the same as for Consequence 2 of Proposition 2.

CONSEQUENCE 2. *Let the function $\hat{\varphi}(\hat{a})$ be determined by the series as given in Eqn. (6.65)*

$$\hat{\varphi}(\hat{a}) = \sum_{n=0}^{\infty} \alpha_n \hat{a}^n$$

and let

$$\hat{\varphi}^{(l)}(\hat{a}) = \sum_{n=l}^{\infty} \alpha_n \frac{n!}{(n-l)!} \hat{a}^{n-l}. \tag{A2.38}$$

Let, further, $\hat{f}(\hat{b})$ be infinitely differentiable with respect to $\hat{b}$. Then

$$\hat{\varphi}(\hat{a})\hat{f}(\hat{b}) = \sum_{l=0}^{\infty} \frac{c^l}{l!} \hat{f}^{(l)}(\hat{b})\hat{\varphi}^{(l)}(\hat{a}) \tag{A2.39}$$

provided the series on the right-hand side of Eqn. (39) is convergent.

The correctness of Eqn. (39) follows from that of Eqn. (37) when $m \to \infty$. A similar equality

$$\hat{\varphi}(\hat{b})\hat{f}(\hat{a}) = \sum_{l=0}^{\infty} (-1)^l \frac{c^l}{l!} \hat{f}^{(l)}(\hat{a})\hat{\varphi}^{(l)}(\hat{b}) \tag{A2.40}$$

is obtained from Eqn. (39) by permutation of $\hat{a}$ and $\hat{b}$, in view of Eqn. (6.54′).

When deriving Eqns. (39) and (40) the functions $\hat{\varphi}$ and $\hat{f}$ were supposed to obey different requirements: $\hat{f}$ was to be infinitely differentiable, whereas $\hat{\varphi}$ had to satisfy Eqns. (6.65) and (38). This was caused by the fact that the deduction proceeded from the first of Eqns. (6.66). If we proceed from the second of Eqns. (6.66), we would come again to Eqns. (40) and (39), but the requirements to the functions $\hat{\varphi}$ and $\hat{f}$ would be just the opposite. This shows that Eqns. (39) and (40) should be valid for any functions $\hat{\varphi}$ and $\hat{f}$, which are anyhow defined and infinitely differentiable with respect to the corresponding operators, provided the series on the right-hand sides of these equations are convergent.

Equations (6.67) and (6.67′) obviously follow from Eqns. (39) and (40), respectively.

We now turn to the consideration of commutators of the form $[\hat{\varphi}(\hat{a}, \hat{b})\hat{f}(\hat{a}, \hat{b})]$. Suppose that $\hat{\varphi}$ is representable in the form

$$\hat{\varphi}(\hat{a}, \hat{b}) = \sum_i \hat{X}_i(\hat{a})\hat{Y}_i(\hat{b}). \tag{A2.41}$$

Then

$$[\hat{\varphi}\hat{f}] = \sum_i \,[\hat{X}_i(\hat{a})\hat{Y}_i(\hat{b}), \hat{f}(\hat{a},\hat{b})]\,. \tag{A2.42}$$

In view of

$$[\hat{a}\hat{b},\hat{c}] = \hat{a}[\hat{b}\hat{c}] + [\hat{a}\hat{c}]\hat{b}, \tag{A2.43}$$

we obtain from Eqn. (42)

$$[\hat{\varphi}\hat{f}] = \sum_i \,\{-\hat{X}_i(\hat{a})[\hat{f}\hat{Y}_i(\hat{b})] + [\hat{X}_i(\hat{a})\hat{f}]\,\hat{Y}_i(\hat{b})\}. \tag{A2.44}$$

Supposing that $\partial^l\hat{f}/\partial\hat{a}^l$ and $\partial^l\hat{f}/\partial\hat{b}^l$ are determined, we apply Eqn. (6.67) to $[\hat{f}\hat{Y}_i]$ and $[\hat{X}_i\hat{f}]$ and find

$$[\hat{\varphi}\hat{f}] = \sum_i \left\{ -\hat{X}_i(\hat{a}) \sum_{l=1} \frac{c^l}{l!}\,\hat{Y}_i^{(l)}(\hat{b})\,\frac{\partial^l\hat{f}}{\partial\hat{a}^l} + \right.$$
$$\left. + \sum_{l=1} \frac{c^l}{l!}\,\frac{\partial^l\hat{f}}{\partial\hat{b}^l}\,\hat{X}_i^{(l)}(\hat{a})\,\hat{Y}_i(\hat{b}) \right\}. \tag{A2.45}$$

By changing the sequence of summation we obtain

$$[\hat{\varphi}\hat{f}] = \sum_{l=1} \frac{c^l}{l!} \sum_i \left\{ \frac{\partial^l\hat{f}}{\partial\hat{b}^l}\,\hat{X}_i^{(l)}(\hat{a})\,\hat{Y}_i(\hat{b}) - \hat{X}_i(\hat{a})\,\hat{Y}_i^{(l)}(\hat{b})\,\frac{\partial^l\hat{f}}{\partial\hat{a}^l} \right\}. \tag{A2.46}$$

But

$$\sum_i \hat{X}_i^{(l)}(\hat{a})\,\hat{Y}_i(\hat{b}) = \frac{\partial^l\hat{\varphi}}{\partial\hat{a}^l}, \tag{A2.47}$$

$$\sum_i \hat{X}_i(\hat{a})\,\hat{Y}_i^{(l)}(\hat{b}) = \frac{\partial^l\hat{\varphi}}{\partial\hat{b}^l}. \tag{A2.48}$$

Hence, we have thereby obtained Eqn. (6.68).

We can, however also represent $\hat{\varphi}$ by means of Eqn. (39) in the form

$$\hat{\varphi}(\hat{a},\hat{b}) = \sum_i \hat{U}_i(\hat{b})\,\hat{V}_i(\hat{a}). \tag{A2.49}$$

Then Eqn. (42) goes over into

$$[\hat{\varphi}\hat{f}] = \sum_i \,[\hat{U}_i(\hat{b})\hat{V}_i(\hat{a}), \hat{f}(\hat{a},\hat{b})] \tag{A2.50}$$

which yields, in view of Eqn. (43),

$$[\hat{\varphi}\hat{f}] = \sum_i \{\hat{U}_i(\hat{b})[\hat{V}_i(\hat{a})\hat{f}] - [\hat{f}\hat{U}_i(\hat{b})]\hat{V}_i(\hat{a})\}. \tag{A2.51}$$

Applying Eqn. (6.67′) to $[\hat{V}_i\hat{f}]$ and $[\hat{f}\hat{U}_i]$, we obtain, after changing the sequence of summation, and by making use of equations similar to Eqns. (47) and (48), the sought after Eqn. (6.68′).

Equations (6.68) and (6.68′) have been obtained on the supposition that $\hat{\varphi}$ is representable in the form as given in Eqns. (41) or (49), respectively, whereas the function $\hat{f}$ has not been supposed to obey such a restriction. However, the functions $\hat{\varphi}$ and $\hat{f}$ are, as a matter of fact, equivalent in this sense. Therefore, Eqns. (6.68) and (6.68′) should be valid for any functions $\hat{\varphi}$ and $\hat{f}$, provided that the right-hand sides of these equations exist.

Equations (6.70) and (6.70′) are the pertinent approximations to the equations

$$[\hat{\varphi}\hat{f}] = \sum_{l=1} \frac{c^l}{l!}\left(\frac{\partial^l\hat{f}}{\partial\hat{b}^l}\,\frac{\partial^l\hat{\varphi}}{\partial\hat{a}^l} - \frac{\partial^l\hat{\varphi}}{\partial\hat{b}^l}\,\frac{\partial^l\hat{f}}{\partial\hat{a}^l}\right) \tag{A2.52}$$

and

$$[\hat{\varphi}\hat{f}] = \sum_{l=1} (-1)^{l+1}\frac{c^l}{l!}\left(\frac{\partial^l\hat{\varphi}}{\partial\hat{a}^l}\,\frac{\partial^l\hat{f}}{\partial\hat{b}^l} - \frac{\partial^l\hat{f}}{\partial\hat{a}^l}\,\frac{\partial^l\hat{\varphi}}{\partial\hat{b}^l}\right) \tag{A2.53}$$

respectively, which are valid for the functions $\hat{\varphi}$ and $\hat{f}$ of n pairs of canonically conjugate operators $\hat{a}_j$ and $\hat{b}_j$ satisfying Eqns. (6.69). Here

$$\frac{1}{l!}\,\frac{\partial^l\hat{f}}{\partial\hat{b}^l}\,\frac{\partial^l\hat{\varphi}}{\partial\hat{a}^l} = \sum_{l_1,\ldots,l_n}\frac{1}{l_1!\ldots l_n!}\,\frac{\partial^l\hat{f}}{\partial\hat{b}_1^{l_1}\ldots\partial\hat{b}_n^{l_n}}\,\frac{\partial^l\hat{\varphi}}{\partial\hat{a}_1^{l_1}\ldots\partial\hat{a}_n^{l_n}}$$

$$\left(\sum_{j=1}^{n} l_j = l\right), \text{ etc.} \tag{A2.54}$$

Equation (52) is derived as follows. Suppose that Eqn. (52) is valid for some n, that is, for functions of the type $\hat{\varphi} = \hat{\varphi}(\hat{a}_1, \ldots, \hat{a}_n, \hat{b}_1, \ldots, \hat{b}_n)$ and $\hat{f} = \hat{f}(\hat{a}_1, \ldots, \hat{a}_n, \hat{b}_1, \ldots, \hat{b}_n)$. Then it is also valid for $n + 1$, that is, for functions of the type $\hat{\Phi} = \hat{\Phi}(\hat{a}_1, \ldots, \hat{a}_n, \hat{a}_{n+1}, \hat{b}_1, \ldots, \hat{b}_n, \hat{b}_{n+1})$ and $\hat{F} = \hat{F}(\hat{a}_1, \ldots, \hat{a}_n, \hat{a}_{n+1}, \hat{b}_1, \ldots, \hat{b}_n, \hat{b}_{n+1})$.

Indeed, let $\hat{\Phi}$ be of the form

$$\hat{\Phi} = \hat{X}(\hat{a}_{n+1})\hat{Y}(\hat{b}_{n+1})\hat{\varphi}(\hat{a}_1, \ldots, \hat{a}_n, \hat{b}_1, \ldots, \hat{b}_n) \tag{A2.55}$$

Then, in view of Eqns. (43) and (6.69),

$$[\hat{\Phi}\hat{F}] = [\hat{X}\hat{F}]\,\hat{Y}\hat{\varphi} + \hat{X}[\hat{\varphi}\hat{F}]\,\hat{Y} - \hat{X}\hat{\varphi}[\hat{F}\hat{Y}]. \tag{A2.56}$$

We now have, in view of Eqn. (6.67),

$$[\hat{X}\hat{F}]\,\hat{Y}\hat{\varphi} = \sum_{l=1} \frac{c^l}{l!} \frac{\partial^l \hat{F}}{\partial \hat{b}_{n+1}^l} \frac{\partial^l \hat{\Phi}}{\partial \hat{a}_{n+1}^l}, \tag{A2.57}$$

$$\hat{X}\hat{\varphi}[\hat{F}\hat{Y}] = \sum_{l=1} \frac{c^l}{l!} \frac{\partial^l \hat{\Phi}}{\partial \hat{b}^l} \frac{\partial^l \hat{F}}{\partial \hat{a}_{n+1}^l}, \tag{A2.58}$$

while Eqns. (52) and (39) yield

$$\hat{X}[\hat{\varphi}\hat{F}]\,\hat{Y} = \sum_{l=1} \frac{c^l}{l!} \sum_{k=0} \frac{c^k}{k!} \left(\frac{\partial^{l+k} \hat{F}}{\partial \hat{b}^l \, \partial \hat{b}_{n+1}^k} \frac{\partial^{l+k} \hat{\Phi}}{\partial \hat{a}^l \, \partial \hat{a}_{n+1}^k} - \frac{\partial^{l+k} \hat{\Phi}}{\partial \hat{b}^l \, \partial \hat{b}_{n+1}^k} \frac{\partial^{l+k} \hat{F}}{\partial \hat{a}^l \, \partial \hat{a}_{n+1}^k} \right). \tag{A2.59}$$

Hence, in view of Eqn. (54),

$$[\hat{\Phi}\hat{F}] = \sum_{l=1} \frac{c^l}{l!} \left(\frac{\partial^l \hat{F}}{\partial \hat{b}^l} \frac{\partial^l \hat{\Phi}}{\partial \hat{a}^l} - \frac{\partial^l \hat{\Phi}}{\partial \hat{b}^l} \frac{\partial^l \hat{F}}{\partial \hat{a}^l} \right) \tag{A2.60}$$

which is Eqn. (52) for $n + 1$. Since Eqn. (52) coincides with Eqn. (6.68) for $n = 1$, the above result demonstrates that it holds for any $n \geqslant 1$.

Due to the linearity of any derivatives, Eqn. (52) also holds when functions $\hat{\Phi}$ are of the type

$$\hat{\Phi} = \sum_j \hat{X}_j(\hat{a}_{n+1}) Y_j(\hat{b}_{n+1}) \varphi_j(\hat{a}_1, \ldots, \hat{a}_n, \hat{b}_1, \ldots, \hat{b}_n). \tag{A2.61}$$

Finally, because of the equivalence of the function $\hat{\Phi}$ and $\hat{F}$ in this derivation, Eqn. (52) is valid for any functions Φ and $\hat{F}$, provided its right-hand side exists.

Equation (53) is obtained in a similar way.

Appendix 3

A General Rotation-Vibration Hamiltonian

As has been indicated in Section 10.2.3, in order to form a Hamiltonian for a mechanical system consisting of n particles in coordinates other than Cartesian coordinates, it is sufficient to express its kinetic part as given in Eqn. (10.22) in terms of the new coordinates. For a molecule, the feasibility of this task is stipulated by the fact that any molecule has an equilibrium configuration (in the Born-Oppenheimer approximation which we strictly confine ourselves to; cf. also Section 11.6). This allows us to choose the new coordinates as follows: $X^0 = R_x^0$, $Y^0 = R_y^0$, and $Z^0 = R_z^0$ are the Cartesian coordinates of the center of mass of the molecule in the initial rest coordinate system; $\varphi = \Omega_1$, $\theta = \Omega_2$, and $\chi = \Omega_3$ are Euler's angles determining the spatial orientation of the molecule equilibrium configuration (more precisely, of the undistorted molecule) and, hence, the spatial orientation of the rotating coordinate system rigidly attached to it; q_j ($j = 1, 2, \ldots, N$) are some independent internal coordinates describing a deformation of the molecule relative to its equilibrium configuration, and $N = 3n - 6$ is the number of these coordinates.

In case of linear molecules $N = 3n - 5$, and the spatial orientation of the molecule is determined by the two angles φ and θ only, the angle χ disappearing. Only the number of independent internal coordinates is essential for this stage of the deduction, since we must get a reversible transformation from the initial coordinates to the final ones. The particular choice of the internal coordinates is of no significance here, and they can be meant, for instance, as changes in some interatomic distances, whereas the real calculation of vibrations can be performed in other appropriate coordinates.

In the new coordinates chosen

$$p^{(i)} = -i\hbar\nabla_i = -i\hbar \frac{\partial}{\partial r^{(i)}} \tag{A3.1}$$

yields

$$\nabla_i = \sum_\alpha \nabla_i R_\alpha^0 \frac{\partial}{\partial R_\alpha^0} + \sum_j \nabla_i \Omega_j \frac{\partial}{\partial \Omega_j} + \sum_j \nabla_i q_j \frac{\partial}{\partial q_j}. \tag{A3.2}$$

By introducing

$$-i\hbar\frac{\partial}{\partial R^0_\alpha}=\hat{P}^0_\alpha;\qquad -i\hbar\frac{\partial}{\partial \Omega_j}=\hat{\Pi}_j;\qquad -i\hbar\frac{\partial}{\partial q_j}=\hat{P}_j,\tag{A3.3}$$

we get from Eqns. (1) and (2)

$$\hat{p}^{(i)}=\sum_\alpha \nabla_i R^0_\alpha \hat{P}^0_\alpha+\sum_j \nabla_i\Omega_j\hat{\Pi}_j+\sum_j \nabla_i q_j\hat{P}_j.\tag{A3.4}$$

It only remains now to find all the $\nabla_i R^0_\alpha$, $\nabla_i\Omega_j$, and $\nabla_i q_j$, and to substitute them in Eqn. (4), and then to replace the $\hat{p}^{(i)}$ in equation

$$\hat{K}=2^{-1}\sum_{i=1}^{n}\epsilon_i(\hat{p}^{(i)},\hat{p}^{(i)})\tag{A3.5}$$

equivalent to Eqn. (10.22), where $\epsilon_i=m_i^{-1}$, by making use of the results obtained. It is very simple to calculate $\nabla_i R^0_\alpha$. Indeed, by definition

$$\mathbf{R}^0=M^{-1}\sum_i m_i\mathbf{r}^{(i)},\tag{A3.6}$$

where $\mathbf{R}^0$ is the radius vector of the center of mass of the molecule and $M=\Sigma_i\, m_i$ is the total mass of it. Therefore,

$$\nabla_i\tilde{R}^0=M^{-1}E_3 m_i,\tag{A3.7}$$

where E_3 is a 3 by 3 unit matrix.

But it is not so simple to find $\nabla_i\Omega_j$ ($j=1,2,3$) and $\nabla_i q_j$ ($j=1,2,\ldots,N$), because it is impossible to represent φ, θ, χ, and all the q_j as explicit functions of the x_i, y_i, and z_i ($i=1,2,\ldots,n$). To find these quantities, we calculate, first, the $3n$ by $3n$ matrix

$$D=\overset{i}{\downarrow}\left\|\frac{\partial r^{(i)}}{\partial R^0}\;\vdots\;\frac{\partial r^{(i)}}{\partial\Omega}\;\vdots\;\frac{\partial r^{(i)}}{\partial q}\right\|,\tag{A3.8}$$

where

$$\frac{\partial r^{(i)}}{\partial R^0}=\overset{\alpha}{\downarrow}\left\|\frac{\partial\alpha_i}{\partial R^0_\beta}\right\|_{\beta\rightarrow};\qquad \frac{\partial r^{(i)}}{\partial\Omega}=\overset{\alpha}{\downarrow}\left\|\frac{\partial\alpha_i}{\partial\Omega_j}\right\|_{j\rightarrow};\qquad \frac{\partial r^{(i)}}{\partial q}=\overset{\alpha}{\downarrow}\left\|\frac{\partial\alpha_i}{\partial q_j}\right\|_{j\rightarrow}\tag{A3.9}$$

($\alpha,\beta=x,y,z$). Then we construct the inverse matrix

$$D^{-1}=\left\|\begin{array}{c}\tilde{\nabla}R^0\\ \hline \tilde{\nabla}\Omega\\ \hline \tilde{\nabla}q\end{array}\right\|=\left\|\begin{array}{c|c|c|c|c}\tilde{\nabla}_1R^0&\cdots&\tilde{\nabla}_iR^0&\cdots&\tilde{\nabla}_nR^0\\ \hline \tilde{\nabla}_1\Omega&\cdots&\tilde{\nabla}_i\Omega&\cdots&\tilde{\nabla}_n\Omega\\ \hline \tilde{\nabla}_1q&\cdots&\tilde{\nabla}_iq&\cdots&\tilde{\nabla}_nq\end{array}\right\|\tag{A3.10}$$

from which we can obviously get all the quantities we need, since

$$\tilde{\nabla}_i R^0 = \overset{\alpha}{\downarrow}\left\| \frac{\partial R^0_\alpha}{\partial \beta_i} \right\|_{\beta\rightarrow}; \qquad \tilde{\nabla}_i \Omega = \overset{j}{\downarrow}\left\| \frac{\partial \Omega_j}{\partial \alpha_i} \right\|_{\alpha\rightarrow}; \qquad \tilde{\nabla}_i q = \overset{j}{\downarrow}\left\| \frac{\partial q_j}{\partial \alpha_i} \right\|_{\alpha\rightarrow}. \tag{A3.11}$$

In view of Eqns. (10.42), the radius vectors $\mathbf{r}^{(i)}$ and $\mathbf{r}^{(i)\prime}$ of ith atom in the rest and rotating coordinate systems, respectively, are related by

$$\mathbf{r}^{(i)} = \mathbf{R}^0 + \tilde{A}_0^{(\Omega)}\mathbf{r}^{(i)\prime}. \tag{A3.12}$$

The $3n$ Cartesian components $\delta\alpha_i$ ($i = 1, \ldots, n; \alpha = x, y, z$), in the rotating coordinate system, of the atomic displacement vectors

$$\delta\mathbf{r}^{(i)} = \mathbf{r}^{(i)\prime} - \mathbf{R}^{(i)} \tag{A3.13}$$

are generally nonlinear functions of any internal coordinates q_j [$\mathbf{R}^{(i)}$ is the equilibrium radius vector of ith atom in the rotating coordinate system]. In virtue of Eqn. (6), the equation

$$\sum_i m_i \mathbf{r}^{(i)\prime} = 0 \tag{A3.14}$$

is valid for any arbitrary $\delta\mathbf{r}^{(i)}$. At the same time, the well-known equation

$$\sum_i m_i (R_\alpha^{(i)}\,\delta\beta_i - R_\beta^{(i)}\,\delta\alpha_i) = 0 \qquad (\alpha \neq \beta; \alpha, \beta = x, y, z) \tag{A3.15}$$

is valid only for small enough $\delta\mathbf{r}^{(i)}$, as we shall see later on.

Equation (12) immediately yields

$$\partial r^{(i)}/\partial R^0 = E_3. \tag{A3.16}$$

From Eqn. (12) we also obtain, in view of Eqn. (10.48),

$$\frac{\partial r^{(i)}}{\partial \Omega} = \tilde{A}_0^{(\Omega)} A_i^{(r)\prime} C^{(\Omega)}. \tag{A3.17}$$

For $\partial r^{(i)}/\partial q$ we obtain from Eqn. (12)

$$\frac{\partial r^{(i)}}{\partial q} = \tilde{A}_0^{(\Omega)} \left\| \frac{\partial r^{(i)\prime}}{\partial q_1} \,\vdots \cdots \vdots\, \frac{\partial r^{(i)\prime}}{\partial q_j} \,\vdots \cdots \vdots\, \frac{\partial r^{(i)\prime}}{\partial q_N} \right\|. \tag{A3.18}$$

Introducing

$$\partial r^{(i)\prime}/\partial q_j = A_{ij}^{(q)} \tag{A3.19}$$

where $A_{ij}^{(q)}$ is a one-by-three column matrix whose elements generally depend on the internal coordinates, we obtain from Eqn. (18)

$$\partial r^{(i)}/\partial q = \tilde{A}_0^{(\Omega)} \left\| A_{i1}^{(q)} \,\vdots\, A_{i2}^{(q)} \,\vdots\, \ldots \,\vdots\, A_{ij}^{(q)} \,\vdots\, \ldots \,\vdots\, A_{iN}^{(q)} \right\| = \tilde{A}_0^{(\Omega)} A_i^{(q)}. \tag{A3.20}$$

We can now rewrite matrix D as given in Eqn. (8) in the form

$$D = \left\| E_3' \,\vdots\, \tilde{A}^{(\Omega)} A^{(r)'} C^{(\Omega)} \,\vdots\, \tilde{A}^{(\Omega)} A^{(q)} \right\|. \tag{A3.21}$$

Here E_3' is the 3 by $3n$ matrix of the form

$$E_3' = (E_3, \ldots, E_3, \ldots, E_3), \tag{A3.22}$$

$\tilde{A}^{(\Omega)}$ is the $3n$ by $3n$ diagonal block matrix of the form

$$\tilde{A}^{(\Omega)} = \left\| \tilde{A}_0^{(\Omega)} \delta_{ij} \right\| \quad (i, j = 2, 3, \ldots, n), \tag{A3.23}$$

$A^{(r)'}$ is the 3 [or 2, in case of linear molecules (LM-case)] by $3n$ matrix of the form

$$A^{(r)'} = \overset{i}{\downarrow} \left\| A_i^{(r)'} \right\| = (A_1^{(r)'}, \ldots, A_i^{(r)'}, \ldots, A_n^{(r)'}), \tag{A3.24}$$

and $A^{(q)}$ is the N by $3n$ matrix of the form

$$A^{(q)} = \overset{i}{\downarrow} \left\| A_i^{(q)} \right\| = (A_1^{(q)}, \ldots, A_i^{(q)}, \ldots, A_n^{(q)}). \tag{A3.25}$$

The matrix D^{-1} should obviously satisfy the relation $D^{-1}D = E_{3n}$ which yields, in view of Eqn. (10),

$$\tilde{\nabla} R^0 E_3' = E_3, \tag{A3.26a}$$

$$\tilde{\nabla} \Omega \tilde{A}^{(\Omega)} A^{(r)'} C^{(\Omega)} = E_3 \quad (\text{or } E_2 \text{ in LM-case}), \tag{A3.26b}$$

$$\tilde{\nabla} q \tilde{A}^{(\Omega)} A^{(q)} = E_N, \tag{A3.26c}$$

$$\tilde{\nabla} R^0 \tilde{A}^{(\Omega)} A^{(r)'} C^{(\Omega)} = 0, \tag{A3.26d}$$

$$\tilde{\nabla} R^0 \tilde{A}^{(\Omega)} A^{(q)} = 0, \tag{A3.26e}$$

$$\tilde{\nabla} \Omega E_3' = 0, \tag{A3.26f}$$

$$\tilde{\nabla} \Omega \tilde{A}^{(\Omega)} A^{(q)} = 0, \tag{A3.26g}$$

$$\tilde{\nabla} q E_3' = 0, \tag{A3.26h}$$

$$\tilde{\nabla} q \tilde{A}^{(\Omega)} A^{(r)'} C^{(\Omega)} = 0. \tag{A3.26i}$$

From Eqn. (7) we immediately obtain

$$\tilde{\nabla} R^0 = M^{-1} \tilde{E}_3' m, \tag{A3.27}$$

where

$$m = \| m_i E_3 \, \delta_{ij} \| \quad (i, j = 1, 2, \ldots, n) \tag{A3.28}$$

is the diagonal $3n$ by $3n$ matrix of atomic masses, and it remains only to find $\tilde{\nabla}\Omega$ and $\tilde{\nabla} q$. In case of linear dependence of atomic displacements on internal coordinates, that is, for diatomic molecules and for small enough displacements of any molecules, this can be done by making use of Eqns. (26), since elements of the matrix $A^{(q)}$ are then constant. But in the general case we have to do as follows.

We calculate, first, the matrix

$$T^{(c)} = \tilde{D} m D. \tag{A3.29}$$

We find, then, the inverse matrix

$$T^{(c)-1} = (\tilde{D} m D)^{-1} \tag{A3.30}$$

and obtain the sought after matrix D^{-1} by the equation

$$D^{-1} = T^{(c)-1} \tilde{D} m. \tag{A3.31}$$

We obtain from Eqn. (29), in view of Eqns. (21) and (14),

$$T^{(c)} = \begin{Vmatrix} ME_3 & 0 & 0 \\ 0 & \tilde{C}^{(\Omega)} I'' C^{(\Omega)} & \tilde{C}^{(\Omega)} \tilde{A}^{(r)\prime} m A^{(q)} \\ 0 & \tilde{A}^{(q)} m A^{(r)\prime} C^{(\Omega)} & T^{(q)} \end{Vmatrix}, \tag{A3.32}$$

where I'' is determined by Eqn. (10.52) and

$$T^{(q)} = \tilde{A}^{(q)} m A^{(q)}. \tag{A3.33}$$

In order to calculate

$$T^{(c)-1} = \begin{Vmatrix} M^{-1} E_3 & 0 & 0 \\ 0 & L^{(r)} & L^{(r,q)} \\ 0 & \tilde{L}^{(r,q)} & L^{(q)} \end{Vmatrix}, \tag{A3.34}$$

we make use of the Frobenius relation (cf., for instance, E. Bodewig, *Matrix Calculus*, New York, Interscience publishers, 1959), which reads: Let a nonsingular square matrix F be of the block form

$$F = \left\| \begin{array}{c|c} A & B \\ \hline C & D \end{array} \right\|. \tag{A3.35}$$

where D is also a nonsingular square matrix. Then

$$F^{-1} = \left\| \begin{array}{c|c} K^{-1} & -K^{-1}BD^{-1} \\ \hline -D^{-1}CK^{-1} & D^{-1} + D^{-1}CK^{-1}BD^{-1} \end{array} \right\|, \tag{A3.36}$$

where

$$K = A - BD^{-1}C. \tag{A3.37}$$

In view of Eqn. (32), Eqn. (34) yields,

$$L^{(r)} = C^{(\Omega)'}\Gamma^{(q)}\widetilde{C}^{(\Omega)'}, \tag{A3.38}$$

$$L^{(r,q)} = -C^{(\Omega)'}\Gamma^{(q)}\widetilde{\kappa}^{(q)}, \tag{A3.39}$$

$$L^{(q)} = G^{(q)} + \kappa^{(q)}\Gamma^{(q)}\widetilde{\kappa}^{(q)}, \tag{A3.40}$$

where

$$\Gamma^{(q)} = (I'' - \widetilde{\kappa}^{(q)}T^{(q)}\kappa^{(q)})^{-1}, \tag{A3.41}$$

$$\kappa^{(q)} = B^{(q)}A^{(r)'}, \tag{A3.42}$$

$$G^{(q)} = T^{(q)-1}, \tag{A3.43}$$

and

$$B^{(q)} = G^{(q)}\widetilde{A}^{(q)}m \tag{A3.44}$$

is the matrix connecting internal coordinates q with the Cartesian coordinates $r' = (r^{(1)'}, \ldots, r^{(i)'}, \ldots, r^{(n)'})$ by the relation

$$B^{(q)} = \partial q/\partial r' \tag{A3.45}$$

inverse to Eqn. (20); $C^{(\Omega)'}$ is as given in Eqn. (10.58).

In view of Eqns. (10), (21), and (31), we now have

$$\widetilde{\nabla}\Omega = C^{(\Omega)'}\Gamma^{(q)}\widetilde{A}^{(r)'}mA^{(\Omega)} - C^{(\Omega)'}\Gamma^{(q)}\widetilde{\kappa}^{(q)}T^{(q)}B^{(q)}A^{(\Omega)}, \tag{A3.46}$$

$$\widetilde{\nabla}q = (G^{(q)} + \kappa^{(q)}\Gamma^{(q)}\widetilde{\kappa}^{(q)})T^{(q)}B^{(q)}A^{(\Omega)} - \kappa^{(q)}\Gamma^{(q)}\widetilde{A}^{(r)'}mA^{(\Omega)}. \tag{A3.47}$$

By introducing Eqns. (27), (46), and (47) into Eqn. (4) rewritten as

$$\hat{p} = \nabla\widetilde{R}^{0}\hat{P}^{0} + \nabla\widetilde{\Omega}\hat{\Pi} + \nabla\widetilde{q}\hat{P}, \tag{A3.48}$$

where

$$\hat{p} = (\hat{p}^{(1)}, \ldots, \hat{p}^{(n)}); \qquad \hat{P}^{0} = (\hat{P}^{0}_{x}, \hat{P}^{0}_{y}, \hat{P}^{0}_{z}),$$

$$\hat{\Pi} = (\hat{\pi}_1, \hat{\pi}_2, \hat{\pi}_3), \quad \text{and} \quad \hat{P} = (\hat{P}_1, \ldots, \hat{P}_N), \tag{A3.49}$$

we obtain, in view of Eqn. (10.94)

$$\hat{p} = M^{-1} mE_3'\hat{P}^0 + \tilde{A}^{(\Omega)}[(mA^{(r)'} - \\ - \tilde{B}^{(q)}T^{(q)}\kappa^{(q)})\Gamma^{(q)}(\hat{\mathcal{M}}' - \hat{\mu}^{(q)}) + \tilde{B}^{(q)}\hat{P}], \tag{A3.50}$$

where

$$\hat{\mu}^{(q)} = \tilde{\kappa}^{(q)}\hat{P} \tag{A3.51}$$

is the operator for the so-called 'vibrational angular momentum'.

When calculating the right-hand side of Eqn. (5) by making use of Eqn. (4), we will generally get terms both quadratic and linear in $\hat{P}^0$, $\hat{\Pi}$, and $\hat{P}$. It is easy to see that no terms linear in any $\hat{P}_\alpha^0$, $\hat{\Pi}_j$, or $\hat{P}_j$ can be due to the term containing $\hat{P}^0$ in Eqn. (48), because all the terms in $\hat{p}$ are independent of R^0. On the other hand, the only nonvanishing quadratic term originating from $M^{-1} mE_3'\hat{P}^0$ is

$$\hat{K}^0 = 2^{-1}M^{-2}\tilde{\hat{P}}^0\tilde{E}_3' m\epsilon mE_3'\hat{P}^0 = 2^{-1}M^{-1}(\hat{P}_x^{02} + \hat{P}_y^{02} + \hat{P}_z^{02}), \tag{A3.52}$$

which is of no interest, since it refers to translations [other quadratic terms containing $\hat{P}^0$ vanish due to Eqns. (26f) and (26h)]. Thus we should deal only with

$$\hat{K}' = \hat{K} - \hat{K}^0 = 2^{-1}\sum_{i=1}^{n} \epsilon_i(\hat{p}^{(i)'}, \hat{p}^{(i)'}) \tag{A3.53}$$

where, in view of Eqns. (4) and (3),

$$\hat{p}^{(i)'} = -i\hbar\left(\sum_{j=1}^{3} \nabla_i\Omega_j \frac{\partial}{\partial\Omega_j} + \sum_{j=1}^{N} \nabla_i q_j \frac{\partial}{\partial q_j}\right) \tag{A3.54}$$

which can also be represented as

$$\hat{p}' = \tilde{A}^{(\Omega)}[(mA^{(r)'} - \tilde{B}^{(q)}T^{(q)}\kappa^{(q)})\Gamma^{(q)}(\hat{\mathcal{M}}' - \hat{\mu}^{(q)}) + \tilde{B}^{(q)}\hat{P}]. \tag{A3.55}$$

When calculating the right-hand side of Eqn. (53) we get terms both quadratic and linear in $\hat{\mathcal{M}} - \hat{\mu}^{(q)}$ and $\hat{P}$. The former ones are obtained from $\tilde{\hat{p}}'\epsilon\hat{p}'$, yielding, in view of Eqns. (55), (26i), (10.52), (44), (45), and (20),

$$2^{-1}\tilde{\hat{p}}'\epsilon\hat{p}' = 2^{-1}\sum_{\alpha,\beta}\gamma_{\alpha\beta}^{(q)}(\hat{\mathcal{M}}_\alpha' - \hat{\mu}_\alpha^{(q)})(\hat{\mathcal{M}}_\beta' - \hat{\mu}_\beta^{(q)}) + 2^{-1}\sum_{j,l} g_{jl}^{(q)}\hat{P}_j\hat{P}_l \tag{A3.56}$$

where $\gamma_{\alpha\beta}^{(q)}$ and $g_{jl}^{(q)}$ are the elements of the matrices $\Gamma^{(q)}$ and $G^{(q)}$, respectively.

To get the terms of $\hat{K}'$ linear in $\hat{\mathcal{M}}' - \hat{\mu}^{(q)}$ and $\hat{P}$ it is convenient to keep the form of $\hat{p}^{(i)'}$ in Eqn. (53) at right from comma as represented by Eqn. (55), namely,

$$\hat{p}^{(i)'} = \tilde{A}_0^{(\Omega)}[(m_iA_i^{(r)'} - \tilde{B}_i^{(q)}T^{(q)}\kappa^{(q)})\Gamma^{(q)}(\hat{\mathcal{M}}' - \hat{\mu}^{(q)}) + \tilde{B}_i^{(q)}\hat{P}] \tag{A3.57}$$

($B_i^{(q)}$ is the submatrix of $B^{(q)}$ corresponding to ith atom), and at left from comma to use $\hat{p}^{(i)\prime}$ as given in Eqn. (54). It is obvious that the linear terms are of the form

$$\text{Linear Terms} = 2^{-1} i\hbar \left[\sum_j s_j^{(q)} \hat{P}_j + \sum_\alpha s_\alpha^{(q)} (\hat{\mathcal{M}}'_\alpha - \hat{\mu}_\alpha^{(q)}) \right], \tag{A3.58}$$

but we shall not calculate the s_j and s_α here for the general case, since for every particular molecule there can be certain simplifications.

Thus, the final form of a general rotation-vibration Hamiltonian is, in view of Eqn. (10.19) and of the fact that U may depend on internal coordinates only,

$$\hat{H}' = 2^{-1} \sum_{\alpha,\beta} \gamma_{\alpha\beta}^{(q)} (\hat{\mathcal{M}}_\alpha - \hat{\mu}_\alpha^{(q)}) (\hat{\mathcal{M}}_\beta - \hat{\mu}_\beta^{(q)}) + 2^{-1} \sum_{j,l} g_{jl}^{(q)} \hat{P}_j \hat{P}_l + \\ + 2^{-1} i\hbar \left[\sum_j s_j^{(q)} \hat{P}_j + \sum_\alpha s_\alpha^{(q)} (\hat{\mathcal{M}}'_\alpha - \hat{\mu}^{(q)}) \right] + U(q)\hat{I}. \tag{A3.59}$$

The terms linear in $\hat{P}_j$ can be reduced, if needed, by a pertinent similarity transformation which cancels them in case of small atomic displacements.

It is easy to see that Eqns. (26) are satisfied identically by $\tilde{\nabla} R^0$, $\tilde{\nabla}\Omega$, and $\tilde{\nabla} q$ as given in Eqns. (27), (46), and (47), respectively. In particular, Eqn. (26g) yields

$$\tilde{A}^{(r)\prime} m A^{(q)} - \tilde{\kappa}^{(q)} T^{(q)} = 0, \tag{A3.60}$$

which, in view of Eqns. (42) and (44), is indeed an identity satisfied for any internal coordinates and any atomic displacements.

For infinitesimal atomic displacements, identity (60), in view of Eqns. (10.49), (13), and (18), yields approximately Eqn. (15).

Thus, Eqn. (15) is actually also an identity, which is indeed valid merely for small enough atomic displacements, and even then only approximately. Since in the deduction of the conventional rotation-vibration Hamiltonian the Eckart conditions, which follow from Eqn. (15), are necessarily used, that Hamiltonian can be in principle valid solely for small enough atomic displacements. It is necessary to stress this fact because wrong claims that the conventional deduction of the rotation-vibration Hamiltonian leads in principle to one valid for any atomic displacements can be found in literature (cf., for instance: E. B. Wilson, Jr., J. C. Decius, P. C. Cross, *Molecular Vibrations*, McGraw-Hill, 1955; J. K. G. Watson, *Mol. Phys.* **15**, 479, 1968).

Answers to Problems

Chapter 1

1.1. The property t_0, which determines that the people involved belong to the family A under consideration, has one value: f_0 = 'a member of the family A'. The property t_1 has two values: $f_1^{(a)}$ = 'male' and $f_1^{(b)}$ = 'female'. The combined property t_{12} has 8 values: $f_{12}^{(aa)}$ = (male, child), $f_{12}^{(ab)}$ = (male, adolescent), $f_{12}^{(ac)}$ = (male, adult), $f_{12}^{(ad)}$ = (male, mature), $f_{12}^{(ba)}$ = (female, child), $f_{12}^{(bb)}$ = (female, adolescent), $f_{12}^{(bc)}$ = (female, adult), $f_{12}^{(bd)}$ = (female, mature). Thus family A is divided into the following classes.

With respect to t_1: $A(f_1^a)$ = {Bob, Stan, John}, $A(f_1^b)$ = {Ann, Lola, Mary}.

With respect to t_2: $A(f_2^a)$ = {Ann}, $A(f_2^b)$ = {Bob}, $A(f_2^c)$ = {Lola, Stan}, $A(f_2^d)$ = {Mary, John}.

With respect to t_{12}: $A(f_{12}^{ab})$ = {Bob}, $A(f_{12}^{ac})$ = {Stan}, $A(f_{12}^{ad})$ = {John}, $A(f_{12}^{ba})$ = {Ann}, $A(f_{12}^{bc})$ = {Lola}, $A(f_{12}^{bd})$ = {Mary}. The classes $A(f_{12}^{aa})$ and $A(f_{12}^{bb})$ are empty.

1.2. The ages of the members of family A will become: John – 64, Mary – 59, Stan – 38, Lola – 34, Bob – 17, and Ann – 13. The classes with respect to t_1 will not change, whereas with respect to t_2 and t_{12} they will, namely: $A(f_2^b)$ = {Ann, Bob}, $A(f_2^c)$ = {Lola}, $A(f_2^d)$ = {Stan, Mary, John}; $A(f_{12}^{ab})$ = {Bob}, $A(f_{12}^{ad})$ = {Stan, John}, $A(f_{12}^{bb})$ = {Ann}, $A(f_{12}^{bc})$ = {Lola}, $A(f_{12}^{bd})$ = {Mary}. The classes $A(f_2^a)$, $A(f_{12}^{aa})$, $A(f_{12}^{ac})$, and $A(f_{12}^{ba})$ will be empty.

1.3. From the standpoint of probabilistics they are different because the concrete people with the same names have different values of age in these two problems.

1.5. The complete set θ of abstract objects in problems 1 and 2 is the same, namely, $\theta = \{f_0, f_{01}^a, f_{01}^b, f_{02}^a, f_{02}^b, f_{02}^c, f_{02}^d, f_{012}^{aa}, f_{012}^{ab}, f_{012}^{ac}, f_{012}^{ad}, f_{012}^{ba}, f_{012}^{bb}, f_{012}^{bc}, f_{012}^{bd}\}$. Their names are: f_0 – 'a member of family A', f_{01}^a – 'a male member of family A', f_{01}^b – 'a female member of family A', f_{02}^a – 'a child of family A', f_{02}^b – 'an adolescent of family A', f_{02}^c – 'an adult of family A', f_{02}^d – 'a mature member of family A', f_{012}^{aa} – 'a male child of family A', f_{012}^{ab} – 'a male adolescent of family A', f_{012}^{ac} – 'a male adult of family A', f_{012}^{ad} – 'a mature man of family A', f_{012}^{ba} – 'a female child of family A', f_{012}^{bb} – 'a female adolescent of family A', f_{012}^{bc} – 'a female adult of family A', f_{012}^{bd} – 'a mature woman of family A'.

1.6. The abstract object f_{02}^d is more abstract than f_{012}^{ad} because the former does not have a definite value of the property t_1 ('sex') whereas the latter does. The most abstract object in these problems is f_0 because it has no definite value of both 'sex' and 'age'. The least abstract objects are f_{012}^{xy} ($x = a, b; y = a, b, c, d$) because in the complete set θ of abstract objects there are no abstract objects having definite values of some additional properties of the concrete members of family A.

1.7. The measures are given by the numbers of people in the classes. They are –

For problem 1: $mA(f_1^a) = 3$, $mA(f_1^b) = 3$; $mA(f_2^a) = 1$, $mA(f_2^b) = 1$, $mA(f_2^c) = 2$, $mA(f_2^d) = 2$; $mA(f_{12}^{aa}) = 0$, $mA(f_{12}^{ab}) = 1$, $mA(f_{12}^{ac}) = 1$, $mA(f_{12}^{ad}) = 1$, $mA(f_{12}^{ba}) = 1$, $mA(f_{12}^{bb}) = 0$, $mA(f_{12}^{bc}) = 1$, $mA(f_{12}^{bd}) = 1$.

For problem 2: $mA(f_1^a) = 3$, $mA(f_1^b) = 3$; $mA(f_2^a) = 0$, $mA(f_2^b) = 2$, $mA(f_2^c) = 1$, $mA(f_2^d) = 3$; $mA(f_{12}^{aa}) = 0$, $mA(f_{12}^{ab}) = 1$, $mA(f_{12}^{ac}) = 0$, $mA(f_{12}^{ad}) = 2$, $mA(f_{12}^{ba}) = 0$, $mA(f_{12}^{bb}) = 1$, $mA(f_{12}^{bc}) = 1$, $mA(f_{12}^{bd}) = 1$.

1.9. In both cases, such a set $A^{(a)}$ can consist of any concrete objects having two properties t_1 and t_2, of which t_1 has two values, and t_2 four values. For instance, the concrete objects can be balls; t_1 is the material they are made of with the values $f_1^{(a)}$ = 'wooden' and $f_1^{(b)}$ = 'plastic'; t_2 is their color with the values $f_2^{(a)}$ = 'red', $f_2^{(b)}$ = 'orange', $f_2^{(c)}$ = 'yellow', and $f_2^{(d)}$ = 'green'.

For problem 1, a set $A_1^{(a)}$ may consist of 1 orange wooden, 1 yellow wooden, 1 green wooden, 1 red plastic, 1 yellow plastic, and 1 green plastic balls.

For problem 2, a set $A_1^{(a)}$ may have 1 orange wooden, 2 green wooden, 1 orange plastic, 1 yellow plastic, and 1 green plastic balls.

We now replace the 6 balls by 6 disks made of any material, and mark them with the initials of the names of the materials and colors related to the balls. We obtain thereby another set $A_2^{(a)}$ adequate to family A in each of problems 1 and 2. For problem 1, $A_2^{(a)}$ may consist of 1 'wo'-, 1 'wy'-, 1 'wg'-, 1 'pr'-, 1 'py'-, and 1 'pg'-disks. For problem 2, $A_2^{(a)}$ may have 1 'wo'-, 2 'wy'-, 1 'po'-, 1 'py'-, and 1 'pg'-disks.

1.10. By making use of Eqn. (11), we obtain, in view of the results of the solution of the problems 5 and 7: for problem 1 – $P(f_{01}^b; f_2^b) = mA(f_{12}^{bb})/mA(f_1^b) = 0$; for problem 2 – $P(f_{01}^b; f_2^b) = 1/3$.

A female member of family A corresponds in the sets $A_1^{(a)}$ and $A_2^{(a)}$ to a plastic ball and a disk marked with the letter 'p', respectively. A female adolescent of family A corresponds in these sets to an orange plastic ball and a 'po'-disk, respectively. By making use of the sets $A_1^{(a)}$ or $A_2^{(a)}$, we obviously obtain the same results for the probabilities sought as those obtained above.

1.11. The previous properties of the concrete members of family A have no direct relation to the problem under consideration. We now have to consider a new set of concrete objects, namely, the set of concrete pairs of members of family A. The only property of such a pair, which now is of interest to us, is 'relationship'. This property has a number of values, for instance, 'father–son', 'mother–daughter', 'couple', etc. The set of concrete pairs involved is divided into classes with respect to these values. To solve the particular problem posed it is sufficient to know the measures of the class of concrete pairs corresponding to the abstract object 'a couple' and of the whole set of concrete pairs corresponding to the abstract object 'a pair of members of family A'. The latter, mA, equals the total number of the pairs, that is, $\binom{6}{2} = 15$. The former $mA(f_1^{\text{couple}})$ is the number of couples in family A, that is 2 (Mary–John and Lola–Stan). We thus have from Eqn. (11): $P(f_0; f_1^{\text{couple}}) = mA(f_1^{\text{couple}})/mA = 2/15$.

1.12. 'A point in a circle of radius R' is the abstract object f_0 corresponding to all the points of this circle. 'A point inside the concentric circle of radius R/n' is the abstract object f_{01} corresponding to all the points of this concentric circle. Hence, in view of Eqn. (11), $P(f_0; f_1) = mA(f_1)/mA = \pi(R/n)^2/\pi R^2 = n^{-2}$.

1.13. The position of a separate gas molecule is definite at every time instant. Hence, in accordance with the concept of a concrete object, a separate molecule at a definite time instant is a concrete molecule. The problem posed should concern the abstract molecule corresponding to the set of concrete molecules formed by the separate molecule in the vessel with the time. This set of concrete molecules is equivalent to the set of points within the vessel determining their positions. Since the only property t_1 of a concrete molecule, which we should take into consideration, is its position, the set of concrete molecules we are interested in is adequate to the set of points inside the vessel. The abstract object $f_0^{(a)}$ 'a point inside the vessel' corresponds to the abstract object f_0 'a molecule inside the vessel'. The abstract object $f_{01}^{(a)'}$ 'a point inside the fixed volume V'' corresponds to the abstract object f'_{01} 'a molecule inside the fixed volume V''. Therefore, the probability sought is, in view of Eqn. (11), $P(f_0; f'_1) = mA(f_1^{(a)'})/mA^{(a)} = V'/V$.

1.14. (10) What are the probabilities that in problems 1 and 2 a female member of the family A chosen at random will be an adolescent?

(11) What is the probability that a pair of members of family A (in the problems 1 and 2) chosen at random will be a couple?

(12) What is the probability that a point chosen at random in a circle of radius R will prove to be in the concentric circle of radius R/n?

(13) What is the probability that at a time instant chosen at random a separate gas molecule in a closed vessel of volume V will prove to be in a fixed volume V'?

Chapter 2

2.1. They are $(f_0; f_{12}^{aa})$, $(f_0; f_{12}^{ab})$, $(f_0; f_{12}^{ac})$, $(f_0; f_{12}^{ad})$, $(f_0; f_{12}^{ba})$, $(f_0; f_{12}^{bb})$, $(f_0; f_{12}^{bc})$, $(f_0; f_{12}^{bd})$.

2.2. The set A under consideration consists of all members of the 100 families. Every concrete member of this set (f_0) has two more properties: t_1 = 'the belonging to a family of a definite type' and t_2 = 'sex'. The property t_1 has in this problem three values denoted, in accordance with the structure of the families, by $f_1^{(a)}$, $f_1^{(b)}$, and $f_1^{(c)}$. The property t_2 has two values: $f_2^{(a)}$ = 'male' and $f_2^{(b)}$ = 'female'. The 6 elementary events are: $(f_0; f_{12}^{aa})$, $(f_0; f_{12}^{ba})$, $(f_0; f_{12}^{ca})$, $(f_0; f_{12}^{ab})$, $(f_0; f_{12}^{bb})$, $(f_0; f_{12}^{cb})$. The event we are interested in is $Q = (f_0; \theta'_{12})$, where $\theta'_{12} = \{f_{12}^{ab}, f_{12}^{bb}, f_{12}^{cb}\}$. It can be represented by Eqn. (22) as $Q = QQ_1 + QQ_2 + QQ_3$, where $Q_1 = (f_0; \{f_{12}^{aa}, f_{12}^{ab}\})$, $Q_2 = (f_0; \{f^{ba}, f^{bb}\})$, $Q_3 = (f_0; \{f^{ca}, f^{cb}\})$, the relation $Q_1 + Q_2 + Q_3 = U$ being obviously satisfied. Therefore, we can calculate the probability sought by Eqn. (23). From the conditions of the problem we have: $P(Q_1) = 25/100$, $P(Q_2) = 50/100$, $P(Q_3) = 25/100$, $P(Q/Q_1) = 2/4$, $P(Q/Q_2) = 4/6$, $P(Q/Q_3) = 5/8$. Hence, $P(Q) = (25/100)(2/4) + (50/100)(4/6) + (25/100)(5/8) = 0.615$.

2.3. The probabilities of the appearance of 'head' (p) and of 'tail' (q) are equal for a 'good' coin: $p = q = 1/2$. The most probable number of the appearances of 'head' is $10{,}000/2 = 5000$. By making use of Eqn. (39), we find: $P_{10{,}000}(5000) = (2/\pi)^{1/2} \times 10^{-2} = 7.979 \times 10^{-3}$.

2.4. We can estimate P by making use of Eqn. (47). Here $p = q = 1/2$, $n = 10{,}000$, and m is in the interval (4950, 5050). Hence, $v = (5000\text{-}m)/50$ and $P = (2\pi)^{-1/2} \int_{-1}^{1} \exp(-v^2/2)\, dv$. By By using a table containing the values of the integral on the right-hand side of Eqn. (45) as a function of z_0 (cf,, for instance, Ref. [2]), we find: $P = 0.683$.

2.5. We use Eqn. (45), where $p = q = 1/2$ and $\epsilon = 0.005$. From the table mentioned in the preceding problem we find that the right-hand side of this equation equals 0.9 for $z_0 = 1.163$. Since $n = (z_0/\epsilon)^2 2pq$, we finally get: $n = (1.163 \times 200)^2/2 = 27{,}052$.

Chapter 3

3.1. The answer depends on the probabilistic problem to be solved. In the usual problem of the probability of the appearance of a definite number on the upper side of a die cast at random, the concrete objects are the upper sides of a die marked by different numbers. Hence, the only random variable of interest here is a number on the upper side of a die, whose values are: 1, 2, 3, 4, 5, 6.

3.2. They are any coordinates uniquely determining the position of the point in space.

3.3. The only random variables in this experiment are the coordinates of a concrete electron on the screen. If they coincide with those of the hole, behind which the Geiger counter is placed, the cat becomes dead. The cat itself merely serves as an indicator which registers the occurrence of the hit of a concrete electron on the Geiger counter.

3.4. They are: coordinates, time, momenta, energy, mass (we omit some specific ones, such as charge, spin, magnetic moment, etc.).

3.5. Energy, E, and the Cartesian momenta p_x, p_y, and p_z are related by $E = (p_x^2 + p_y^2 + p_z^2)/2m$, where m is the mass of the particle (nonrelativistic case). For a free particle, the direction of its motion is constant, and we can choose it along the positive or negative direction of the x-axis. Then $p_y = p_z = 0$, and $p_x = p$ is connected with E by $p = \pm(2mE)^{1/2}$.

The joint probability density of p and E is given, in view of Eqn. (26), by $\rho(p, E) = \delta[p \mp (2mE)^{1/2}]\rho(E)$. For a free particle, the energy has a constant value, say E'. Hence, in view of Eqn. (15), $\rho(E) = \delta(E - E')$. We now have from Eqn. (29): $\rho(p) = \int_0^\infty \delta[p \mp (2mE)^{1/2}] \times \delta(E - E')\,dE = \delta[p \mp (2mE')^{1/2}]$. This means that $p = \pm(2mE')^{1/2}$ is also constant, and depending on the direction of the motion, we have either $\rho(p) = \delta(p - p')$ or $\rho(p) = \delta(p + p')$, where $p' = (2mE')^{1/2}$.

3.6. Suppose that the particle moves freely in a finite interval $-a < x < a$. Then, in view of Eqn. (17), we can write: $\rho(x) = \delta_{2a}(x)$. When a goes to infinity, we can write $\rho(x) = \lim_{a\to\infty} \delta_{2a}(x) \equiv \delta_\infty(x)$ (cf. also Sect. 10.4).

3.7. For p, we have, in view of Eqns. (48) and (59a), and the results of the solution of problem 5, $Mp^n = \int_{-\infty}^{\infty} p^n\,\delta(p - p')\,dp = p'^n$; $Dp = p'^2 - p'^2 = 0$.

For x, we obtain, by making use of the results of the solution of the preceding problem,

$$Mx^n = \lim_{a\to\infty} \int_{-a}^{a} x^n\,\delta_{2a}(x)\,dx = \lim_{a\to\infty} (2a)^{-1} \int_{-a}^{a} x^n\,dx = \begin{cases} 0 & (n \text{ odd}) \\ \infty & (n \text{ even}) \end{cases}; \quad Dx = \infty.$$

3.8. Let the values of t be z. Then, in view of Eqn. (34), the joint probability density of the random variables t, t_1, and t_2, is $\rho(z, x, y) = \delta(z - x - y)\rho(x, y)$. Hence $\rho(z) = \int_{-\infty}^{\infty}\int_{-\infty}^{\infty} \delta(z - x - y)\rho(x, y)\,dx\,dy$. Integrating over x or over y, we obtain, in view of Eqn. (30), $\rho(z) = \int_{-\infty}^{\infty} \rho(z - y, y)\,dy$ or $\rho(z) = \int_{-\infty}^{\infty} \rho(x, z - x)\,dx$.

3.9. Denoting the values of t by z, from Eqn. (34) we have: $\rho(z, x, y) = \delta(z - x/y)\rho(x, y)$. Hence, $\rho(z) = \int_{-\infty}^{\infty}\int_{-\infty}^{\infty} \delta(z - x/y)\rho(x, y)\,dx\,dy$. Integrating over x, we obtain, in view of Eqn. (30), $\rho(z) = \int_{-\infty}^{\infty} |y|\rho(yz, y)\,dy = \int_0^\infty y\rho(yz, y)\,dy - \int_{-\infty}^{0} y\rho(yz, y)\,dy$.

Chapter 5

5.1. If the concrete cat involved were aware of the mortal danger which threatened it, one would have the right to think of the state of its mood. Fortunately, it was unaware of this, but that is immaterial to the case under consideration, the more so because the notion of a state is irrelevant here for any concrete object, including the concrete cat and the concrete electron that can kill it. In this experiment, the notion of a state is only relevant to the abstract electron involved, and its state is determined by the conditions of the experiment and can be described by the probability distribution of its coordinates.

5.2. The abstract object meant is the 'upper side of a die'. The state under examination is that described by the probability of different values of the random variable t_1 – the number marking that side (cf. also problem 3.1).

5.3. The values of the random variable t_1 are $\lambda_j = j$ ($j = 1, 2, \ldots, 6$). The state under consideration can be described by:

(1) the respective probabilities $p_j = 1/6$ ($j = 1, 2, \ldots, 6$);
(2) the probability density $\rho(\lambda) = 6^{-1} \sum_{j=1}^{6} \delta(\lambda - j)$;
(3) the distribution function $F(\lambda) = 6^{-1}k$ ($k < \lambda$; $0 \leqslant k \leqslant 6$);
(4) the characteristic function $\Phi(\epsilon) = 6^{-1} \sum_{j=1}^{6} \exp(i\epsilon j)$.

5.4. To that end, let us first find a Hermitian matrix whose eigenvalues coincide with the λ_j's of the preceding problem. The simplest matrix of this kind is a diagonal 6 by 6 matrix of the form (40): $\Lambda = \|\delta_{kj} j\|$. Its eigenvector corresponding to $\lambda_j = j$, which is the jth column, e_j, of the unit matrix E_6, describes the unreal jth state of the abstract object involved. The vector describing the real state we are interested in is the linear combination of the e_j: $x = \sum_{j=1}^{6} a_j e_j$.

Since the probabilities of all the λ_j are 1/6, we should set $a_j = (1/6)^{1/2}$. Hence, the real state under examination can also be described by:

(5) the state vector $x = 6^{-1/2}(1, 1, 1, 1, 1, 1)$ and

(6) the probability matrix $W = x\tilde{x} = 6^{-1}\|1\|$, whose elements are 1/6 overall. It is easy to see that the norm of x and the trace of W are unities.

By making use of Eqn. (47), we can describe this state also by:

(7) a state vector $\psi(\lambda) = \sum_{j=1}^{6} 6^{-1/2}\,\delta^{1/2}(\lambda - j)\exp[i\eta_j(\lambda)]$, where η_j are any real functions, and

(8) the density matrix $W(\lambda, \lambda') = \psi(\lambda)\bar{\psi}(\lambda')$.

5.5. By making use of the pertinent equations, in particular Eqns. (109) and (110), we obtain: $Mt_1 = 7/2$.

5.6. $\delta(x - x')$ cannot be normalized. Indeed, in view of Eqn. (135), $(\delta_\epsilon(x - x'), \delta_\epsilon(x - x')) = \epsilon^{-1}$. Since $\delta(x - x') = \lim_{\epsilon\to 0}\delta_\epsilon(x - x')$, we get $(\delta(x - x'), \delta(x - x')) = \infty$.

5.7. In view of Eqns. (130) and (3.17),

$$f(x)\,\delta_\epsilon^{1/2}(x - x') = \begin{cases} 0 & (|x - x'| > \epsilon/2) \\ \epsilon^{-1/2}f(x) & (|x - x'| < \epsilon/2) \end{cases}.$$

This gives, because of

$$f(x) = f(x') + f'(x')(x - x'),$$

$$f(x)\,\delta_\epsilon^{1/2}(x - x') = f(x')\delta_\epsilon^{1/2}(x - x') + \begin{cases} 0 & (|x - x'| > \epsilon/2) \\ \eta & (|x - x'| < \epsilon/2) \end{cases},$$

where

$$|\eta| = |f'(x')(x - x')|\epsilon^{-1/2} < 2^{-1}|f'(x')|\epsilon^{1/2} = O(\epsilon^{1/2}).$$

Hence,

$$f(x)\,\delta_\epsilon^{1/2}(x - x') = f(x')\,\delta_\epsilon^{1/2}(x - x') + O(\epsilon^{1/2}),$$

which yields, in view of

$$\delta^{1/2}(x - x') = \lim_{\epsilon\to 0}\delta_\epsilon^{1/2}(x - x'), \qquad f(x)\,\delta^{1/2}(x - x') = f(x')\,\delta^{1/2}(x - x').$$

Since Eqn. (136) is valid, regardless of the magnitude of ϵ, $\delta^{1/2}(x - x')$ also is normalized to unity.

Chapter 6

6.1. Equation (28) yields, in view of $\tilde{x}^{(1)}\eta = \tilde{\eta}\dot{x}^{(1)}$,

$$\frac{\partial L}{\partial \tilde{x}^{(1)}} = -\frac{\partial \varphi_2}{\partial \tilde{x}^{(1)}} + \frac{\partial \tilde{\eta}}{\partial \tilde{x}^{(1)}}\dot{x}^{(1)} \quad \text{and} \quad \frac{\partial L}{\partial \dot{\tilde{x}}^{(1)}} = \frac{\partial \varphi_1}{\partial \dot{\tilde{x}}^{(1)}} + \eta.$$

Hence,

$$\frac{\mathrm{d}}{\mathrm{d}x_n}\frac{\partial L}{\partial \dot{\tilde{x}}^{(1)}} = \frac{\partial^2 \varphi_1}{\partial \dot{\tilde{x}}^{(1)}\,\partial \dot{x}^{(1)}}\ddot{x}^{(1)} + \frac{\partial \eta}{\partial x^{(1)}}\dot{x}^{(1)}$$

and, in view of Eqn. (27),

$$\frac{\partial L}{\partial \tilde{x}^{(1)}} - \frac{\mathrm{d}}{\mathrm{d}x_n}\frac{\partial L}{\partial \dot{\tilde{x}}^{(1)}} = -Q - A_1\dot{x}^{(1)} - A_0\ddot{x}^{(1)}.$$

Thus, Eqns. (1) and (2) are identical in the case under consideration.

6.2. No, because a ψ-function of this type does not satisfy Eqn. (84) and, hence, Eqn. (91).

6.3. These intervals are: $0 \leqslant \beta < \beta_1$ and $\beta_2 < \beta \leqslant 1$, where β_1 and β_2 satisfy the equation

$$\frac{\pi^2\beta^2(1-\beta)^2}{3[\beta^2+(1-\beta)^2]^2} + \frac{4\beta^3(1-\beta)^3}{[\beta^2+(1-\beta)^2]^3} = \tfrac{1}{4}.$$

Since the left-hand side of this equation does not change when interchanging β and $(1-\beta)$, $\beta_2 = 1 - \beta_1$. Hence, it is sufficient to calculate β_1. The calculation gives $\beta_1 = 0.205$ and, hence, $\beta_2 = 0.795$.

Chapter 10

10.1. By introducing $\psi_E(r) = \chi^E(r)/r$ we get from Eqn. (262), for $l = 0$, a reduced Schrödinger equation

$$\frac{\hbar^2}{2m}\frac{d^2\chi^E(r)}{dr^2} + [U(r) - E]\chi^E(r) = 0.$$

The function $\chi^E(r)$ should satisfy this equation in all the three intervals, 1, 2, and 3. It can be represented in the form $\chi^E(r) = \chi_1^E(r) + \chi_2^E(r) + \chi_3^E(r)$, where each χ_1^E, χ_2^E, and χ_3^E satisfies the reduced Schrödinger equation in the pertinent interval, being zero outside it. Both the probability density $\rho(r)$ and its derivative with respect to r may be supposed to be finite and continuous in the entire interval $0 \leqslant r < \infty$. Therefore, $\chi_1^E(r)$, $\chi_2^E(r)$, and $\chi_3^E(r)$ should obey the conditions: $\chi_1^E(r_1) = \chi_2^E(r_1)$, $\chi_1'^E(r_1) = \chi_2'^E(r_1)$; $\chi_2^E(r_2) = \chi_3^E(r_2)$, $\chi_2'^E(r_2) = \chi_3'^E(r_2)$ (the prime denotes a derivative with respect to r). These functions can be written as:

$$\chi_1^E(r) = 2^{-1}A\exp(ikr) + 2^{-1}A_1\exp(-ikr);$$

$$\chi_2^E(r) = AB\exp[ik''(r-r_1)] + AC\exp[-ik''(r-r_1)];$$

$$\chi_3^E(r) = AD\exp[ik(r-r_2)] + AF\exp[-ik(r-r_2)].$$

In order for them to satisfy the corresponding Schrödinger equations, the following relations must be fulfilled: $k = \hbar^{-1}(2mE)^{1/2}$; $k'' = \hbar^{-1}[2m(E-U_1)]^{1/2} = -ik'$, where $k' = \hbar^{-1}[2m(U_1-E)]^{1/2}$. Hence,

$$\chi_2^E(r) = AB\exp[k'(r-r_1)] + AC\exp[-k'(r-r_1)].$$

Since the probability density of r is finite at $r = 0$, $\chi_1^E(0) = 0$, and hence, $\chi_1^E(r) = A\sin kr$. The above conditions yield: $\sin kr_1 = B + C$; $k\cos kr_1 = k'B - k'C$, whence $B = (2k')^{-1}(k'\sin kr_1 + k\cos kr_1)$ and $C = (2k')^{-1}(k'\sin kr_1 - k\cos kr_1)$. Since $\chi_2^E(r)$ is real, so is $\chi_3^E(r)$. Therefore, $F = \bar{D}$, and from the above conditions we obtain:

$$D = 2^{-1}(1 - ik'/k)B\exp(k'L) + 2^{-1}(1 + ik'/k)C\exp(-k'L),$$

where $L = r_2 - r_1$. The factor A should be obtained from a normalization condition.

10.2. Since $\rho(\infty)$ should be finite, the coefficient B in $\chi_2^E(r)$ must vanish. Hence, $k'\sin kr = -k\cos kr$, whence $k'\tan kr = -k$. This is an equation in E because both k and k' are functions of E. Solution of this transcendental equation determines the energy values E_j of the corresponding discrete stationary states. Because $B = 0$ and $L = \infty$, $D = 0$ as well. Hence $\chi_3^E(r) = 0$. For $\chi_1(r)$ and $\chi_2(r)$, we now have:

$$\chi_1^j(r) = A\sin k_jr, \qquad \chi_2^j(r) = A\sin k_jr_1\exp[-k_j'(r-r_1)],$$

where k_j and k_j' are the discrete values of k and k' corresponding to E_j.

10.3. The operator for $\dot{r} = v$ is, in view of Eqn. (4), $\hat{v} = -(i\hbar/m)(\partial/\partial r + r^{-1}\hat{I})$. Its eigenfunctions are $\varphi(v', r) = r^{-1} \sin(mv'r/\hbar)$. To find $\rho(v')$ in a state described by $\psi_E(r)$ of problem 10.1, one is supposed to normalize both this function and $\varphi(v', r)$, and then to expand the $\psi_E(r)$ normalized in terms of $\varphi(v', r)$ normalized. The magnitude squared of the coefficients in this expansion are proportional to the pertinent values $\rho(v')$ of the probability density of v. But we can avoid both normalizations and find the necessary factor at the last stage of the calculation. Thus, we have to calculate

$$4\pi \int_0^\infty \psi_E(r)\varphi(v', r) r^2 \, dr = 4\pi \int_0^\infty \chi^E(r) \sin(mv'r/\hbar) \, dr.$$

If we normalized $\chi^E(r)$, we would see that in the integral

$$\int_0^\infty |\chi^E(r)|^2 \, dr = \int_0^{r_1} |\chi_1^E(r)|^2 \, dr + \int_{r_1}^{r_2} |\chi_2^E(r)|^2 \, dr + \int_{r_2}^\infty |\chi_3^E(r)|^2 \, dr,$$

the third term is infinite, whereas the first two are finite. Hence, they can be omitted, and we may expand merely $\chi_3^E(r)$ in terms of $\sin(mv'r/\hbar)$. Since $k = mv/\hbar$, we obtain:

$$a(v') = \int_{r_2}^\infty \chi_3^E(r) \sin(mv'r/\hbar) \, dr \propto$$

$$\propto \int_{r_2}^\infty [D \exp(imvr/\hbar) + \bar{D} \exp(-imvr/\hbar)] \sin(mv'r/\hbar) \, dr.$$

Because

$$\sin(mv'r/\hbar) \propto [\exp(imv'r/\hbar) - \exp(-imv'r/\hbar)]$$

and

$$\int_{-\infty}^\infty \exp(i\alpha r) \, dr \propto \delta(\alpha),$$

the integration yields terms of the form $\delta(v' - v)$ and $\delta(v' + v)$ with equal factors before them. Thus, we get: $a(v') = C[\delta(v' - v) + \delta(v' + v)]$ and, hence, we should have $\rho(v') = C^2[\delta^2(v' - v) + \delta^2(v' + v)]$, where $C^2 = 1/\int_{-\infty}^\infty [\delta^2(v' - v) + \delta^2(v' + v)] \, dv'$. All this may seem meaningless to one, since $\delta^2(v' \pm v)$ is known not to exist at all. Nevertheless, $\rho(v')$ does exist and can be calculated in the following way. Recall that $\delta(x - x') = \lim_{\epsilon \to 0} \delta_\epsilon(x - x')$. We have, thus, in view of Eqn. (5.130), $\delta^2(v' - v) = \lim_{\epsilon \to 0} \delta_\epsilon^2(v' - v) = \lim_{\epsilon \to 0} \epsilon^{-1} \delta_\epsilon(v' - v)$ and $\delta^2(v' + v) = \lim_{\epsilon \to 0} \epsilon^{-1} \delta_\epsilon(v' + v)$. Therefore,

$$\rho(v') = \lim_{\epsilon \to 0} \left\{ \frac{\epsilon^{-1}[\delta_\epsilon(v' - v) + \delta_\epsilon(v' + v)]}{\epsilon^{-1}\int_{-\infty}^\infty [\delta_\epsilon(v' - v) + \delta_\epsilon(v' + v)] \, dv} \right\}.$$

This yields, in view of Eqn. (3.19), the final result:

$$\rho(v') = 2^{-1} \lim_{\epsilon \to 0} [\delta_\epsilon(v' - v) + \delta_\epsilon(v' + v)] = 2^{-1}[\delta(v' - v) + \delta(v' + v)].$$

10.4. The energy values in this case are those of a linear harmonic oscillator, namely, $E_n = (n + \frac{1}{2})\hbar\nu$ $(n = 0, 1, \ldots)$, where ν is the vibrational frequency. Hence, the vibrational partition function for one molecule is, in view of Eqn. (289),

$$Z = \exp\left(-\frac{\hbar\nu}{2kT}\right) \sum_{n=0}^\infty \exp\left(-\frac{n\hbar\nu}{kT}\right) = \frac{\exp(-\hbar\nu/2kT)}{1 - \exp(-\hbar\nu/kT)},$$

and, in view of Eqn. (9.68), $F = \hbar\nu/2 + kT \ln[1 - \exp(-\hbar\nu/kT)]$, for one molecule. Therefore, for N molecules, $F = N\hbar\nu/2 + NkT \ln[1 - \exp(-\hbar\nu/kT)]$.

Chapter 11

11.1. There are two variants of ammonia. Variant I is the old molecule of ammonia having the old equilibrium configuration, variant II is the new ammonia molecule with the new equilibrium configuration. The transformation path coordinate q is a change in the height of the pyramid of NH_3 whose top is at atom N. The transitional point is determined by the value $q_0 = -l$, where l is the height of the equilibrium NH_3-pyramid. A transitional configuration is any plane configuration of the molecule.

11.2. At $T = 0°$ K, Eqns. (94) and (53) coincide and yield

$$\kappa = \kappa_s = 3.088 \times 10^{13} \exp\left[-\frac{(0.36/1.09)^2}{2 \times 0.0824^2}\right] \sec^{-1} \cong 1 \times 10^{10} \sec^{-1}.$$

Chapter 12

12.1. It follows from Newton's gravitation law and special relativity that the equation $-\gamma Mm/r + mc^2 = \text{const.}$ is valid for the system under consideration, where γ is the gravitation constant, m is given by Eqn. (13), and r is the distance from the ICS origin to the particle. We can rewrite this equation, in view of Eqn. (18), as $(1 - r^{-1}R_0)/\eta = (1 - r_0^{-1}R_0)/\eta_0$, where $R_0 = \gamma M/c^2$ and $\eta_0 = (1 - v_0^2/c^2)^{1/2}$.

As r increases, so does η, and the velocity of the particle diminishes. But neither $1 - R_0/r$ nor η can exceed 1. Hence, if (a) $1 - R_0/r_0 > \eta_0$, then $1 - R_0/r = 1$ at $r = \infty$, while $\eta_\infty < 1$, that is $v_\infty \neq 0$; and if (b) $1 - R_0/r_0 < \eta_0$, then $\eta = 1$ for a finite value r_{max} of r, that is $v(r_{max}) = 0$. In case (a) the particle goes to infinity with a nonzero velocity, whereas in case (b) it reaches a definite point ($r = r_{max}$) with a zero velocity and then moves backwards.

12.2. The solution of the preceding problem does not depend on the rest mass of the particle which can even be a weighty brick. Hence, for a nonrelativistic emon, for which $\eta_0 \cong 1$, the case (b) should generally be materialized. On the contrary, for a photon, for which $\eta_0 \cong 0$, case (a) generally occurs. In order that for a proton case (b) were relevant, it is obviously necessary that $1 - R_0/r_0 < \eta_0 \cong 0$ would have to be valid. But this means that the massive body would have to be practically a 'black hole'.

Index